Springer Atmospheric Sciences

More information about this series at http://www.springer.com/series/10176

Boris M. Smirnov

Microphysics of Atmospheric Phenomena

Springer

Boris M. Smirnov
Institute for High Temperatures
Russian Academy of Sciences
Moscow
Russia

ISSN 2194-5217 ISSN 2194-5225 (electronic)
Springer Atmospheric Sciences
ISBN 978-3-319-30812-8 ISBN 978-3-319-30813-5 (eBook)
DOI 10.1007/978-3-319-30813-5

Library of Congress Control Number: 2016940329

Printed on acid-free paper

This Springer imprint is published by Springer Nature
The registered company is Springer International Publishing AG Switzerland

Preface

The goal of this book is to give a reliable description of various global atmospheric properties and some phenomena including photoprocesses, processes of atmospheric electricity with participation of aerosols, the atmospheric chemistry with processes involving ions, oxygen atoms, excited atomic particles and ozone molecules, as well as processes in an ionospheric plasma. In the course of this consideration, we are based on observational data, existing atmospheric concepts, and measured rate constants of elementary processes in atmospheric air. Fundamentals of global atmospheric phenomena are based on the standard atmosphere model with averaged atmospheric parameters over the globe and time, the global electric circuit, the energetic balance of the Earth, and other global concepts. As a result, one can give a qualitative description of atmospheric phenomena and estimate their parameters. This analysis is based on processes involving elementary atmospheric particles and gives a simple and reliable understanding of the Earth processes and phenomena. In addition, this analysis allows one to glance at future trends of topical atmospheric problems, such as ozone problem and climate change.

Moscow, Russia Boris M. Smirnov

Contents

Chapter 1
Introduction

The goal of this book is the analysis of some aspects of atmospheric physics and chemistry on the basis of elementary processes in the Earth's atmosphere, as well as global atmospheric properties and phenomena that are based on these processes. This book has common features with [1] for the kinetics of excited air. Here we consider mostly other aspects of this general problem including radiative and aerosol processes, and also respective atmospheric phenomena. In consideration of atmospheric phenomena, we are grounded on observational and measured data using simple reliable concepts and models. Taking elementary processes in excited air as a basis, we use appropriate global models and concepts including the global circuit model [2] for electric processes in the atmosphere and the model of standard atmosphere [3–6]; that is, we use an average of atmospheric parameters over the globe. This allows us to obtain detailed information about atmospheric properties or phenomena in a simple form. Following is a list of such problems to deepen understanding on the basis of this analysis.

1. Collisions of neutral aerosols in different aggregate states leads to their charging, and the subsequent fall of positive and negative aerosols with different falling velocities leads to charge separation in the atmosphere and the creation of atmospheric electric fields as a source of atmospheric electrical phenomena. Charged aerosols constitute cumulus clouds, and due to a charge, growth of aerosols in cumulus clouds proceeds with a lower rate compared with that for neutral aerosols. Then from rates of the coagulation process and the lifetime of a cumulus cloud it follows that a typical charge of aerosols in cumulus clouds includes $Z = 25-30$ electron charges. Ionization of atmospheric air is realized by seconadary Mev-energy particles which are formed in nuclear rections involving Gev-energy protons or neutrons.

2. Note a significant role of cosmic rays in atmospheric processes. Atmospheric ionization by cosmic rays is important for aerosol discharging and creating a plasma that does not allow for clouds to expand in a surrounding space. In addition, showers of cosmic rays initiate the beginning of lightning in a thunderstorm.

3. The atmospheric optical depth u for infrared radiation is approximately equal to $u = 2.7$, as follows from the analysis of the energetic balance of the Earth and its atmosphere. The greenhouse atmospheric effect is determined mostly by atmospheric

B.M. Smirnov, *Microphysics of Atmospheric Phenomena*,
Springer Atmospheric Sciences, DOI 10.1007/978-3-319-30813-5_1

water, and, as follows from the subsequent analysis, the above optical depth may be created by atmospheric aerosols if $(1-2)$ % of atmospheric water is converted in aerosols. This may cause anxiety because of an atmospheric instability for the global climate that can be changed by a weak action. Next, according to NASA data [7], the average Earth temperature has increased by (0.8 ± 0.1)°C from 1880 up to now. Keeping in mind the huge effort to obtain this value which requires monitoring by thousands of meteorological stations because temperature fluctuations are tens of degrees, one can note that this change corresponds to an increase of the total mass of atmospheric water by approximately 10 %. The press and TV convince us that the main reason for the above temperature increase is an accumulation of atmospheric CO_2 because its concentration in the atmosphere has increased by about 30 % during the twentieth century. In this book we analyze the contribution of atmospheric CO_2 in the greenhouse effect step by step and find that doubling of the CO_2 concentration in the atmosphere leads to an increase of the Earth's global temperature by (0.4 ± 0.2)°C and at the contemporary rate of change of the CO_2 concentration the doubling of its concentration will take 130 years. Thus, the contribution of atmospheric CO_2 to the greenhouse effect of the atmosphere is several times less than that from atmospheric water. This contribution also depends on the manner of CO_2 generation, so that deforestation acts more strongly on the balance of atmospheric CO_2 than that due to combustion of fossil fuels.

4. In spite of a small ozone concentration (<0.01 %) in the stratosphere, it is important both for absorption of ultraviolet solar radiation and for thermal balance of the stratosphere. But seasonal and daily fluctuations of the ozone concentration are compared with its concentration, as well as its average concentration in various parts of the globe. The latter is reflected by the term "ozone hole" for poles where the ozone concentration at the poles is less than approximately twice that at equatorial regions, but this does not characterize an absorption of ultraviolet radiation in atmospheric air. For example, the probability of reaching the Earth's surface for ultraviolet solar photons, which mostly are absorbed by ozone molecules, is $\exp(-200)$ in an equatorial zone and $\exp(-100)$ at the poles. The same changes relate to daily and seasonal ozone concentrations, so that the term "ozone hole" does not correspond to a meaning of these words.

Although above we enumerate some atmospheric problems that are distorted in the press or are not taken into account in atmospheric science, it is not the main goal of this book, which consists of an academic description of elementary atmospheric processes and their kinetics. This, together with other atmospheric reviews and books, allows us to formulate the physical picture of atmospheric processes and to correct some positions of this description. Let us demonstrate this for electrical atmospheric processes. One can consider processes of atmospheric electricity as a secondary phenomenon with respect to water circulation through the atmosphere [8]. The central process of atmospheric electricity is the charging process as a result of collisions of aerosols that are located in different aggregate states. This characteristic of the charging process follows from the experiment [9] which, in the author's opinion, is underestimated.

Cosmic rays are of importance for atmospheric electricity and electrical atmospheric processes. Air ionization by cosmic rays in the troposphere and above the tropopause is important for the Earth's discharge, which is charged negatively [10, 11]. In reality, the study of cosmic rays starts from investigation of air ionization in the troposphere [12]. But the role of cosmic rays in processes of atmospheric electricity is not restricted by ion formation in the troposphere. One of the problems of atmospheric electricity is the origin of lightning. Because a typical electrical field strength in a thunderstorm due to a cloud charge is two orders of magnitude lower than the breakdown of electrical field strength in air, the development of lightning is impossible from one seed electron. It is possible as a result of streamer propagation, but approximately 10^8 electrons must be gathered in a small volume for streamer realization [13] in order to create electrical field strengths above the breakdown. The analysis [14] shows that known mechanisms do not explain this stage of lightning evolution. Evidently, the answer for this follows from the specific investigations in the Tien Shan station of cosmic rays [15] where the correlation was observed between showers of cosmic rays and radio emission from lightning that is evidence of lightning origin as a result of atmospheric ionization by cosmic showers.

Lightning or electrical breakdown of the atmosphere is a stage of atmospheric electricity that leads to negative charging of the Earth. The contemporary state of lightning physics is represented in Uman's books [16–19], where experimental methods are given and observational parameters of lightning are obtained in detail. The theoretical analysis of lightning development is contained in the Bazelyan and Raizer books [20–23] where models are used on the basis of electric schemes and a uniform plasma. Therefore we describe this phenomenon briefly, referring to the above books and mentioning only some aspects of this phenomenon where physics of elementary and other processes are able to give a new standpoint.

However, we correct the appropriate atmospheric electricity parameters if they contradict the results of the special analysis. In particular, the first stage of lightning, the stepwise leader, is an ionization wave that forms a conductive channel for subsequent propagation of a lightning electric current. In order to conserve a plasma in this channel during leader propagation, it is necessary to suppress attachment of electrons to oxygen molecules, and this is realized due to a heightened channel temperature. This temperature is expected above 5000 K according to [20–23]. On the basis of rate constants of the electron attachment process in the following we prove that this boundary temperature is about 1200 K.

Next, it is supposed on the basis of measurements that the lightning leader transfers on average a charge of 5 C that corresponds to a negative charge gradient along the leader channel $\sim 10^{-3}$C/m [14]. One can see that at a channel radius of 10 cm this corresponds to the number density of excess electrons on the order of 10^{11}cm^{-3}. At this excess number density of electrons the electric potential of 50 MV, a typical potential of a thundercloud, arises at a length of 1 m; that is, the ionization wave is locked at such distances. This means that the statement about a charge transfer at the stepwise leader stage is wrong, and an indicated transferred charge of 5 C relates to other stages of lightning evolution.

The above examples for lightning demonstrate the possibilities of using rates of microphysical processes in the analysis of atmospheric processes and phenomena. Returning to lightning, note that such an approach allows us to compare the contribution of various mechanisms to energy release in the lightning channel during propagation of an electric current through it, the lifetime of the conductive channel in the absence of an electric current in it, the character of expansion of the conductive channel in time, and so on. All this may be added to a contemporary understanding of the lightning physics.

The analysis of the Earth's ionosphere as one of the objectives of this book is based, in the first place, on elementary processes involving its atomic particles. According to H.S.W. Massey, the classic atmospheric physics and physics of atomic collisions, "A detailed understanding of the ionospheres of the Earth and planets requires knowledge of the rates of many ionic reactions as well as of electron recombination coefficients and photoionization cross sections" [24]. Indeed, the Earth's ionosphere contains nonequilibrium excited air; that is, its properties are determined by processes involving its atomic particles, and this air nonequilibrium starts from photoionization and photodissociation of atmospheric air under the action of solar radiation. Study of the ionosphere develops in this manner (e.g., [25–28]). Below we collect information about elementary processes that is useful for ionospheric properties.

Thus, the goal of this book is to give a simple and reliable analysis of atmospheric processes and corresponding properties of the Earth's atmosphere on the basis of physical laws and rate constants of elementary processes involving atmospheric particles. This analysis allows us to ground some existing concepts of atmospheric physics and chemistry with additional information. As a result, the analysis gives a new perspective on some aspects of processes and phenomena under consideration, and some additions to the contemporary physical picture of the Earth's atmosphere. Briefly, we consider the Earth's atmosphere as a physical object, and this approach allows us to determine definitely the mechanisms of some atmospheric phenomena with estimations for their parameters.

References

1. M. Capitelli, C.M. Ferreira, B.F. Gordiets, A.I. Osipov, *Plasma Kinetics in Atmospheric Gases* (Springer, Berlin, 2000)
2. C.T.R. Wilson, J. Frankl. Inst. **208**, 1 (1929)
3. U.S. Standard Atmosphere (U.S. Government Printing Office, Washington, 1976)
4. K.S.W. Champion, A.E. Cole, A.J. Kantor, in *Chemical Dynamics in Extreme Environments*, ed. by R.A. Dressler (World Scientific Publishing, Singapore, 2001)
5. M. Davies, *The Standard Handbook for Aeronautical and Astronautical Engineers* (McGraw-Hill, New York, 2003)
6. D.R. Lide (ed.), *Handbook of Chemistry and Physics*, 86th edn. (CRC Press, London, 2003–2004)
7. http://climate.nasa.gov
8. R. Braham, J. Meteorol. **9**, 227 (1952)

9. S.E. Reynolds, M. Brook, M.F. Gourley, J. Met. **14**, 426 (1957)
10. C.T.R. Wilson, Proc. R. Soc. **92A**, 555 (1916)
11. C.T.R. Wilson, Philos. Trans. R. Soc. **221A**, 73 (1921)
12. V.F. Hess, Phys. Zs. **113**, 1084 (1912)
13. H. Raether, Arch. Electrotech. **34**, 49 (1940)
14. J.R. Dwyer, M. Uman, Phys. Rep. **534**, 147 (2014)
15. A.V. Gurevich, A.N. Karashtin, V.A. Ryabov, Usp. Phys. Nauk **179**, 779 (2009)
16. M.A. Uman, *Lightning* (McGraw Hill, New York, 1969)
17. M.A. Uman, *About Lightning* (Dover, New York, 1986)
18. M.A. Uman, *The Lightning Discharge* (Academic Press, New York, 1987)
19. V.A. Rakov, M.A. Uman, *Lightning, Physics and Effects* (Cambridge University Press, Cambridge, 2003)
20. E.M. Bazelyan, Yu.P. Raizer, Usp. Fiz. Nauk **170**, 753 (2000) (Phys. Uspekhi **43**, 753 (2000))
21. E.M. Bazelyan, Yu.P. Raizer, *Lightning Physics and Lightning Protection* (IOP Publishing, Bristol, 2000)
22. E.M. Bazelyan, Yu.P. Raizer, *Physics of Lightning and Lightning Protection* (Fizmatlit, Moscow, 2001) (in Russian)
23. Yu.P. Raizer, *Physics of Gas Discharge* (Intellect, Dolgoprudnyi, 2009) (in Russian)
24. H.S.W. Massey, J. de Physique **40**, C7 (1979)
25. J.A. Rateliffe, *Physics of the Upper Atmosphere* (Academic Press, New York, 1960)
26. A.D. Danilov, *Chemistry of Ionosphere* (Plenum Press, New York, 1970)
27. M.J. McEwan, L.N. Phillips, *Chemistry of the Atmosphere* (Edward Arnold, London, 1975)
28. M. McIven, L. Philips, *Chemistry of Atmospheres* (Mir, Moscow, 1978)

Part I
Fundamentals of Atmosphere and Atmospheric Processes

Chapter 2
Properties of Neutral Atmosphere

Abstract Some atmospheric parameters as a function of altitude are represented for the atmospheric standard model that operates with averaging over time and season parameters at the latitude of 45°N. As a result of air circulation due to air convective motion, molecules and small aerosols are captured by vortices, and their concentration in air does not vary with altitude, whereas micron-size aerosols do not penetrate altitudes above those of their formation. Therefore concentration of water molecules in the stratosphere is small compared to that in the troposphere. The water and carbon balances in the atmosphere are analyzed, and accumulation of CO_2 in the atmosphere is considered from various standpoints. The atmospheric energetic balance of the Earth and its atmosphere is represented and is compared with energetics of other processes.

2.1 Standard Atmosphere

In order to analyze global properties of the Earth's atmosphere, it is convenient to deal with average atmospheric parameters. A widespread average over time and the globe leads to the so-called standard atmosphere model. The standard atmosphere parameters are averaged over longitude, season, and time, so that they depend on the altitude only. Instead of averaging over the latitudes, these data are taken for the latitude of 45°N. Atmospheric parameters depend on radiative fluxes from the Sun, and we use a basis of atmospheric parameters at moderate solar activity. Below we use atmospheric parameters taken from [1–4]. In this case basic atmospheric parameters at sea level are equal as follows. The air pressure is $p = 101.3\,\mathrm{kPa}$, the air temperature is $T = 288.2\,\mathrm{K}$, the air density is $\rho = 1.225\,\mathrm{kg/m^3}$, and the number density of air molecules is $2.547 \times 10^{19}\mathrm{cm}^{-3}$. In addition, the mean free path of air molecules is $\lambda = 66.33\,\mathrm{nm}$, the mean molecular weight in atomic units of masses is $M = 28.96$, the sound speed is $u = 340.3\,\mathrm{m/s}$, the freefall acceleration is $g = 980.7\,\mathrm{cm/s^2}$, the heat capacity at constant pressure is $C_p = 1.005\ \mathrm{J/(g \cdot K)}$, the heat capacity at constant volume $C_V = 0.718\ \mathrm{J/(g \cdot K)}$, the dynamic viscosity coefficient of air is $\eta = 1.789\,\mathrm{P}$ (the dynamic viscosity unit P, Poiseuille, is equal to $1P = 0.1 Pa \cdot s$) and corresponds to the kinematic viscosity coefficient of air $\nu = 0.146\,\mathrm{cm^2/s}$,

B.M. Smirnov, *Microphysics of Atmospheric Phenomena*,
Springer Atmospheric Sciences, DOI 10.1007/978-3-319-30813-5_2

and the thermal conductivity coefficient of air is $\kappa = 2.596 \times 10^{-4} W/(\mathrm{cm} \cdot \mathrm{K})$ corresponds to the thermal diffusivity coefficient $\chi = \kappa/(C_p \cdot \rho) = 0.21\,\mathrm{cm}^2/\mathrm{s}$. Note that we express the temperature in energetic units, and a transfer from Kelvin to other energetic units is given in Appendix A.3.

Figure 2.1 represents the temperature distributions over altitudes. As shown, the atmosphere temperature has nonmonotonic dependence on the altitude. At sea level the US standard atmosphere has a temperature of 288 K [1]. The first temperature minimum $T = 217$ K is located above the tropopause in the altitude range h = 11–20 km. Another temperature minimum $T = 187$ K is found in a range of the mesopause at altitudes h = 86–93 km. One can repeat that current average values of the atmospheric temperature may differ from a given value; that is, the data represented have a qualitative character.

Figures 2.2, 2.3, and 2.4 give the altitude dependencies for the pressure, air density, and density of air molecules, respectively, under conditions of standard atmosphere in accordance with [1–3]. It is convenient to introduce the typical size (scale) with respect to pressure Λ_p, density Λ_ρ, and number density of air molecules Λ_N on the basis of relations

$$\Lambda_p = \left(\frac{d\ln p}{dh}\right)^{-1}, \quad \Lambda_\rho = \left(\frac{d\ln \rho}{dh}\right)^{-1}, \quad \Lambda_N = \left(\frac{d\ln N}{dh}\right)^{-1}, \tag{2.1}$$

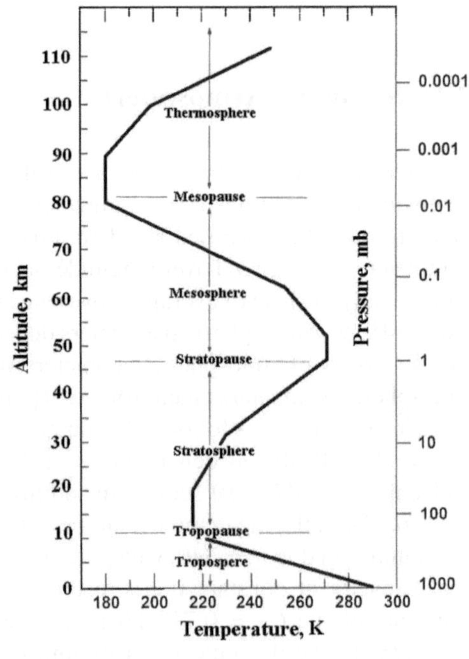

Fig. 2.1 Temperature distribution over altitude in the atmosphere with average parameters and notations of low atmosphere regions

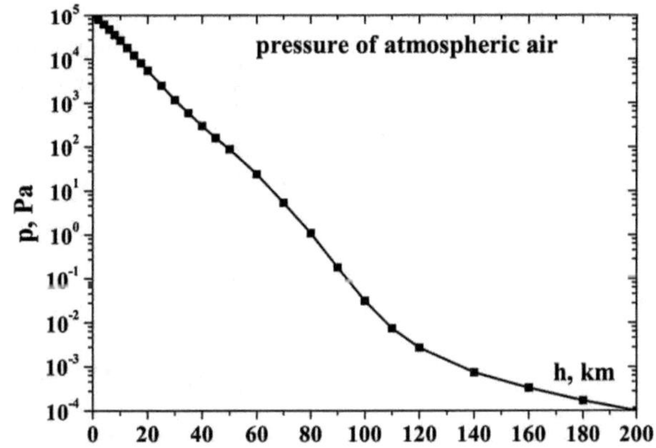

Fig. 2.2 Dependence of the air pressure p on the altitude for standard atmosphere

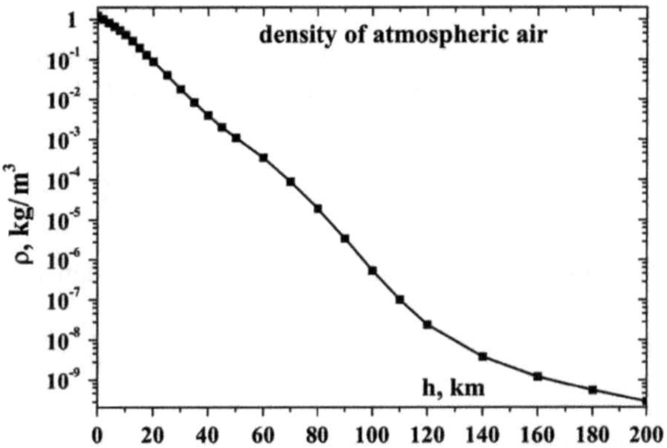

Fig. 2.3 Dependence of the air density ρ on the altitude h for standard atmosphere

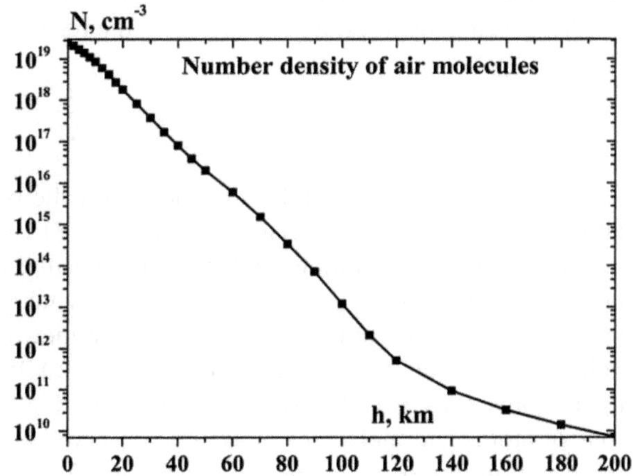

Fig. 2.4 Dependence of the number density of air molecules N on the altitude h for standard atmosphere

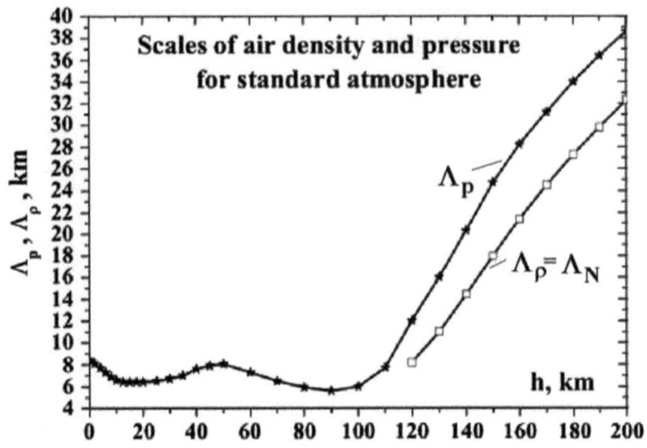

Fig. 2.5 Dependence of scales of the air pressure $\Lambda_p = (d \ln p/dh)^{-1}$ and the air density $\Lambda_N = (d \ln N/dh)^{-1}$ on the altitude h for standard atmosphere

These quantities describe the character of variation of corresponding parameters, in particular, for the air pressure near the altitude h_o, where it is $p(h_o)$, we have

$$p(h) = p(h_o) \cdot \exp\left(-\frac{h - h_o}{\Lambda_p}\right) \tag{2.2}$$

The same relation corresponds to the number density of molecules or atoms $N(h)$ at an altitude h

$$N(h) = N(h_o) \cdot \exp\left(-\frac{h - h_o}{\Lambda_N}\right) \tag{2.3}$$

In addition, one can use the definition of the mass density $\rho = \mathrm{Nm}$, where m is the air molecule mass, and we have $\Lambda_\rho = \Lambda_N$. The state equation for an ideal gas $p = NT$ given in appropriate units leads to the relation:

$$\frac{1}{\Lambda_p} = \frac{1}{\Lambda_N} + \frac{1}{\Lambda_T} \tag{2.4}$$

Figure 2.5 contains the altitude dependence for the parameter Λ_p and also for the parameter $\Lambda_N = \Lambda_\rho$ in accordance with data [1–3].

In the above consideration we define air molecules that have the average parameters of nitrogen and oxygen molecules. We now take into account that basic components of the Earth's atmosphere within the framework of the standard atmosphere include at sea level 78.08 % nitrogen molecules, 20.95 % oxygen molecules, 0.93 % argon atoms, 0.04 % CO_2 molecules, and 0.0018 % neon atoms. These concentrations of atoms and molecules in air are conserved at not high altitudes because of mixing of air components due to convection. The atmospheric composition varies at altitudes above 100 km, in the first place, due to dissociation of oxygen molecules under the action of solar radiation. Figure 2.6 represents the altitude dependencies for the number densities of nitrogen molecules $N(N_2)$, oxygen molecules

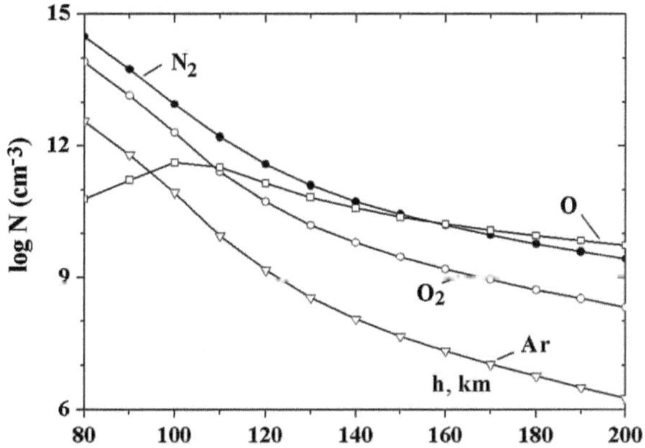

Fig. 2.6 Altitude dependence for the density logarithm ($\log_{10} N$) of the number densities of nitrogen and oxygen molecules, as well as the number densities of oxygen and argon atoms

$N(O_2)$, oxygen atoms $N(O)$, and argon atoms $N(Ar)$ in the upper atmosphere. As shown, the number density of oxygen atoms becomes equal to the number density of oxygen molecules at an altitude of about 100 km and to the number density of nitrogen molecules at an altitude of 155 km.

2.2 Air Transport in the Atmosphere

Heat transport in the lower atmospheric layers proceeds at altitudes 3–10 km in the form of air convection which is accompanied by processes of water condensation with formation of water aerosols [5]. Simultaneously, micron-size liquid aerosols and solid aerosols-ice particles-may be formed at such altitudes. In addition, atmospheric convection is of importance for these altitudes because of large atmospheric sizes and high atmospheric pressure. Convective motion is important for transport of air and water in the atmosphere and for atmospheric electric phenomena.

In the first place, transport of atmospheric air and water is important for the weather. Describing these processes by corresponding equations, one can expect the possibility to obtain atmospheric parameters at each point and a given time on the basis of values of these parameters at the beginning. Then one can predict the weather by solving the corresponding transport equations, given initial conditions on the basis of data at the beginning for a wide region of the globe. This concept of weather prediction at a given locality and in a given time as a result of numerical solution of transport equations has existed for almost a century (e.g., [6]), and the contemporary forecast is based on power computer codes. The authenticity of such a forecast decreases with increasing prediction time, and the maximum time is accepted as one week, a time during which airstreams on the troposphere boundary encircle the globe.

The uncertainty in the local weather forecast follows from the random character of atmospheric processes due to transport of heat and moisture. Indeed, dynamics of the Earth's atmosphere is determined by mass and energy transport [7–9], and a turbulent character of motion results from collision of air streams [10, 11], creating a random factor in the atmospheric development at a given locality. Such a character of air evolution in the atmosphere determines the climate [12, 13] and its change in time, as well as other atmospheric phenomena including the electric ones [14–20].

Thus, in predicting the weather and climate we are based on the values of initial atmospheric parameters and on the deterministic character of atmospheric processes. An error in such a description of atmospheric parameters due to random processes in the atmosphere is accumulated in time. As a result, deterministic weather prediction is not reliable for large periods of time. Empiric methods may be used for longer times. As for electric phenomena, their predictions may be more rough. For example, it is impossible to predict at what time a thunderstorm will be observed in a certain locality, and to which house a thunderbolt will come. Nevertheless, the above problems of variation of atmospheric parameters, including electric ones, are the object of meteorological investigations. Our analysis in this book related to global atmospheric phenomena is based on the concepts of standard atmosphere and the global electric circuit. This means that we use averaged parameters over the globe and time. In this consideration the information which is of interest for meteorology is lost. (In particular, the Earth's temperature equals 288 K and does not vary in time and over the globe). But such a description is convenient in the analysis of global processes that are the object of this book. Such a simplification allows us to extract the processes and mechanisms that determine a certain atmospheric property. In addition, analysis on the basis of averaged parameters has a qualitative character. Therefore in this analysis one can use estimations for some atmospheric parameters.

Convective air motion may be represented as a sum of individual vortices, and below we estimate the size of an elementary vortex in a middle troposphere assuming that convection results from the Rayleigh–Taylor instability [21–25] under the action of the gravitational force and temperature gradient. If conditions of the convective instability are fulfilled in a region between two heated plates, structures such as Bennard cells [26–28] are formed near the threshold of this instability. If a convection motion proceeds in an infinite space, such structures are mixed, leading to a random convective air motion in the form of vortices of different sizes. Let us estimate the minimal vortex size on the basis of the criterion that the Rayleigh–Taylor instability arises and develops, if the Rayleigh number is $Ra \sim 1000$ [21, 22, 29].

The Rayleigh number for a vortex of size l, that is, a size of the elementary air cell, has the form [21]

$$Ra = \frac{dT}{Tdh}\frac{gl^4}{\nu\chi} = \frac{gl^4}{\Lambda\nu\chi} \tag{2.5}$$

Here $g = 980\,\text{cm/s}^2$ is the freefall acceleration at the Earth's surface, $\nu = \eta/\rho = 0.3\,\text{cm}^2/\text{s}$ is the air kinematic viscosity, $\rho = 0.6\,\text{kg/m}^3$ is the air density, $\Lambda = Tdh/dT \approx 44\,\text{km}$ is a distance that characterizes a temperature variation in the troposphere, and we are guided by altitudes where the air pressure is $p = 0.5$ atm. Next, $\chi = \kappa/Nc_p = 0.4\,\mathrm{cm}^2/\mathrm{s}$ is the air thermal diffusivity, so that N is the number density of air molecules, and $c_p = 7/2$ is the heat capacity per molecule. This gives for a typical vortex size

$$l \approx \left(\frac{Ra \cdot \nu\chi\Lambda}{g}\right)^{1/4}, \qquad (2.6)$$

From this one can estimate the minimal vortex size in the atmosphere that is equal to $l \sim 20$ cm under the above conditions at altitudes where the pressure is $p = 0.5$ atm. This quantity depends on the number density of air molecules N as $l \sim 1/\sqrt{N}$, so that near the Earth's surface $l \sim 10$ cm.

One can estimate the velocity v_l of motion for a vortex of a size l from the Navier–Stokes equation

$$\eta\frac{\partial^2 v}{\partial h^2} \sim \eta\frac{v_l}{l^2} \sim PN', \qquad (2.7)$$

where η is the air viscosity, $P = mg$ is the molecule weight (m is the molecule mass, g is the free fall acceleration), $N' \sim Nl/\Lambda$ is the difference of the molecule number density at a distance l between two air layers, and N is the number density of air molecules ($N' \ll N$). From this we have for the vortex velocity v_l of a size l

$$v_l \sim \frac{mgNl^3}{\Lambda\eta}, \qquad (2.8)$$

and the diffusion coefficient for transfer of such vortices is estimated as

$$D_l \approx \frac{v_l l}{3} \sim \frac{mgNl^4}{3\Lambda\eta} \qquad (2.9)$$

Applying these results for the troposphere at altitudes approximately 5 km, where the air pressure is $p = 0.5$ atm, one can obtain there $v_l \sim 20$ cm/s, and the diffusion coefficient for air transport through vortices of a minimal size equals $D_l \sim 100\mathrm{cm}^2/\mathrm{s}$ for the troposphere. The Reynolds number is equal for this vortex size $Re = v_l l/\nu \sim 10^3$ that has the same order of magnitude for gasses as the Rayleigh number Ra [25].

Air transport over large distances that significantly exceed a size L of the largest elementary vortex proceeds through turbulent motion. Transition to the turbulent character of motion corresponds to the critical Reynolds number; that is, $Re_{cr} = v_L L/\nu = 2 \times 10^5$ [21]. From this one can obtain for the diffusion coefficient of atmospheric air through large vortices

$$D_L \approx \frac{Re_{cr}\nu}{3} \sim 2 \times 10^4\mathrm{cm}^2/\mathrm{s} \qquad (2.10)$$

In addition, Formulas (2.8) and (2.9) allow one to estimate the size of large vortices in accordance with

$$L \approx \left(\frac{3\Lambda \eta D_L}{mgN} \right)^{1/4} \tag{2.11}$$

Applying these formulas to the troposphere, as before, we obtain $L \sim 60$cm, $v_L \sim 10^3$ cm/s. Thus, we obtain roughly for the ratio of sizes for large and small vortices in the course of air convection in the troposphere $L/l \approx 3$. One can see that transport of atmospheric air and its components over large distances is realized through vortices of a maximum size, and the appropriate diffusion coefficient for tropospheric air is estimated as $D_L \sim 2 \times 10^4 \text{cm}^2/\text{s}$.

Thus, mass and energy transport in atmospheric air over large distances has a random character [7–9] that is created by collisions of air fluxes resulting from the mixing of air layers [10, 11]. Because of the random character of the transport process, one can characterize this transport by the diffusion coefficient (2.10). This diffusion coefficient varies with altitude, and Fig. 2.7 gives the turbulent diffusion coefficient for air of the standard atmosphere, where observational data for air transport are treated on the basis of the diffusion character of its motion.

In the regime realized in the troposphere we have the following dependence of parameters of the largest vortices on the number density N of air molecules

$$L \sim \frac{1}{\sqrt{N}}, \; D_L \sim \frac{1}{N}, \; v_L \sim \frac{D_L}{L} \sim \frac{1}{\sqrt{N}} \tag{2.12}$$

This similarity law holds true until the velocity v_L of vortices is small compared to the thermal velocity of air molecules or the sound speed c_s. This takes place in the low atmosphere

$$N \gg N_o, \tag{2.13}$$

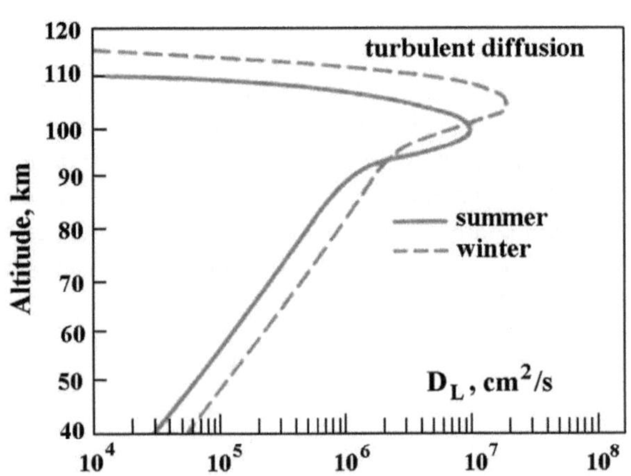

Fig. 2.7 Altitude dependence for the coefficient of turbulent diffusion [30, 31]

where $N_o \sim 10^{16}\text{cm}^{-3}$ which corresponds to altitudes $h < 55\,\text{km}$; that is, the regime under consideration is fulfilled in the mesosphere and lower layers.

In considering transport of air at higher altitudes, one can assume the velocity of the largest vortices to be equal to the sound speed $v_L \approx c_s \approx 3 \times 10^4\,\text{cm/s}$. Because the critical value of the Reynolds number $Re_{cr} = v_L L/\nu = 2 \times 10^5$ [21] corresponds to the largest vortices, one can obtain the similarity law instead of (2.12)

$$L \sim \frac{1}{N}, \quad D_L \sim \frac{1}{N}, \quad v_L = c_s, \tag{2.14}$$

and the largest vortex size is

$$L = \frac{Re_{cr}\nu}{c_s} \tag{2.15}$$

In particular, for the altitude $h = 100\,\text{km}$ this formula gives ($\nu = 30\text{m}^2/\text{s}$ [1]), which corresponds to $L \sim 10\,\text{km}$ as it follows from treatment of observational data (see Fig. 2.7). Next, Formula (2.10)

$$D_L \approx \frac{Re_{cr}\nu}{3} \tag{2.16}$$

gives at the altitude $h = 100\,\text{km}$ $D_L \sim 2 \times 10^6\text{cm}^2/\text{s}$. This value is less than that following from treatment of observational data $D_L \sim 1 \times 10^7\text{cm}^2/\text{s}$ [32]. Note that in this case $L \sim \Lambda$; that is, the diffusion character of air transport for large distances is violated.

In the above consideration convection is created by air molecules, that is, nitrogen and oxygen molecules. Therefore the average mass of these molecules ($m = 29$ in units of atomic mass) is included in Formulas (2.8) and (2.9). Other components and admixtures do not contribute to convective air motion. Collisions between molecules are important; for example, the mean free path of air molecules at the altitude 100 Km is 0.14 m [1], that is, small compared with values L and Λ responsible for convection. Therefore components of air are captured by its motion and are mixed with air. As a result, the concentration of air admixture does not depend on the altitude h up to $h \sim 100\,\text{km}$. One can add to this that transport of air in this consideration consists of two parts, namely, diffusion motion of air under the action of convection and drift motion of molecules due to gravitational forces. In this consideration displacement of molecules owing to gravitational forces is weak, so that one can neglect the action of the molecule weight on the convection process.

In order to give the criterion of the convection regime under consideration, where air vortices lead to mixing of molecules of air and its components, let us introduce an effective drift velocity w_L of a given component due to convection, which follows from the expression for a flux j of this component:

$$j = -D_L \nabla N = w_L N, \tag{2.17}$$

where N is the number density of particles for this component, and its drift velocity is

$$w_L \sim \frac{D_L}{\Lambda}, \tag{2.18}$$

For mixing of this component with air as a result of convective air motion, this drift velocity w_L must be large compared to the drift velocity w_g under the action of the gravitational force (particle weight). The latter is equal according to the Einstein relation [33–35]

$$w_g = \frac{Dmg}{T} \tag{2.19}$$

Here m is the mass of an admixture particle, $g \approx 980\text{cm}/\text{s}^2$ is the freefall acceleration, T is the air temperature given in energetic units, and D is the diffusion coefficient of a given atomic particle in air. Within the framework of the Chapman–Enskog approximation [36–38], the hard-sphere model for colliding atoms and molecules is given by [38, 39]

$$D = \frac{3\sqrt{\pi T}}{8\sqrt{2\mu} N_a \sigma_g}, \tag{2.20}$$

where $\mu = m_1 m_2/(m_1 + m_2)$ is the reduced mass of colliding particles of masses m_1 and m_2, N_a is the number density of air molecules, and σ_g is the gas-kinetic cross section for colliding atoms or molecules. Let us calculate the drift velocity for oxygen atoms located in the atmosphere. In this case from Formulas (2.13) and (2.14) we have (the gas-kinetic cross section at room temperature is $\sigma_g = 25\,\text{Å}^2$ [39])

$$N_a w_g = 1.5 \times 10^{13} \text{cm}^{-2} \text{s}^{-1}; \tag{2.21}$$

From this we have for the troposphere $w_L \sim 0.03$ cm/s, and $w_g \sim 10^{-6}$ cm/s, whereas at the altitude ~ 100 km which is roughly the boundary of the convective air motion, we obtain for air molecules $w_L \sim 3 \times 10^4$ cm/s, and $w_g \sim 1$ cm/s. Hence, the criterion

$$w_L \gg w_g, \tag{2.22}$$

that allows us to neglect the action of gravitational forces on the convective character of motion of air molecules in the Earth's atmosphere, holds true for air molecules.

2.3 Atmospheric Water

Water is an important component of the Earth's atmosphere, in the first place, because electrical atmospheric processes accompany water circulation in the Earth's atmosphere [40]. Within the framework of the global circuit model for atmospheric electricity [41], where the Earth and its atmosphere are considered as a whole in some electric scheme, we represent parameters of water and transport through the atmosphere to be averaged over time and to be summed over the globe. The rate of

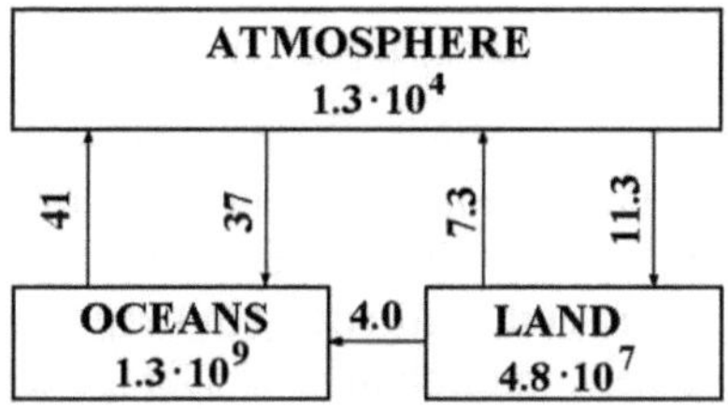

Fig. 2.8 Rates of water transitions between oceans, Earth's land, and the atmosphere expressed in units $10^4 km^3/year$ are indicated at corresponding *arrows*. An amount of water in these species is given in km^3 inside *rectangles*

water evaporation from the Earth's surface and its passage through the atmosphere is on average $4.8 \times 10^5 km^3/year$ [13, 42, 43] and the same water amount returns to the Earth in the form of precipitation. Here we take the volume of liquid water as a unit, as usual. Transferring to standard units, we obtain for the rate of water evaporation from the Earth's surface to the atmosphere as well as the rate of water precipitation on the Earth's surface from the atmosphere as $4.8 \times 10^{20} g/year$ or 1.5×10^{13} g/s. An average content of water in the atmosphere is $1.3 \times 10^{19} g$ [44–47], compared to the total air mass $5.1 \times 10^{21} g$ which is located in atmospheric air on average. From this one can find the residence time of water in the atmosphere which is approximately 9 days. If atmospheric water in the liquid state is precipitated uniformly over the Earth's surface, it forms a layer of the depth of 2.5 cm [48]. In addition, the Earth contains approximately $1.4 \times 10^{24} g$ of water, and 96 % of this water is saline. This includes approximately $1.0 \times 10^{22} g$ of fresh groundwater. Figure 2.8 shows the rates of exchange by water between the atmosphere, oceans, and land, as well as the water amount inside each component. Note also that the rate of snow precipitation on the Earth's surface is $1.0 \times 10^3 km^3/year$ [49] compared to $4.8 \times 10^5 km^3/year$ for liquid water. The accuracy of the above data is 20 %, which follows from comparison of data from different sources.

As follows from the above parameters, the average water vapor mass density in the atmosphere near the Earth's surface is $3g/m^3$ in comparison with the air mass density $1.2 kg/m^3$ for dry air. An indicated water-specific density in air corresponds to the partial water pressure $2Torr$ near the Earth's surface. The saturated vapor pressure of water at the temperature 0^oC is $4.7Torr$ [4, 50]. One can conclude from this that water is located in the Earth's atmosphere in the form of molecules, and condensation of water vapor in regions with a low temperature leads to formation of aerosols, micron-size water particles. Because the condensation process proceeds at temperatures near the water melting point, water may be present in the atmosphere in the form of aerosols where water is found in any aggregate state.

In considering the atmospheric electric current as a secondary phenomenon of water circulation in the atmosphere, we find the ratio of the atmospheric electric current due to lightning ($I = 1700A$ [17, 51]) to the transferring mass through the atmosphere per unit time $dM/dt = 1.5 \times 10^{13}$ g/s, that gives the following specific transferring mass [52, 53]

$$\frac{I}{dM/dt} = 1.2 \times 10^{-10} \frac{C}{g} \tag{2.23}$$

We are guided by this value in the analysis of atmospheric electric processes.

The temperature of atmospheric air decreases with increasing altitude up to the tropopause [47] that is located at the altitude 9 km for poles and 17 km for the equator where the temperature has a minimum. The average temperature of the Earth's surface is $15^o C$, the tropopause temperature is $-55^o C$, and the temperature gradient is −6.5 K/km [54]. From this one can see that water condensation proceeds at altitudes, at least, above 2–3 km. Forming micron-size aerosols are liquid, snow, or ice microparticles, and, possibly, a mixture of these forms.

Table 2.1 [4] contains the temperature dependence for the water-saturated vapor pressure and other water parameters, so that water microparticles are formed at pressures which exceed the saturated vapor pressure at a given temperature. Because only water aerosols are responsible for electrical atmospheric phenomena, these processes take place if the water content exceeds a certain threshold. In particular, it is usually accepted that thunderstorm weather is possible if the water vapor mass mixing ratio exceeds 7 g/kg (grams of water per kilogram of air) [55]. This water content corresponds to the saturated vapor pressure at the temperature 8°C.

If the partial water pressure exceeds the saturated vapor pressure, its excess is transformed in a condensed phase. This process in a space starts from formation of nano-size water clusters as a result of attachment of water molecules to condensation nuclei, that is, to ions or chemically active atomic particles, in particular, to radicals. Subsequent growth of small water particles by attachment of water molecules leads to formation of water aerosols, that is, micron-size water particles. Water aerosols may be found in different aggregate states depending on the air temperature where their growth takes place. Namely, water aerosols may be liquid drops, ice particles, snow particles, and a mixture of snow and water, that is, graupels. Aerosols in atmospheric air constitute clouds of various types. In considering processes involving aerosols, we are guided by aerosols in cumulus clouds which take part in atmospheric electricity and finally cause the origination of lightning. A mature cumulus cloud contains aerosols with the following averaged parameters [56–59].

$$r = 8\mu m, \; N_p = 10^3 \mathrm{cm}^{-3}, \tag{2.24}$$

where r is the average aerosol radius, and N_p is the average number density of aerosols in the cloud. Note that the average density of water in the atmosphere is $2g/m^3$, and this corresponds to the specific water mass 1.6 g/kg under standard conditions (the temperature $0^o C$ and pressure 1 atm). If we take these aerosols from the atmosphere and transform them in liquid water, the ratio of the thickness of the water layer h to the cloud thickness L is $h/L = 2$mm/km.

Table 2.1 Saturated water vapor in atmospheric air [4]: p_{sat} is the water-saturated vapor pressure at an indicated temperature, N_{sat} is the number density of water molecules in a saturated vapor, and A_s is the ratio of a mass of saturated water vapor to an air mass located in a given volume at atmospheric pressure

$T, ^oC$	T, K	p_{sat}, KPa	N_{sat}, 10^{17}cm^{-3}	A_s, g/kg
–20	253	0.103	0.295	0.631
–15	258	0.165	0.463	1.01
–10	263	0.260	0.716	1.59
–5	268	0.402	1.09	2.46
0	273	0.611	1.62	3.74
5	278	0.873	2.27	5.35
10	283	1.23	3.14	7.52
15	288	1.71	4.29	10.4
20	293	2.34	5.78	14.3
25	298	3.17	7.70	19.4
30	303	4.25	10.2	26.0

It is convenient to introduce a number of molecules in a spherical aerosol; that is,

$$n = \left(\frac{r}{r_W}\right)^3, \tag{2.25}$$

where r_W is the Wigner-Seits radius [60, 61] which in the water case is equal to $r_W = 1.92$Å [62]. Correspondingly, the number of molecules in a typical cumulus cloud (2.24) is equal to $n = 7 \times 10^{13}$. This corresponds to the number density of bound water molecules in aerosols of $7 \times 10^{16} cm^{-3}$ and to the concentration of bound water molecules in aerosols of 0.27 % at zero temperature, or to the water content in the atmosphere of 1.7 g/kg at zero temperature. From this it follows that equilibrium between aerosols of clouds with parameters (2.24) and ambient wet air at zero temperature 0^oC takes place at the total water content in air 5.4 g/kg. In this case approximately 31 % of water molecules are bonded and 69 % of water molecules are free.

In considering the character of aerosol motion in the atmosphere, note that the criterion (2.22) may be violated for micron-size aerosols. This means the absence of the total mixing of water aerosols with air that can lead to a stronger decrease of the aerosol number density with increasing altitude compared to that of air molecules. We first analyze the character of aerosol motion in air vortices. Guided by micron-size water aerosols, determine the relaxation time τ_{rel} of the aerosol velocity to that of air. This relaxation time is given by [63]

$$\tau_{rel} = \frac{2\rho r_o^2}{9\eta}, \tag{2.26}$$

where $\rho = 1g/\mathrm{cm}^3$ is the mass water density, and $\eta = 1.85g/(\mathrm{cm}\cdot \mathrm{s})$ is the air viscosity coefficient at room temperature. One can compare it with a typical time τ_L of velocity variation in large vortices. Restricting ourselves to large vortices that determine the air transport due to convection, we have $\tau_L \sim L/v_L \sim 0.06s$ for large vortices in tropospheric air. On the other hand, we have $\tau_{rel}/r_o^2 = 1.2 * 10^3 s/\mathrm{cm}^2$ on the basis of formula (2.26). From comparison it follows that the criterion $\tau_{rel} \ll \tau_L$ holds true for aerosol sizes $r_o \ll 70\mu m$. Because in the analysis of atmospheric clouds, the size range $r_o \sim 1 - 10\mu m$ is of interest, such aerosols are captured by air vortices in the troposphere. In addition, the mean free path of air molecules $\lambda \sim 0.1\mu m$ is small compared to an aerosol radius r_o, that justifies the viscosity character of relaxation of the aerosol velocity used above.

We now analyze the distribution of the aerosol number density over altitudes. The velocity of gravitational fall of aerosols is given by [21]

$$w_g = g\tau_{rel} = \frac{2\rho g r_o^2}{9\eta}, \tag{2.27}$$

and for water aerosols this formula has the form $w_g/r_o^2 = 1.2 \times 10^6 s^{-1}\mathrm{cm}^{-1}$. Criterion (2.22) is fulfilled for aerosols of a size $r_o \ll 2\mu m$, and the altitude distribution function (2.3) is fulfilled for such aerosols. Let us introduce the altitude distribution of the number density of aerosols $N_a(h)$ near an altitude $h_o$in the form

$$N_a(h) = N_a(h_o)\cdot \exp\left(-\frac{h-h_o}{\lambda_a}\right) \tag{2.28}$$

Evidently, in the limiting case $w_L \ll w_g$ we have $\lambda_a \sim \Lambda w_L/w_g$. One can approximate a typical altitude variation λ_a for the aerosol number density

$$\lambda_a = \frac{\Lambda}{1 + w_g/w_L} \tag{2.29}$$

In particular, from this it follows that λ_a ~2 km for aerosols of a size $r_o = 8\mu m$, that is, for an average aerosol size (2.24) in cumulus clouds. As shown, such aerosols do not penetrate in high atmospheric layers. This means that the concentration of free water molecules drops with increasing altitude. In particular, Fig. 2.9 contains the altitude dependence of the concentration of water molecules which confirms the above statement. Note that the concentration of water molecules at the Earth's surface in the measurements used [64] exceeds several times that of the standard atmosphere.

2.4 Carbon Dioxide in the Earth's Atmosphere

The greenhouse effect for the gas layer located over a solid surface takes place if this layer is transparent for a visible spectrum range and is not transparent for infrared radiation. Therefore solar radiation goes through this layer, whereas infrared is absorbed

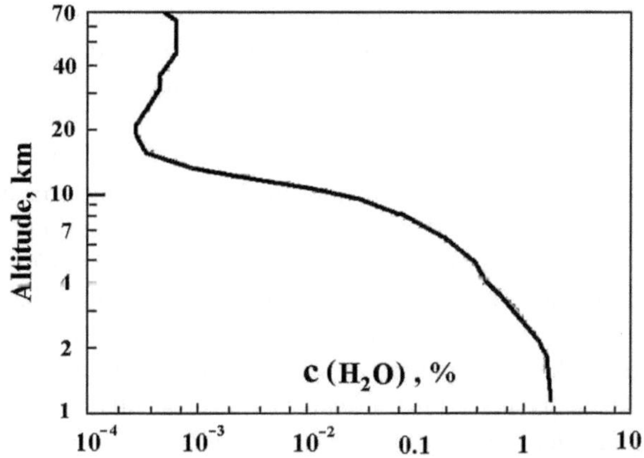

Fig. 2.9 Concentration of free water molecules (the ratio of the number density of water molecules to the number density of air molecules) as an altitude function according to measurements [64]

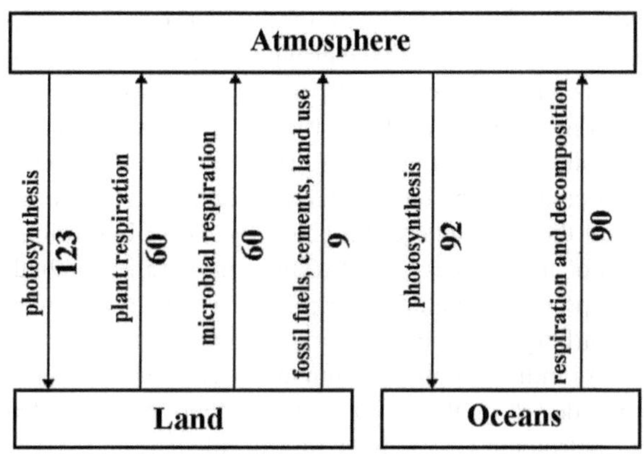

Fig. 2.10 Carbon fluxes through the Earth's atmosphere in the carbon cycle in the form of carbon dioxide. The fluxes are given in 10^9 tons of carbon per year

by it. The atmospheric greenhouse effect is determined by atmospheric water in the form of free water molecules and aerosols. In addition to this, the greenhouse effect is determined by carbon dioxide because the most intense absorption band of the CO_2 molecule at $15\mu m$ corresponds to the absorption center for thermal emission of the atmosphere. Absorption by water in the atmosphere takes place at altitudes below 10 km, and then the concentration of water molecules drops strongly with increasing altitude. On the contrary, CO_2 molecules are mixed with those of air because of their large lifetime in the atmosphere, and their concentration is almost constant over a space far from sources and absorbers of carbon dioxide. Below we consider some aspects of accumulation of carbon dioxide in the atmosphere and its influence on the Earth's energetic balance.

Figure 2.10 gives the channels of the carbon cycle in the atmosphere on the basis of data [65–67]. Note that values of carbon fluxes of some channels are different for different sources, but in any case the total carbon flux into the atmosphere is larger than that towards the Earth. But in ignoring carbon fluxes due to human activity, we obtain the opposite relation between these fluxes. Hence there is accumulation of carbon dioxide in the atmosphere, but the content of atmospheric CO_2 would be decreased if fossil burning would be stopped. One can add to Fig. 2.10 data that now the atmosphere contains approximately $8 \times 10^{17} g$ of carbon in the form of CO_2 molecules. For comparison, the average mass of atmospheric water is 1.3×10^{19} g, and the mass of atmospheric air equals $5.1 \times 10^{21} g$. Atmospheric CO_2 molecules are mixed with air molecules, and therefore the concentration of these molecules in air is almost constant far from sources and absorbers of carbon dioxide.

As follows from Fig. 2.10, the accumulation of CO_2 in the atmosphere results from human activity. If this channel of CO_2 injection will be stopped, the CO_2 content in the atmosphere would be decreased. Photosynthesis is the main process of CO_2 removal from the atmosphere. This is a complex process where chloroplasts of plants are the catalysts [68] that transfer CO_2 in various forms of bound carbon. The complex character of photosynthesis [69–74] follows from a specific character of the process depending on the plant. One can divide plants in $C3$ and $C4$ groups [75] in accordance with a number of carbon atoms that partake of an elementary photosynthesis process. In the case of plants of the group $C3$, such as wheat, rice, and haricot, a CO_2 molecule reacts with a intermediate product, whereas in the case of plants of the group $C4$, a CO_2 molecule is captured by a cell, and subsequently bound molecules partake in the photosynthesis process. This leads to different dependencies on conditions, in particular, on the CO_2 concentration [76]. Another channel of the carbon transition from the atmosphere to oceans by CO_2 dissolution leads to formation of calcium carbonates such as $CaCO_3$. The carbon transition in the atmosphere results from natural processes including respiration of plants and microbes, as well as decomposition of carbonates in oceans.

As indicated above, human activity leads to an increase of atmospheric carbon dioxide. Roughly, 35 % of CO_2 ejection results from coal burning, 36 % relates to petroleum, 20 % falls to natural gas, and 3 % of CO_2 ejections proceeds in cement production [77]. Deforestation, that is, annihilation of tropic forests in South America, Africa, and Asia is of importance for climate change [78–80], inasmuch as approximately $1.5 \times 10^5 \mathrm{km}^2$ of forests are lost each year in tropical regions. But it is difficult to determine the character of the influence of deforestation on the climate. If an unfettered land is used for agriculture, the photosynthesis process takes place here, but in the other form. On the other hand, the deforestation process changes evaporation of water for an unfettered area and its influence on the climate is not unambiguous. For this reason in the subsequent analysis we are not based on carbon fluxes through the atmosphere.

One can see from the diagram of Fig. 2.10 that if we exclude human activity in the carbon balance, the amount of atmospheric CO_2 will be decreased, whereas in reality human activity leads to CO_2 accumulation in the atmosphere. Indeed, if we exclude from the carbon balance the channels due to human activity, the equilibrium content of atmospheric CO_2 molecules would be lower than its contemporary value. In particular, prior to 1750 where human activity made little contribution to the carbon cycle, the concentration of CO_2 molecules in the atmosphere ranged from 260 ppm up to 280 ppm for approximately 10,000 years [81]. We note also that now the Earth's atmosphere contains approximately $8 \times 10^{17} g$ of carbon in the form of

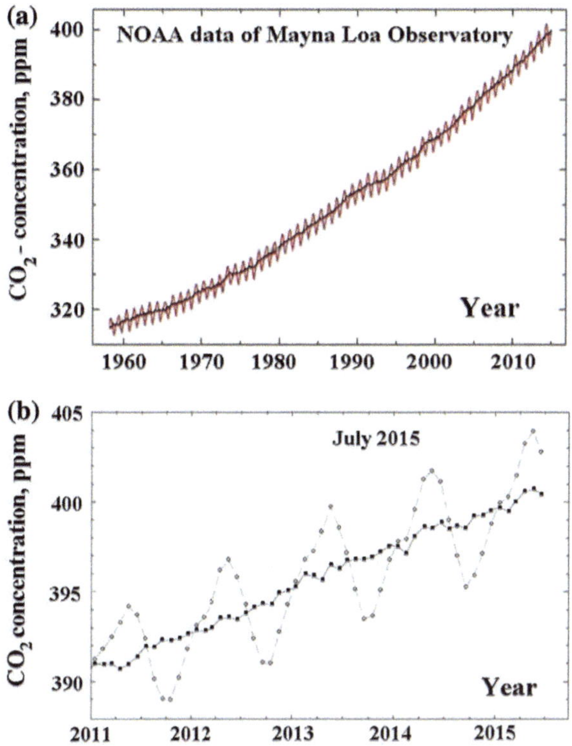

Fig. 2.11 Concentration of carbon dioxide in atmospheric air according to measurements at the Mauna Loa Observatory [82] **a** during the last 50 years, **b** during the last few years. *Open circles* relate to averaging during a month; *closed squares* correspond to the year averaging

CO_2 molecules. As follows from the diagram in Fig. 2.10, the residence time for CO_2 molecules in the atmosphere is estimated as approximately (4–5) years. During this time CO_2 molecules are mixed with air molecules in the atmosphere.

The basic information about the contemporary concentration of atmospheric carbon dioxide follows from measurements at the Mauna Loa Observatory (Hawai, USA) [83]. This observatory is located at an altitude of 3400 m above sea level and gives a stable concentration of CO_2 molecules. The monitoring of the atmospheric content of carbon dioxide has taken place since 1958 and the results of this monitoring are given in Fig. 2.11. According to the data of Fig. 2.11a, the concentration of CO_2 molecules varies from $316ppm$ in 1959 up to $402ppm$ in 2015 ($1ppm \equiv 10^{-6}$). In addition, the rate of CO_2 accumulation in the atmosphere varies from $0.7ppm$ per year in 1959 up to $2.1ppm$ per year in 2015. Seasonal variations of the CO_2 concentration results from a more intense photosynthesis process in the northern hemisphere. As a result, from May to September where the photosynthesis process proceeds in the northern hemisphere, the concentration of atmospheric carbon dioxide decreases.

The analysis of carbon dioxide content in glacier deposits for the concentration of atmospheric CO_2 molecules represented in Fig. 2.12, gives additional information along with the monitoring of the contemporary evolution of atmospheric carbon dioxide. Leaving aside the periodic variation of this value in time, one can estimate

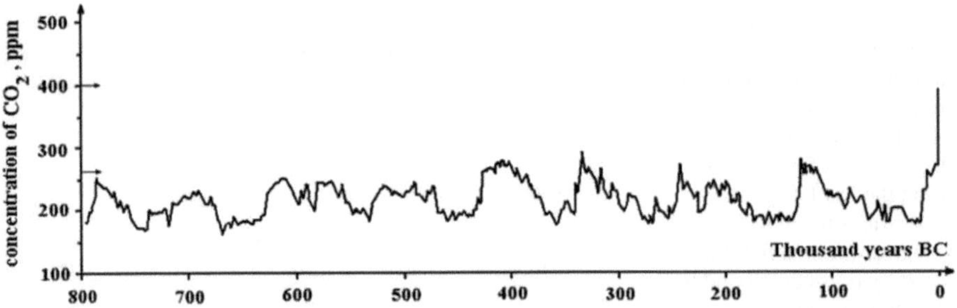

Fig. 2.12 The concentration of atmospheric CO_2 that is restored from the analysis of the glacier ice [84]

the carbon dioxide concentration in preindustrial times to be below 260*ppm*. Variation of conditions on Earth, including that under human activity, leads to a change of the Earth's parameters and its climate. In particular, since 1880 the global Earth temperature increased by 0.8 K [85]. Often this is connected with an accumulation of CO_2 molecules in the atmosphere. One can check this statement because CO_2 is a simple molecule with a simple spectrum. In addition, atmospheric CO_2 molecules are mixed with air molecules in the atmosphere allowing one to measure their concentration as well as the concentration variation for CO_2 molecules in time. Therefore one can determine reliably the variation of the radiation power that is emitted by CO_2 atmospheric molecules. Hence, one can connect an accumulation of CO_2 in the atmosphere with the greenhouse effect [86, 87]. We consider this problem below.

Let us make some estimations on the basis of the above data. According to Fig. 2.11b, the contemporary increase 2.1*ppm*/*year* for the atmospheric concentration of CO_2 molecules means that the contemporary increase of carbon amount in the atmosphere is 4.4×10^9 ton per year, that is less than the manmade amount (see Fig. 2.10). Next, from Fig. 2.11a it follows that the concentration of atmospheric CO_2 molecules decreases from 402*ppm* to 395*ppm* from May to September 2014 and then increases from 395*ppm* to 404*ppm* from September 2014 to May 2015. This corresponds to the transfer of 15×10^9 ton of carbon from the atmosphere in the first case and the transfer of 19×10^9 ton of carbon to the atmosphere in the second case. These amounts exceed remarkably those accumulated in the atmosphere as a result of human activity. Thus, contemporary measurements allow one to determine reliably a current concentration of atmospheric CO_2 molecules and evolution of this value in time.

One can consider two limiting cases in the photosynthesis process. In the first limit that corresponds to the $C3$ group the photosynthesis rate is proportional to the CO_2 atmospheric content; the second limit is typical of the $C4$ group where this rate is determined by the capacity of plants for capture of CO_2 molecules. Ejection of carbon in the atmosphere in the form of CO_2 as a result of combustion of fossil fuels (oil, coal, natural gas) gives 9–10 billion tons of carbon. As can be seen, the second limit takes place in reality. This means that the photosynthesis rate depends on the mass of Earth plants. Therefore deforestation is of importance for carbon balance.

2.5 Energetics of Earth and Atmospheric Processes

Energetics of the atmosphere is one of its global properties. In its analysis we consider the Earth and its atmosphere as individual objects that exchange by energy both with each other and with a surrounding space. Operating with the total powers obtained by the Earth and its atmosphere due to various processes, we base the following on the data [88] for powers of these processes. Note that energetic parameters of these processes on the basis of sources [89–92] differ from those by 1–2 %. Table 2.2 and Fig. 2.13 contain the total powers of basic energetic processes related to the Earth and its atmosphere which start from solar radiation that is absorbed by the Earth and its atmosphere or is reflected back by them. This solar radiation in a visible spectrum range is transformed into thermal radiation in a infrared spectrum range that is absorbed and emitted by the Earth and its atmosphere. It should be noted that the atmosphere in this consideration may be separated as an independent object because its optical thickness for infrared radiation is large. This allows one to represent the energetic balance independently for the Earth, the Earth's atmosphere, or the object consisting of the Earth and its atmosphere, as represented in Table 2.3.

As shown, the basis of the energy budget of the Earth and its atmosphere is solar radiation. The Sun emits radiation of the power $3.86 \times 10^{26} W$ on average [93] and this radiation corresponds to the visible and infrared spectrum, whereas ultraviolet radiation, radio waves, and X-rays make a contribution below 1 % to the Sun's power. The radiation flux at the Earth level (named the solar constant) or the total solar irradiance is equal on average to $1365 W/m^2$ [94–97] that corresponds to the effective temperature $T_{ef} = 5777$ K of solar radiation. This value is accepted as a constant, however, it varies in time [98, 99] because of a nonstationarity of solar processes, although these variations are less than percent. For example, at solar

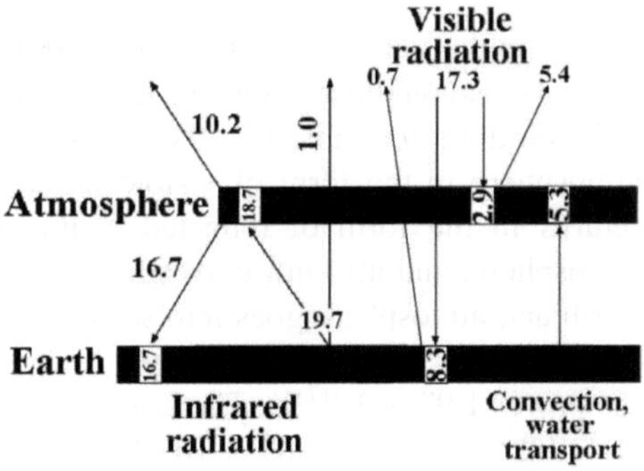

Fig. 2.13 Powers obtained and lost by the Earth as a whole and its atmosphere in various channels. All the powers are expressed in $10^{16} W$. Absorbed powers are given inside corresponding *rectangles*; emitted or scattered powers are indicated near *arrows*

Table 2.2 Powers of the Earth's processes P summarized planetwide

Form of Energy	Process	Power, P, W
Sun's radiation	reached Earth's atmosphere	1.73×10^{17}
	reflected by the Earth	7×10^{15}
	absorbed by the Earth	8.3×10^{16}
	reflected by the atmosphere	5.4×10^{16}
	absorbed by the atmosphere	2.9×10^{16}
Infrared radiation	atmospheric emission to space	1.02×10^{17}
	atmospheric emission to Earth	1.67×10^{17}
	absorbed by the atmosphere	1.87×10^{17}
	absorbed by the Earth	1.67×10^{17}
	emitted by the Earth	1.97×10^{17}
	transmitted through the atmosphere	1.0×10^{16}
Air motion	air convection in troposphere	1.3×10^{16}
Release of water molecules	water evaporation from Earth surface	3.5×10^{16}
Ionization by solar hard radiation	ionization of molecules and atoms of the upper atmosphere	7×10^{10}
Heat transfer	thermal conductivity of air in troposphere	2×10^{11}
Kinetic energy of particles	solar wind outside atmosphere	5×10^{10}
Electric energy	lightning	1×10^{11}
Tropospheric ionization	ionization of air molecules by cosmic rays	2×10^{8}
Electric energy	ion drift in troposphere	5×10^{8}

Table 2.3 Summary Powers (in $10^{16}\,W$) in the power balance of the Earth and the Earth atmosphere as a whole

	Income	Loss
The Earth and atmosphere	$8.3 + 2.9 + 0.7 + 5.4 = 17.3$	$0.7 + 5.4 + 1.0 + 10.2 = 17.3$
The Earth	$8.3 + 16.7 = 25.0$	$8.3 + 16.7 = 25.0$
The atmosphere	$2.9 + 18.7 + 5.3 = 26.9$	$16.7 + 10.2 = 26.9$

minimum this value was $1361\,W/m^2$ [100]. Of course, the value of the solar constant depends also on the accuracy of its radiation [101].

According to Table 2.2, the power of $1.73 \times 10^{17}\,W$ penetrates the Earth's atmosphere in the form of visible radiation. A part of this power, $6.1 \times 10^{16}\,W$, returns in the form of reflected visible radiation from the Earth's surface and atmosphere, and also infrared radiation of the power $1.02 \times 10^{17}\,W$ emitted by the Earth and atmosphere goes into surrounding space. Next, the Earth absorbs visible and infrared radiation of the power of $2.5 \times 10^{17}\,W$ and emits infrared radiation of the power $1.97 \times 10^{17}\,W$. The Earth's atmosphere emits infrared radiation of powers 1.02×10^{17} and $2.69 \times 10^{17}\,W$ towards the Earth and outside it. In addition, let us divide the power of $5.3 \times 10^{16}\,W$ which is transmitted from the Earth to its atmosphere in the form of convection and water evaporation. One can find the power P_{ev} of this value due to water evaporation from the Earth's surface according to the formula

$$P_{ev} = \frac{\varepsilon}{m_w}\frac{dM}{dt}, \tag{2.30}$$

where $\varepsilon = 0.43eV$ is the energy consumed on evaporation of one molecule, and m_w is the mass of a water molecule, and $dM/dt = 1.5 \times 10^{13}$ g/s is the evaporation rate of water from the Earth's surface. From this one can obtain for the power $P_{ev} = 3.5 \times 10^{16} W$ that provides the observed rate of water evaporation from the Earth's surface. This power may be smaller if evaporated water molecules form bonds with salt molecules and a part of water returns to the Earth in the form of free water molecules rather than of bound water molecules in rain, mist, haze, or fog. From this we have for the power $P_h = 1.9 \times 10^{16} W$ due to heat flux from the Earth through the atmosphere. One can define the thermal conductivity coefficient κ_{con} by the formula

$$P_h = -S\kappa_{con}\frac{dT}{dh}, \tag{2.31}$$

where $S = 5.1 \times 10^{18}\text{cm}^2$ is the area of the Earth's surface, and the average observed temperature gradient for a low atmosphere is $dT/dh = 6.5$ K/km. From this we obtain for the thermal conductivity coefficient $\kappa_{con} = 40\text{W}/(\text{cm}^2 \cdot \text{K})$ of the atmosphere due to convection compared with that of motionless air $\kappa_{air} = 2.6 \times 10^{-4} W/(\text{cm}^2 \cdot K)$ [102]. We conclude once more that the transport of mass and heat results from the convection motion of air.

Solar energy heats the Earth. Assuming that the Earth emits radiation as a blackbody of a temperature T, one can connect it with the radiation power $P_{em} = 1.97 \times 10^{17} W$ of the Earth in accordance with the Stephan-Boltzmann law

$$P_{em} = S \cdot \sigma T^4, \tag{2.32}$$

where S is the area of the Earth's surface and $\sigma = 5.67 \times 10^{-12} W/(\text{cm}^2 \cdot K^4)$ is the Stephan-Boltzmann constant. From this one can find the Earth's temperature $T = 287$ K which is averaged over time and geographic coordinates. Note that the Earth's temperature within the framework of standard atmosphere conditions [1] is equal to $T = 288$ K (see also Fig. 2.1). This analysis also testifies to the accuracy of using the averaged data over the globe.

Along with the basic processes that are responsible for energetics of the Earth and its atmosphere, weak energetic processes may be of importance because they determine some atmospheric properties. Powers of some processes of such a type are given in Table 2.2. Some atmospheric processes, such as air transport in the form of winds and convection, and also evaporation of water from the Earth's surface and water reservoirs, are induced by solar radiation. Only a small part of solar radiation is emitted in the vacuum ultraviolet spectrum with wavelengths of 100 nm and below. This radiation does not reach the Earth's surface, but it is important because it creates ionization of atmospheric molecules and atoms. The average photon flux with wavelengths below 100 nm is equal approximately to $2.4 \times 10^{10} \mathrm{cm}^{-2} s^{-1}$. This photon flux corresponds to the power $7 \times 10^{10} W$ for solar radiation in the vacuum ultraviolet spectrum that penetrates the upper layers of the Earth's atmosphere. Of course, this power is very small compared to that of radiation in the visible spectrum. Note the vertical gradient of the atmospheric temperature. For a standard atmosphere it is equal to $dT/dh = -6.5$ K/km (h is the altitude), as follows from Fig. 2.1. In a quiet atmosphere this creates the heat flux q due to air thermal conductivity that is equal to $q = \kappa |dT/dz| = 3 \times 10^{-8} W/\mathrm{cm}^2$. Summarized over the globe, this flux corresponds to the heat power $P_{th} \approx 1 \times 10^{11} W$. This is smaller than the power of air convection by three orders of magnitude, if we assume that air convection results from the heat transfer.

We also include in Table 2.2 some electrical processes in which the power is small compared to the power of radiative processes. In particular, in considering ionization of the troposphere under the action of cosmic rays, one can assume the energetic cost of one electron-ion pair to be 30 eV. Because the total rate of formation of electrons and ions per unit of an atmospheric column under the action of cosmic rays is $4.5 \times 10^7 \mathrm{cm}^{-2} s^{-1}$ [103], one can estimate from this the energy flux due to atmospheric ionization by cosmic rays as $2 \times 10^{-10} W/\mathrm{cm}^2$. Assuming this flux to be created by the Sun, one can find from this the total power of cosmic rays causing this ionization over the globe as $\sim 1 \times 10^9 W$.

The solar wind, a plasma flux from the solar corona [104–106], may be considered as one form of cosmic rays. Note that charged particles of the solar wind, electrons and protons, cannot penetrate the Earth's atmosphere (except in the polar regions) due to the action of the Earth's magnetic field. A typical velocity of the solar wind is $v = 4 \times 10^7$ cm/s, the energy per charged particle is $\varepsilon = 1 keV$, and the number density of electrons or ions is $N_i = 7 \mathrm{cm}^{-3}$. This gives for the power P_{sw} of the solar wind per the Earth's cross section $\pi R_\oplus^2$ ($R_\oplus$ is the Earth radius) as

$$P_{sw} = 2 N_i v \varepsilon \pi R_\oplus^2 \sim 10^{11} W \quad (2.33)$$

Of course, it is small compared to the power of solar radiation in the Earth's atmosphere.

The power of electrical processes as secondary processes of water circulation are small compared to radiative processes and processes involving water transport in the atmosphere. As a result of electrical processes, the electric current occurs with a strength $I = 1700 A$ with transfer of a negative charge to the Earth's surface. Transport of electricity through lightning gives the main contribution to the electric energy of the Earth's atmosphere. The electric potential of the lower edge of a cumulus cloud with respect to the Earth's surface is 20–100 MV [107], and taking its average value $U = 60 MV$, one can obtain for the power contained in lightning $P_l = UI = 1 \times 10^{11} W$. This value is given in Table 2.2. Another type of electrical energy is responsible for ion transport that provides the Earth's discharge. Because the electric potential of the upper layers of the troposphere is $U_o = 240$–300 kV [108–110], the power consumed on ion transport is $P_{dr} = U_o I = 5 \times 10^8 W$, and this value is represented in Table 2.2. One can add to this the power due to contemporary energetics. The mass of fossil fuels in 2015 is equivalent to approximately 13 billion of oil per year. Combustion of these fossil fuels corresponds to the power of approximately 1.8×10^{10}W, and this is small compared to the powers of natural channels.

References

1. U.S. Standard Atmosphere (U.S. Government Printing Office, Washington, 1976)
2. K.S.W.Champion, A.E.Cole, A.J.Kantor, in *Chemical dynamics in extreme environments.*, ed. by R.A. Dressler (World Scientific Publishing, Singapore, 2001)
3. M. Davies, *The Standard Handbook for Aeronautical and Astronautical Engineers* (McGraw-Hill, New York, 2003)
4. Handbook of Chemistry and Physics, ed. by 86, D.R. Lide (CRC Press, London, 2003–2004)
5. http://en.wikipedia.org/wiki/Thunderstorm
6. L.F. Richardson, *Weather Prediction by Numerical Processes* (Cambridge University Press, Cambridge, 1922)
7. E. Lorenz, *The Nature and Theory of General Circulation of the Atmosphere* (World Metheorological Organization, Geneva, 1967)
8. D.G. Andrews, J.R.Holton, C.B.Leovy, *Middle Atmospheric Dynamics.* (Academic Press, 1987)
9. D.G. Andrews, *An Introduction to Atmospheric Physics* (Cambridge University Press, Cambridge, 1999)
10. G.T. Csanady, *Turbuletnt Diffusion in Environment* (Holland, Reidel, Dordrecht, 1973)
11. D.J. Tritton, *Physical Fluid Dynamics* (Claredon Press, Oxford, 1988)
12. L.T.Matveev, *Fundamentals of General Metheorology: Physics of the Atmosphere.* (Israel Programme for Scientific Translations, Jerusalem, 1967)
13. J.P. Peixoto, A.H. Oort, *The Physics of Climate* (The American Institute of Physics, New York, 1992)
14. B.F.J. Schonland, *Atmospheric Electricity* (Methuen, London, 1932)
15. J.A. Chalmers, *Atmospheric Electricity* (Claredon Press, Oxford, 1949)
16. B.F.J. Schonland, *Atmospheric Electricity* (Methuen, London, 1953)
17. H. Israel, *Atmospheric Electricity* (Keter Press Binding, Jerusalem, 1973)
18. R. Reiter, *Phenomena in Atmospheric and Environmental Electricity* (Elsevier, New York, 1992)
19. E.P. Krider, R.G. Roble, *The Earth's Electrical Environment* (National Academy Press, Washington, 1986)
20. D.R. MacGorman, W.D. Rust, *The Electrical Nature of Storms* (Oxford University Press, Oxford, 1998)
21. L.D. Landau, E.M. Lifshits, *Fluid Mechanics* (Pergamon Press, London, 1959)
22. G.K. Batchelor, *An Introduction to Fluid Dynamics* (Cambridge University Press, Cambridge, 1967)
23. P. Drazin, W. Reid, *Hydrodynamic Stability* (Cambridge University Press, Cambridge, 1981)
24. D.J. Acheson, *Elementary Fluid Dynamics* (Oxford University Press, Oxford, 1990)
25. V.P. Krainov, *Qualitative Methods in Physical Kinetics and Hydrodynamics* (American Institute of Physics, New York, 1992)
26. S. Chandrasekhar, *Hydrodynamic and Hydromagnetic Stability* (Oxford University Press, London, 1961)
27. E.L. Koschmieder, *Benard Cells and Taylor Vortices* (Cambridge University Press, Cambridge, 1993)

28. A.V. Getting, *Rayleigh-Taylor Convection Structure and Dynamics* (World Scientific Publishing, London, 1998)
29. G.Z.Gershuni, E.M.Zhukhovitskii. *Convection Stability of Compressed Liquids.* (Moscow, Nauka, 1972)
30. T. Shimazaki, A.R. Laird, Radio Sci. **7**, 23 (1972)
31. M.J. McEwan, L.N. Phillips, *Chemistry of the Atmosphere* (Edward Arnold, London, 1975)
32. D.F. Strobel, Radio Sci. **7**, 1 (1972)
33. A. Einstein, Ann. Phys. **17**, 549 (1905)
34. A. Einstein, Ann. Phys. **19**, 371 (1906)
35. A. Einstein, Zs.für. Electrochem. **14**, 235 (1908)
36. S. Chapman, T.G. Cowling, *The Mathematical Theory of Non-uniform Gases* (Cambridge University Press, Cambridge, 1952)
37. J.H. Ferziger, H.G. Kaper, *Mathematical Theory of Transport Processes in Gases* (North Holland, Amsterdam, 1972)
38. M.Capitelli, D.Bruno, A.Laricchiuta. *Fundamental Aspects of Plasma Chemical Physics. Transport.* (Springer, New York, 2013)
39. B.M. Smirnov, *Reference Data on Atomic Physics and Atomic Processes* (Springer, Heidelberg, 2008)
40. R. Braham, J. Meteorol. **9**, 227 (1952)
41. C.T.R. Wilson, J. Franklin Inst. **208**, 1 (1929)
42. A. Baumgartner, E. Reichel, *The World Water Balance* (Elsevier, Amsterdam, 1975)
43. K.E. Trenberth, L. Smith, T. Qian et al., J. Hydrometeorol. **8**, 758 (2007)
44. I.A.Shiklomanov, In *Water in Crisis : A Guide to the World's Fresh Water Resources.* ed. by P.H.Gleick (Oxford Univ.Press, Oxford, 1993), pp. 13–24
45. I.A. Shiklomanov, J.C. Rodda (eds.), *World Water Resources at the Beginning of the Twenty-First Century* (Cambridge University Press, Cambridge, 2003)
46. R.W. Healy, T.C. Winter, J.W. Labaugh, O.L. Franke, *Water budgets: Foundations for effective water-resources and environmental management.* (Reston, Virginia, U.S.Geological Survey Circular 1308, 2007)
47. http://en.wikipedia.org/wiki/Atmosphere-of-Earth
48. http://water.usgs.gov/edu/watercyrcleatmosphere.html
49. http://en.wikipedia.org/wiki/water-circle
50. http://en.wikipedia.org/wiki/Properties-of-water
51. R.P. Feynman, R.B.Leighton, M.Sands, *The Feynman Lectures of Physics.*, vol. 2(Addison-Wesley, Reading, 1964)
52. B.M. Smirnov, *Cluster Processes in Gases and Plasmas* (Wiley, Berlin, 2010)
53. B.M. Smirnov, Phys. Uspekhi **184**, 1153 (2014)
54. http://en.wikipedia.org/wiki/Troposphere
55. C.B. Moore, B. Vonnegut. *Point-discharge Currents.* in *Lightning*, ed. by R.H. Golde (Acadamic Press, London, 1977), p. 51
56. B.J. Mason, *The Physics of Clouds* (Claredon Press, Oxford, 1971)
57. J. Warner, Tellus **7**, 450 (1955)
58. W.R. Leaitch, G.A. Isaak, Atmos. Environ. **25**, 601 (1991)
59. http://en.wikipedia.org/wiki/Liquid-water-content
60. E.P. Wigner, W.F. Seits, Phys. Rev. **46**, 509 (1934)
61. E.P. Wigner, Phys. Rev. **46**, 1002 (1934)
62. B.M. Smirnov, *Clusters and Small Particles in Gases and Plasmas* (Springer, New York, 1999)
63. B.M. Smirnov, *Nanoclusters and Microparticles in Gases and Vapors* (De Gruyter, Berlin, 2012)
64. B.C. Kindel, P. Pilewskie1, K.S.Schmidt, Atmos. Meas. Tech. **8**, 1147 (2015)
65. T. Pedersen, Y. Rosental, S. Seitzinger et al., Science **290**, 291 (2000)
66. D.S. Schimmel, *The Carbon Cycle.* ed. by T.M. Wigley (Cambridge, Cambridge University Press, 2000)

67. https://en.wikipedia.org/wiki/Carbon-cycle
68. R. Hill, Nature **139**, 881 (1937)
69. H.G. Jones, *Plants and Microclimate: A Quantitative Approach to Environmental Plant Physiology* (Cambridge University Press, Cambridge, 1992)
70. D.O. Hall, K.K. Rao, *Photosynthesis* (Cambridge University Press, Cambridge, 1999)
71. D.W. Lawlor, *Photosynthesis.* (BIOS, Oxford, 2001)
72. J.E. Bidlack, K.R. Stern, S. Jansky, *Introductory Plant Biology* (McGraw-Hill, New York, 2003)
73. Handbook of Photosynthesis (CRC Press, Boca Ratonm, 2005)
74. R.E. Blankenship, *Molecular Mechanisms of Photosynthesis* (Wiley, London, 2014)
75. G.E. Edwards, D.A. Walker, *C3, C4: Mechanisms and Cellular and Environmental Regulation of Photosynthesis* (Blackwell, Oxford, 1983)
76. http://www.ciesin.org/docs/004-038/004-038a.html
77. https://www.wikipedia.org/wiki/Greenhouse-effect
78. J. Lean, P.R. Rowntree, J. Climate **10**, 1216 (1997)
79. A.N. Hahmann, R.E. Dickinson, J. Climate **10**, 1944 (1997)
80. D. Lawrence, K. Vandecar, Nature Climate Change **5**, 27 (2015)
81. https://www.ipcc.ch.publications-and-data/ar4/wg1/en/ch757-3.html
82. http://www.esrl.noaa.gov/gmd/ccgg/trends
83. https://en.wikipedia.org/wiki/Mauna-Loa-Observatory
84. D. Lüthi, M. Floch, B. Bereiter e.a. Nature **453**, 379 (2008)
85. http://climate.nasa.gov
86. J.B.J. Fourier, Annal. Chem. Phys. **27**, 136 (1824)
87. J.B.J. Fourier, Mem. Acad. Roy. Sci. **7**, 569 (1827)
88. B.M. Smirnov, *Energetics of the Atmosphere.* (Znanie, Physical series, Moscow, 1979, N3; in Russian)
89. H.N. Pollack, S.J. Hunter, R. Johnson, Rev. Geophys. **30**, 267 (1997)
90. M.L. Salby, *Physics of the Atmosphere and Climate* (Cambridge University Press, Cambridge, 2012)
91. http://en.wikipedia.org/wiki/Earth's-energy-budget
92. https://en.wikipedia.org/wiki/Greenhouse-effect
93. http://en.wikipedia.org/wiki/solar-constant
94. P.V. Foukal, J. Lean, Science **247**, 556 (1990)
95. R.C. Wilson, H.S. Hudson, Nature **351**, 42 (1991)
96. C. Fröhlich, Space Sci. Rev. **125**, 53 (2006)
97. S. Mekaoui, S. Dewitte, Solar Phys. **247**, 203 (2008)
98. C. Fröhlich, J. Lean. Geophys. Res. Lett. **25**, 4377 (1998)
99. C. Fröhlich, J. Lean. Astron. Astrophys. Rev. **12**, 273 (2004)
100. G. Kopp, J.L. Lean, Geophys. Res. Lett. **38**, L01706 (2011)
101. R.C. Wilson, Solar Phys. **74**, 217 (1981)
102. N.B. Vargaftik, *Tables of Thermophysical Properties of Liquids and Gases.* 2nd edn. (Halsted Press, New York, 1975)
103. M.A. Ruderman, J.W. Chamberlain, Planet Space Sci. **23**, 247 (1975)
104. B.W. Carroll, D.A. Ostlie, *An Introduction to Modern Astrophysics* (Pearson, London, 2013)
105. B.V. Somov, *Cosmic Plasma Physics* (Kluwer, Dordrecht, 2000)
106. N. Meyer-Vernet, *Basic of Solar Wind* (Cambridge University Press, Cambridge, 2007)
107. K. Berger, *The Earth flash.* In *Lightning*, ed. by R.H. Golde (London, Academic Press, 1977), p. 119
108. M. Rycroft et al., Space Sci. Rev. **137**, 83 (2008)
109. M.J. Rycroft et al., in *Planetary Atmospheric Electricity.* ed. by F. Leblanc et al. (Springer, Heidelberg, 2008), p. 83
110. E. Williams, Atmos. Res. **91**, 140 (2009)

Chapter 3
Charged Particles in Atmosphere

Abstract Electrons and ions are present in the atmosphere and determine its conductivities at various altitudes. A negative charge of the atmosphere at altitudes below approximately 70 km comes from negative ions, whereas electrons are present at higher altitudes. Mobilities of various positive and negative ions at atmospheric pressure and room temperature are close to $2\,\mathrm{cm}^2/(\mathrm{V}\cdot\mathrm{s})$. Complex ions are formed at altitudes below 100 km; the type of ion depends on the altitude and local impurities. Collisions of two neutral aerosols in different aggregate states lead to their charging, and formation of an atmospheric charge is determined mostly by collisions of graupels and ice particles. Along with this, cosmic rays ionize air in the troposphere and stratosphere. Motion of these ions under the action of the Earth's electric field determines the discharging of the Earth.

3.1 Ionosphere

Electrons formed as a result of ionization at low altitudes attach to oxygen molecules with formation of the negative ions. Therefore the negative atmospheric charge at low altitudes is connected with negative ions. As the altitude increases, the atmospheric density decreases, and a typical time of electron attachment to oxygen molecules increases also. The presence of solar radiation at high altitudes leads to photodestruction of negative ions. Hence electrons constitute the atmospheric negative charge at high altitudes. Figure 3.1 [1] gives a schematic dependence of the number density of electrons and ions on the altitude. As follows from this figure, electrons become the main component of the negative atmospheric charge at altitudes above 70 km, and this atmospheric region is called the ionosphere. The transition from negative ions to electrons with the altitude increasing at the ionosphere boundary leads to a sharp increase of the air conductivity.

The presence of electrons in the ionosphere prevents long electromagnetic waves(radio waves) from propagating therein. Namely, radio waves reflect from the ionosphere and return to the Earth's surface [2–4]. Usually the ionosphere is divided in several layers [5–8] in accordance with their respective properties to reflection of radio waves. Table 3.1 represents the definition of these ionosphere layers. The lower

B.M. Smirnov, *Microphysics of Atmospheric Phenomena*,
Springer Atmospheric Sciences, DOI 10.1007/978-3-319-30813-5_3

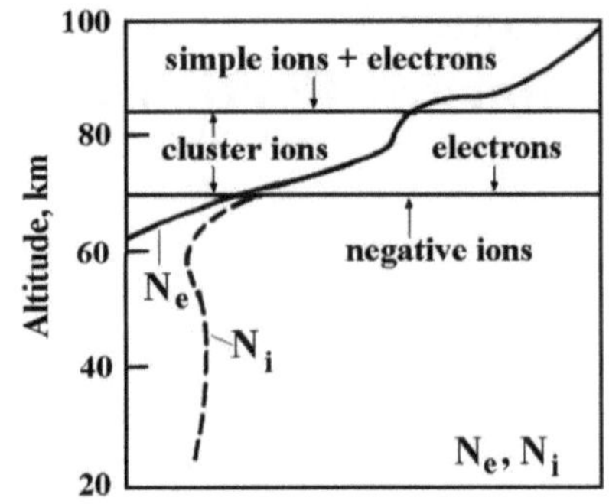

Fig. 3.1 Charged atomic particles in a low atmosphere. Boundaries of existence of different charged particles are indicated [1]

Table 3.1 Layers of the ionosphere

Layer	Altitude, km
D	60–90
E	90–140
F_1	140–300
F_2	300–1000

D-layer of the ionosphere occupies the altitude region from 50 to 90 km. A typical number density of charged particles therein is on the order of 10^3 cm^{-3}. Because electrons are absent at altitudes below approximately 70 km, where the negative charge is connected with negative ions, radio waves propagate through these layers freely.

The next layer of the ionosphere, the E-layer, is located at altitudes from 90 up to 140 km. Charged particles of the E-layer are formed by photoionization of air caused by solar UV radiation. These charged particles drift to lower layers of the atmosphere and are the source of plasma in the D-layer. The electron number density in the E-layer of the ionosphere is on the order of 10^5 cm^{-3}. Negative ions are practically absent in this layer, and the basic types of positive ions are NO^+ and O_2^+. The decay of charged particles in the E-layer results from dissociative recombination of electrons and molecular ions, or by transport of charged particles to lower layers of the atmosphere.

A higher ionospheric layer (the F_1-layer) is located at altitudes from 140 up to 200 km. Above it (up to an altitude of about 400 km) the ionospheric F_2-layer exists. The electron number density in these F-layers ranges between 10^5 and 10^6 cm^{-3}. The basic type of positive ion is O^+. Charged particles of the F-layers of the ionosphere are formed by photoionization of atmospheric oxygen (the basic component of the atmosphere at these altitudes) under the action of solar radiation. Loss of electrons in these layers is caused by photorecombination of electrons and oxygen ions, photoattachment of electrons to oxygen atoms, and by transport of electrons in lower atmospheric layers. Thus, E- and F-layers of the ionosphere, as layers with a high number density of electrons, are responsible for reflection of radio signals, so that these layers are the radio mirror of the atmosphere for radio waves.

3.2 Mobility of Atmospheric Ions

Along with elementary processes of formation and destruction of molecular ions in the lower atmosphere, transport of molecular ions is of interest because decay of molecular ions proceeds in atmospheric layers other than those where ions are formed. The characteristic of ion motion in a gas in an external electric field is its mobility. If ions are moving under the action of an electric field of a small strength E, the electric current density i arises in accordance with Ohm's law

$$i = EN_i(K_+ + K_-), \tag{3.1}$$

where N_i is the number density of positive and negative ions that are identical for a quasineutral plasma, and K_+, K_- are the mobilities of positive and negative ions, respectively.

Table 3.2 contains values of the mobilities for basic negative and positive ions in nitrogen measured at low electrical field strengths. As shown, the mobilities of different ions differ slightly from each other, and below we use the value $K \approx 2\,\mathrm{cm}^2/(\mathrm{V}\cdot\mathrm{s})$ [9] as the ion mobility. Let us use the Einstein relation [10–12]

$$K_i = \frac{eD_i}{T} \tag{3.2}$$

between the mobility K_i and the diffusion coefficient D_i of a charged particle in a buffer gas. For the reduced mobility of molecular ions in air at room temperature this gives:

$$D_i N_{air} = 5.6 \cdot 10^{19}\,\mathrm{cm}^{-1}\mathrm{s}^{-1}$$

In particular, at the air pressure 0.5 atm and room temperature this formula gives $D_i = 0.45\,\mathrm{cm}^2/\mathrm{s}$ for the ion diffusion coefficient, and this diffusion process determines ion transport at not large distances in a motionless atmosphere.

Ions are nuclei of condensation for formation of clusters (nanometer-size particles) and micron-size aerosols in atmospheric air. Below we determine the mobilities of water clusters and aerosols moving in atmospheric air. There are two regimes of ion

Table 3.2 Mobilities K of negative and positive ions in nitrogen [13–16] at room temperature reduced to the normal number density of molecules $N = 2.69 \times 10^{19}\,\mathrm{cm}^{-3}$ and expressed in units $\mathrm{cm}^2/(\mathrm{V}\cdot\mathrm{s})$

Ion	K	Ion	K	Ion	K
NO_2^-	2.5	$CO_3^- \cdot H_2O$	2.1	N_2O^+	2.3
NO_3^-	2.3	N_2^+	1.9	N_4^+	2.3
CO_3^-	2.4	CO_2^+	2.2	$H^+ \cdot H_2O$	2.8
$NO_2^- \cdot H_2O$	2.4	N_2H^+	2.1	$H^+ \cdot (H_2O)_2$	2.3
$NO_3^- \cdot H_2O$	2.2	N_3^+	2.3	$H^+ \cdot (H_2O)_3$	2.1

motion in a gas depending on the relation between the mean free path for gaseous particles in a gas λ and a particle radius r [17]. The ion mobility K_{kin} in a weak electrical field in the kinetic regime is given by

$$K_{kin} = \frac{3e}{8\sqrt{2\pi mT}N_a r^2} = \frac{K_o}{n^{2/3}}, \; K_o = \frac{3e}{8\sqrt{2\pi mT}N_a r_W^2}, \; r \ll \lambda \tag{3.3}$$

Here r_W is the Wigner–Seits radius defined by Formula (2.25) that connects a particle radius r and a number of particle molecules, e is the electron charge that is assumed to be equal to the ion charge, m is the mass of an air molecule, T is the temperature expressed in energetic units, and N_a is the number density of gas molecules. In considering water aerosols in air, we have $r_W = 1.92\,\text{Å}$ [18] and the reduced mobility at room temperature is $K_o = 17\,\text{cm}^2/(\text{V}\cdot\text{s})$.

Kinetic and diffusion regimes of motion of a particle differ by the character of particle interaction with air molecules. Namely, a test particle may collide with one air molecule only at each moment in the kinetic regime. However, in the diffusion regime of particle motion in air many air molecules interact simultaneously with a test particle. If this particle is moving in a gas with a velocity $\mathbf{v} = K\mathbf{E}$, the force $\mathbf{F}$ acted on a particle in the diffusion regime is determined by the Stokes formula $\mathbf{F} = 6\pi r K\mathbf{E}\eta$, where η is the gas viscosity. Equalizing the friction force $\mathbf{F}$ to the force from an electric field $e\mathbf{E}$, one can find the mobility K_{dif} of a singly charged particle in the diffusion regime

$$K_{dif} = \frac{e}{6\pi r\eta} = \frac{K_1}{n^{1/3}}, \; K_1 = \frac{e}{6\pi r_W\eta}, \; r \gg \lambda \tag{3.4}$$

If a water aerosol is located in air at room temperature, this formula gives $K_1 = 0.024\,\text{cm}^2/(\text{V}\cdot\text{s})$. One can obtain that the equality $K_{kin} = K_{dif}$ takes place under consideration at an aerosol radius $r = 0.14\,\mu\text{m}$ ($n = 4\cdot 10^8$). Note that the mean free path of air molecules in air is $\lambda \approx 0.1\,\mu\text{m}$ at room temperature and atmospheric air pressure.

Figure 3.2 gives the dependence of the mobility of water clusters in air depending on a cluster size. This dependence is assumed to be continuous, and $K = 2\,\text{cm}^2/(\text{V}\cdot\text{s})$ at $n = 1$, whereas for large n it is determined by Formula (3.3). Note that the maximum cluster size in Fig. 3.2 ($n = 10^6$) corresponds to the cluster radius $r = 20\,\text{nm}$.

3.3 Formation of Molecular Ions in Earth's Atmosphere

Ions are formed in the Earth's atmosphere under the action of short-length radiation of a Sun corona. Negative ions may result from attachment of electrons to oxygen molecules. The rate of ion formation is larger at higher altitudes because both the

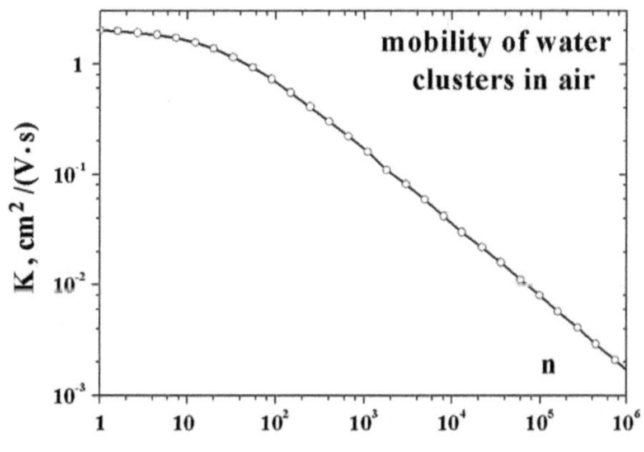

Fig. 3.2 Mobility of water clusters in atmospheric air at room temperature as a function of water molecules of the cluster

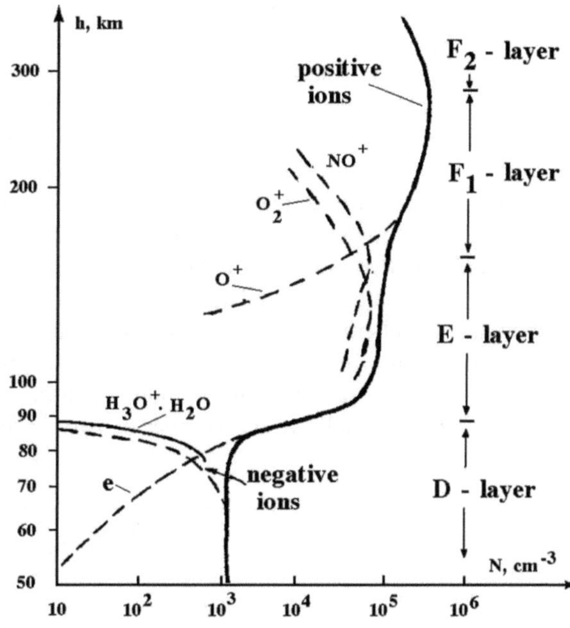

Fig. 3.3 Ions in the upper atmosphere (ionosphere) and notations for ionosphere layers. The average number densities of electrons (e) and indicated ions are given

depth of penetration of X-ray radiation inside the Earth's atmosphere and the intensity of X-ray radiation decrease with a decreasing wavelength.

Figure 3.3 contains the distribution of electrons, as well as atomic and molecular ions, in the middle atmosphere (i.e., in the Earth's atmosphere with averaged parameters). As shown, electrons are located in the ionosphere at altitudes above 80 km, whereas a negative atmospheric charge in the lower atmospheric layers corresponds to negative ions. Therefore an electron-ion plasma exists in E- and F-ionospheric layers at altitudes above 100 km, and the number density of electrons and ions in this plasma is on the order of $10^4 - 10^6\,cm^{-3}$. In the troposphere at altitudes below 15 km the number density of positive and negative ions is on the order of $10^3\,cm^{-3}$.

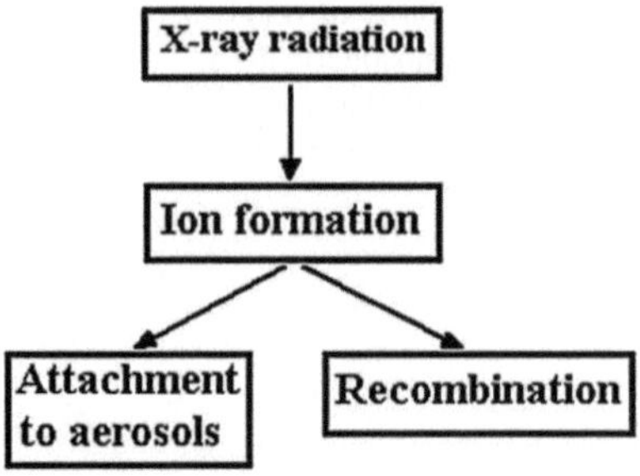

Fig. 3.4 Chain of processes for formation and decay of molecular ions in the troposphere

Only these ions are responsible for discharging of the Earth whose charging proceeds due to the action of lightning. According to the sum of observational data, the average current density of the Earth discharging is $2.4 \times 10^{-16}\,\mathrm{A/cm^2}$ over land and $3.7 \times 10^{-16}\,\mathrm{A/cm^2}$ over oceans [19, 20]. This corresponds approximately to the total discharged current for the Earth's atmosphere $I = 1700\,\mathrm{A}$ [21].

Thus, molecular ions are formed in the troposphere as a result of absorption of X-ray radiation that is emitted by the solar corona creating an atmospheric current in atmospheric dry regions that leads to Earth discharging. The scheme of elementary processes for formation and decay of molecular ions in the troposphere are given in Fig. 3.4. Decay of molecular ions in a dry atmosphere results from recombination of positive and negative molecular ions, and the rate constant of this process is represented in Fig. 3.5 for positive and negative molecular ions that exist in atmospheric air. As shown, atmospheric pressure corresponds to the maximum of the rate constant of ion recombination, where this rate constant is equal approximately to $2 \times 10^{-6}\,\mathrm{cm^3/s}$. At low pressures the rate constant of mutual neutralization of positive and negative ions varies proportionally to the number density of air molecules according to the

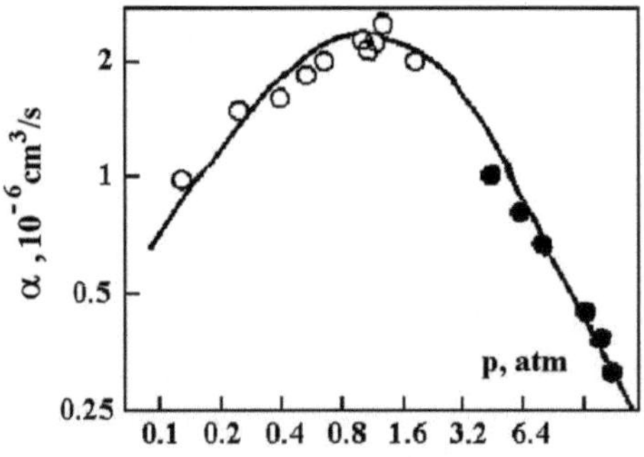

Fig. 3.5 Recombination coefficient of positive and negative ions in air. The *solid curve* is an approximation of experimental data represented by *open* [24] and *closed* [25] circles. The *left part* of the solid curve corresponds to three-body recombination of positive and negative ions in accordance with the Thomson model [22], and recombination at large pressures (the *right part* of the curve) is given by the Langevin formula [23]

Thomson model [22]. At large number densities of air molecules a recombination time is limited by the approach of ions in a dense gas, and the rate constant of this process is inversely proportional to the number density of air molecules as well as the rate of ion approach that is proportional to the total ion mobility in accordance with the Langevin theory [23]. This behavior of the rate constant of ion recombination is confirmed by Fig. 3.5.

3.4 Complex Ions in Atmosphere

Complex ions are ions constructed from simple ions bonded with several molecules; cluster ions may contain a larger number of bonded molecules. Being guided by atmospheric ions, one can see from Fig. 3.1 that complex or cluster ions are the main ion types at altitudes 70–85 km (upper thermosphere and lower ionosphere) because of a low temperature in this region (about 180 K). Complex ions are present also in lower regions of the atmosphere depending on impurities and other local parameters. In particular, below we consider the existence of complex ions in the stratosphere where corresponding mass-spectrometric measurements were made.

Table 3.3 represents ion types that are observed in the stratosphere over an ocean in accordance with observed mass of these ions [26–29, 32]. One can see that these ions are constructed on the basis of sodium and potassium ions from an ocean which

Table 3.3 Mass spectrum of stratospheric positive ions over an ocean at altitudes of 35–40 km

Observed mass [26]	Observed mass [27]	Observed mass [28]	Suggested complex ion [29]
29 ± 2			
42 ± 2	43 ± 3	–	$Na^+ \cdot H_2O$ (41)
60 ± 2	60 ± 2		$K^+ \cdot H_2O$ (57)
			$Na^+ \cdot (H_2O)_2$ (59)
80 ± 2	78 ± 2	78 ± 2	
	82 ± 2		$Na^+ \cdot H_2O \cdot NaOH$ (81)
	96 ± 2	96 ± 1	$Na^+ \cdot (H_2O)_4$ (95)
			$K^+ \cdot (H_2O)_3$ (93)
		100 ± 1	$Na^+ \cdot (H_2O)_2 \cdot NaOH$ (99)
		114 ± 1	$Na^+ \cdot (H_2O)_5$ (113)
			$K^+ \cdot (H_2O)_4$ (113)
		118 ± 1	$Na^+ \cdot (H_2O)_3 \cdot NaOH$ (117)
			$K^+ \cdot H_2O \cdot KOH$ (117)
		136 ± 1	$Na^+ \cdot (H_2O)_4 \cdot NaOH$ (135)
		140 ± 1	$Na^+ \cdot (H_2O)_2 \cdot (NaOH)_2$ (139)

Observed masses expressed in atomic mass allow one to find an appropriate ion, and its mass is given in parentheses

Table 3.4 Mass spectrum of stratospheric positive ions over an ocean at altitude 35–40 km

Observed ion mass	Suggested ion ($X \equiv CH_3CN$)	c, % [30]	c, % [31]
55	$H^+ \cdot (H_2O)_2$	3.4	5.5
73	$H^+ \cdot (H_2O)_4$	26.8	29.6
78	$H^+ \cdot X \cdot (H_2O)_2$	7.2	8.4
91	$H^+ \cdot (H_2O)_5$	8.9	2.8
96	$H^+ \cdot X \cdot (H_2O)_3$	19.6	44
101	$H^+ \cdot X_2 \cdot H_2O$	10.7	3.5
114	$H^+ \cdot X \cdot (H_2O)_4$	-	0.9
119 ± 1	$H^+ \cdot X_2 \cdot (H_2O)_2$	16.1	4.9
136 ± 1	$H^+ \cdot X_2 \cdot (H_2O)_3$	4.4	-
141 ± 1	$H^+ \cdot X_3 \cdot H_2O$	2.9	-

Observed masses (left) expressed in atomic mass allow one to find an appropriate ion (right), and its concentration

penetrate the stratosphere. The ion sort may be changed depending on the underlying Earth surface, air wetness, season, or time of day. Tables 3.4 and 3.5 give the concentration of complex positive and negative ions in the stratosphere over land. One can see that negative complex ions are built on the basis of simple ions bonded with NO_3 and HSO_4. Nitrogen acid, sulfuric acid, and water molecules are bonded with these ions in complex ions. Under conditions of Tables 3.3 and 3.5, the number density of air molecules is on the order of $10^{17}\,\mathrm{cm}^{-3}$, whereas the number density of sulfuric acid H_2SO_4 molecules is $(10^5 - 10^6)\,\mathrm{cm}^{-3}$ and the number density of nitrogen acid HNO_3 molecules is $(10^7 - 10^8)\,\mathrm{cm}^{-3}$.

From this the question occurs of which concentration of an impurity is required for the existence of complex ions. According to Henshen and Arnold [35], such a minimal concentration is 10^{-12} for altitudes below 30 km and 5×10^{-13} at altitudes near 42 km. Calculations carried out by Bates [36] lead to this boundary concentration of 5×10^{-13} at altitudes 20–42 km. Note that the above negative ions in the stratosphere are nuclei of condensation for formation of aerosols of nitrogen and sulfuric acids in the surface layer of the Earth's atmosphere. Namely, nitrogen and sulfur aerosols are the main constituents of charged aerosols near the Earth's surface [37].

Complex ions are formed in ionized gases at low temperatures and not low pressures. Along with the stratosphere, a lower D-layer of the ionosphere with a low temperature is favorable for formation of complex ions. As follows from Fig. 3.3, in contrast to stratospheric layers, for lower layers of the ionosphere complex ions are basic ions. In spite of low pressure at the 80-km altitude, the atmosphere temperature there is below 200 K that is favorable for formation of complex ions. The main types of positive cluster ions are $H^+ \cdot (H_2O)_2$, $H^+ \cdot (H_2O)_3$, and H_3O^+ [38–41], whereas ions $O^- \cdot (H_2O)_n$, $O_2^- \cdot (H_2O)_n$, and $NO_3^- \cdot (H_2O)_n$ with $n = 1 - 5$ are the main type of negative complex ions in this region [42, 43]. These ions are formed under the action of hard radiation emitted by solar corona and not absorbed in upper layers. Absorption of hard radiation causes formation of simple ions and subsequently they

Table 3.5 Concentration of negative complex ions over land in the stratosphere at altitudes near 36 km [33] and 32–34 km [34]

Observed ion mass	Suggested ion	c, % [33]	c, % [34]
62 ± 2	NO_3^-	-	0.6
125 ± 1	$NO_3^- \cdot HNO_3$	3	6.5
143 ± 2	$NO_3^- \cdot HNO_3 \cdot H_2O$	-	0.9
160 ± 1	$HSO_4^- \cdot HNO_3$	5	14
178 ± 2	$HSO_4^- \cdot HNO_3 \cdot H_2O$	-	2
188 ± 1	$NO_3^- \cdot (HNO_3)_2$	65	19
195 ± 3	$HSO_4^- \cdot H_SO_4$	7	10
223 ± 1	$HSO_4^- \cdot (HNO_3)_2$	14	1.4
251 ± 1	$NO_3^- \cdot (HNO_3)_3$	-	0.3
276 ± 1	$HSO_4^- \cdot H_2SO_4 \cdot HNO_3$	3	1
293 ± 1	$HSO_4^- \cdot (H_2SO_4)_2$	3	1.2
311 ± 3	$HSO_4^- \cdot (H_2SO_4)_2 \cdot H_2O$	-	0.3
374 ± 3	$HSO_4^- \cdot (H_2SO_4)_2 \cdot HNO_3$	-	0.3
391 ± 2	$HSO_4^- \cdot (H_2SO_4)_3$	-	14
409 ± 3	$HSO_4^- \cdot (H_2SO_4)_3 \cdot H_2O$	-	13
427 ± 3	$HSO_4^- \cdot (H_2SO_4)_3 \cdot (H_2O)_2$	-	2.2
454 ± 2	$HSO_4^- \cdot (H_2SO_4)_3 \cdot HNO_3$	-	1.5
472 ± 3	$HSO_4^- \cdot (H_2SO_4)_3 \cdot HNO_3 \cdot H_2O$	-	0.4
489 ± 3	$HSO_4^- \cdot (H_2SO_4)_4$	-	1.2
507 ± 3	$HSO_4^- \cdot (H_2SO_4)_4 \cdot H_2O$	-	1.5
587 ± 3	$HSO_4^- \cdot (H_2SO_4)_5$	-	0.4

are transformed in complex ions as a result of appropriate processes [44–46]. Note that because of the irregular character of processes in the solar corona, forming ions and their number density may be strongly varied in time.

At altitudes about 90 km, complex ions with a sodium ion as their basis are formed [47–54]. Surprisingly, this altitude collects sodium, and the number density of sodium atoms is $10^3 - 10^4\,\mathrm{cm}^{-3}$ in a belt of a thickness of approximately 10 km [43, 55, 56]. The chain of processes involving sodium ions makes the ion $Na^+ \cdot H_2O$ the dominated ion in a layer of a thickness of 10 km. It is necessary to restate that the character of processes depends on atmospheric conditions which may be different. The possibility of formation of complex negative ions that are formed in a plasma flow with small admixtures of vapors of water and corresponding acids is confirmed by the data given in Fig. 3.6.

As shown, complex ions are formed effectively in Earth's atmosphere and may be the basic ion type at some atmospheric altitudes with a low temperature. Complex ions may be condensation nuclei that constitute an origin for aerosol growth in the atmosphere. Nevertheless, they may occupy a small part of aerosols and therefore complex ions are not essential for kinetics of aerosols that lead to evolution of charges in clouds.

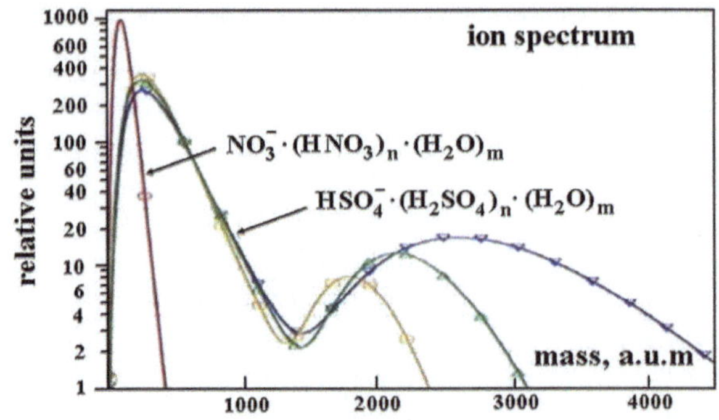

Fig. 3.6 Mass spectrum of negative complex ions formed under laboratory conditions [57–59]

3.5 Processes of Aerosol Charging

The above character of charge creation in the atmosphere was demonstrated experimentally in [60], although this study is underestimated. One can expect that this concept allowed one to measure the aerosol charge depending on aggregate states, size, temperature, and collision velocity. Unfortunately, we only have fragmentary information about this process. Evidently, the rate of this process is sensitive to its mechanism and to external conditions. Charge separation also proceeds at the interface between ice particles and liquid water particles [61, 62], and in the case of growing the ice particle is charged positively, whereas ice sublimation leads to a negative ice charge [63, 64]. This character of charging explains why the attempt [65, 66] to repeat the experiment [60] gave the charging rates several orders of magnitude below those according to experiment [60]. Evidently, a large difference of results follows from different experimental conditions [67], so that aerosol growth takes place in the first case, whereas in the second case water molecules are evaporated from the aerosol surface. Because of a high sensitivity of the charging rate to the aerosol temperature and size, we do not analyze experiments of this type. One more peculiarity of this process is that the Coulomb interaction energy of charged aerosols significantly exceeds their thermal energy. Therefore, if aerosols do not acquire large speed in the experiment, their charge in air is much less than the optimal one.

Thus, in atmospheric regions with temperature about 0° C aerosols in different aggregate states can coexist, and they get a charge as a result of their collisions. The aerosol charging is an important stage of atmospheric electricity, and coexistence of aerosols in different aggregate states is necessary for this. From this it follows that thunderstorms are absent in winter, and polar regions do not contribute to the Earth's charging. Subsequently charged aerosols are included in cumulus clouds which are a source of electric breakdown in the atmosphere.

Subsequently charge separation proceeds due to gravitational aerosol fall in the atmosphere. In considering this process, we assume that aerosols of different charge signs have a different mass, therefore fall velocities for positively and negatively charged aerosols are different. Moreover, because the Earth is charged negatively, one can assume liquid negatively charged aerosols to be heavier. Inasmuch as solid aerosols fall more slowly, one can assume for simplicity that they remain in the charging region. This charging character leads to the dipole structure of a cumulus cloud, although this assumption is rough for a charge structure of a cumulus cloud. One can add to this that convective air motion leads to mixing of atmospheric layers, which in turn gives a charge redistribution over the altitudes. Nevertheless, because convective motion of air has a random character, this character of charge separation is conserved if aerosols take part in convective motion.

The above character of charge separation in the atmosphere allows one to determine the scale of aerosol sizes in this process [9]. This process proceeds in an electric field created both by the charged Earth surface and by the charged layer of a formed cloud. In this process the aerosol weight exceeds the force acted on an aerosol from an electric field. Assuming an aerosol to be singly charged, we have that the force from an electrical field is proportional to an aerosol radius, whereas its weight is proportional to a radius cube. Hence, small aerosols of a radius below $0.1\,\mu m$ move up [9]. In particular, from this it follows that molecular ions cannot take part in the above mechanism of charge separation. Next, the efficiency of charge transfer, that is, the ratio of the aerosol charge to the mass drops with an increasing aerosol size, and the maximum aerosol radius for which Formula (2.23) is fulfilled, is $20-30\mu m$. Below we are guided by an aerosol radius $r = 8-10\,\mu m$ that is found in an indicated range.

Clouds are necessary for atmospheric electric phenomena [68–75] and allow one to focus atmospheric electrical energy in a small region, which may cause the development of thunderstorms. Clouds consist of micron-size water aerosols, and these aerosols are responsible for charge separation in the Earth's atmosphere. Clouds are formed as a result of convective transport in the atmosphere [76–78], so that atmospheric moisture is condensed in cold atmosphere regions. In spite of energy release in the course of water condensation, this process intensifies as a result of transport phenomena, and condensing water is concentrated in restricted atmospheric regions [79–81]. This process is determined by convective transport in the atmosphere, therefore creation of cumulus clouds is a random process accompanied by large gradients in aerosol distribution. According to existing models of transport in clouds (e.g., [76, 82–85]), convective transport leads both to creation and smearing of cumulus clouds. Below, leaving aside processes of cloud development and cloud influence on the Earth's climate, we consider clouds as a medium in which electric currents are formed and lead to charge separation. In considering cumulus clouds as sources of atmospheric electricity, we are guided by aerosol parameters (2.24) in accordance with data [75, 86–88]. These parameters include the aerosol radius $r_o = 8\,\mu m$ and the water density $2\,g/m^3$ in aerosols, although it may reach the value $20\,g/m^3$ [89] in the last stage of development of cumulus clouds before water precipitation. The average water content in aerosols corresponds to the number density $N_p = 1 \times 10^3\,cm^{-3}$ of aerosols.

Based on these parameters, below we analyze the character of formation of charged aerosols and charge separation in the atmosphere. A charge is formed at altitudes where water may be found in different aggregate states. Along with liquid and solid (ice) phases, aerosols may contain amorphous water (hailstones) or a mixture of snow and liquid water (graupels). These types of aerosols may coexist at altitudes of several kilometers with temperatures about 0° C. For this reason thunderstorms are observed in warm seasons if zero temperature is attained at altitudes of several kilometers.

Charging in collisions of water and ice microparticles was observed first in 1957 [60] where microdroplets obtained a negative charge, whereas ice microparticles were charged positively. Various mechanisms of interaction may lead to this charging, but in any case the charging process results from the contact voltage at the interface between different aggregate states of water aerosols [90, 91] leading to charging of colliding particles [92–96]. Only this process determines the aerosol charging in clouds [67, 75, 94–100]. In spite of different standpoints [92–96] about the rate of charging and existence of empiric relations [97, 101, 102] for a certain range of charging parameters, the description of this process has a qualitative character and the basic thesis relates to collisions of ice particles and graupels consisting of a snow and water mixture [100, 103–105]. Therefore we consider below the nature of the process of aerosol charging, where the aerosol charge after charge separation is determined by the growth rate of aerosols at their gravitation association.

In considering aerosol water as a weak electrolyte containing positive ions H_3O^+ and negative ions OH^-, one can describe aerosol charging as ion transfer through the interface during contact of aerosols, so that one aerosol contains an excess of ions H_3O^+, and another aerosol has an excess of ions OH^-. If aerosols are in the solid state, one can connect the aerosol charging with formation of charged defects. Then the aerosol charging proceeds both in aerosol collisions and in processes of aerosol growth and evaporation. Figure 3.7 shows the interface of contacted water aerosols that are characterized by different potentials with respect to free ions (the Workman-Reynolds potential [61]). This potential is not determined by the behavior of internal ions, and only surface ions create this potential. One can connect this surface contact potential with the interaction of positive H_3O^+ and negative OH^- ions located in a surrounding space with the aerosol surface. In particular, the presence of charged defects on the aerosol surface that depend on its aggregate state influences the character of ion attachment to an aerosol surface and hence determines the surface potential. The difference of surface potentials at a contact between two aerosols in different aggregate states creates the contact potential [93, 106]. As a result of contact

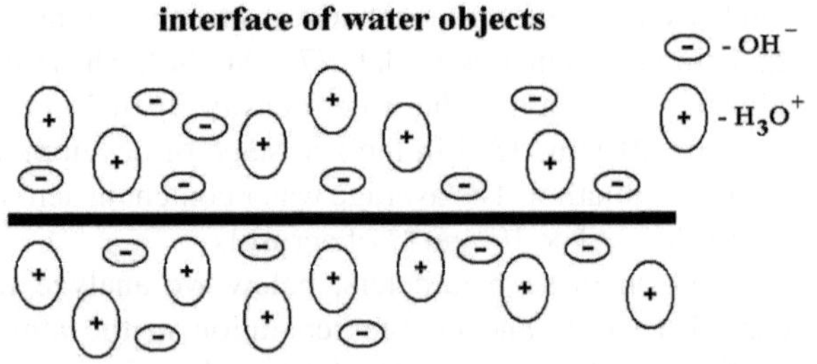

Fig. 3.7 Positive (H_3O^+) and negative (OH^-) ions on the interface, the boundary between two water aerosols in different aggregate states

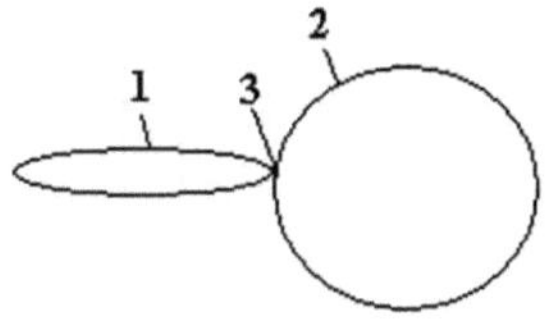

Fig. 3.8 Aerosol charging in collisions of ice (*1*) and liquid water microparticles (*2*). At their contact (*3*) an electric voltage arises and causes the particle charging after their separation

between two water aerosols located in different aggregate states, some ions transfer through their interface, and this equalizes the potentials of these aerosols. According to experiments [92, 107, 108] an excess of ions OH^- corresponds to aerosol growth with the joining of water molecules to an aerosol, whereas an excess of ions H_3O^+ takes place in the sublimation process. Electrolyte properties of water lead to aerosol charging in collisions with a surface. For example, electric discharges are observed in collisions of snowflakes with flight wires [109], and charges of clouds influence flight electrization [110].

The character of aerosol charging is given in Fig. 3.8. Leaving aside the mechanism of this process, note that the interaction energy at the interface significantly exceeds the thermal energy of aerosols, in particular, the specific energy of ice melting 334 J/g [111] or 0.06 eV per water molecule that exceeds a thermal aerosol energy (0.026 eV). Hence, charged aerosols are joined, and their separation is possible if the potential of the Coulomb attraction of aerosols does not exceed their thermal energy $\sim T$, so that a typical charge Z_o of forming aerosols is connected with their radius r_o in accordance with the following estimation [9, 17]

$$\frac{Z_o^2 e^2}{r_o} \sim T \tag{3.5}$$

Hence, a typical charge of forming aerosols Z_o is given by

$$Z_o = b\sqrt{\frac{r_o T}{e^2}}, \tag{3.6}$$

where a numerical factor b is on the order of one. Let us assume that subsequent aerosol collisions conserve the charge distribution of aerosols; that is, the ratio of the aerosol charge to its mass does not vary in the course of aerosol association. Then, based on Formula (3.6) and assuming that the initial charging gives the aerosol charge $Z = 1$, one can find the initial aerosol mass $m = 8 \times 10^{-11}$ g that corresponds to an initial aerosol radius $r_o = 2.7\,\mu$m. Subsequent aerosol growth in a cumulus cloud leads to an approximately 30-fold increase of its charge (Formula (8.14)) and mass.

Note that thermal velocities of aerosols in a cumulus cloud are small and restrict the aerosol charge after collisions. Indeed, a thermal velocity v_T of a water aerosol of a radius $r = 2.7\,\mu$m, as it is at the beginning of aerosol charging in a cumulus cloud, is equal to $v_T = 0.04$ cm/s. Aerosols of an average radius (2.24) contain a

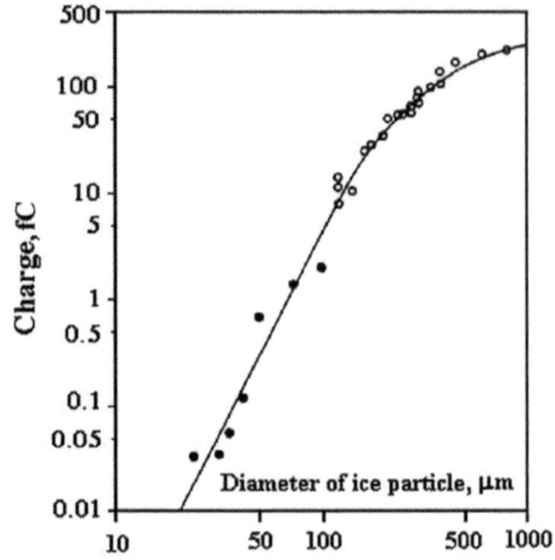

Fig. 3.9 Charge of an ice particle as a result of collisions with a riming metal rod [112, 113]

charge (3.6) resulting from the joining of such aerosols. An increase in the collision velocity leads to an increase of the transferring charge. As an example, Fig. 3.9 gives the dependence on the aerosol radius for a transferring charge to an ice particle that collides with a riming metal rod of a diameter of 5 mm at collision velocities in the range between 2 m/s and 9 m/s. As shown, both the collision velocities and observed charges of ice particles significantly exceed those in cumulus clouds. In particular, according to Formula (3.6) an aerosol charge is equal to 4×10^{-6} fC. An increase of sizes of colliding aerosols leads to an increasing transferring charge in one collision event [114].

An important peculiarity of the charging process in aerosol collisions is that an aerosol may obtain a charge of a different sign depending on external conditions. An example of this dependence is represented in Fig. 3.10. The temperature T_* at which the sign charge of an aerosol after collision is changed depending on the water content in a cumulus cloud. Table 3.6 gives the water density in a cumulus cloud

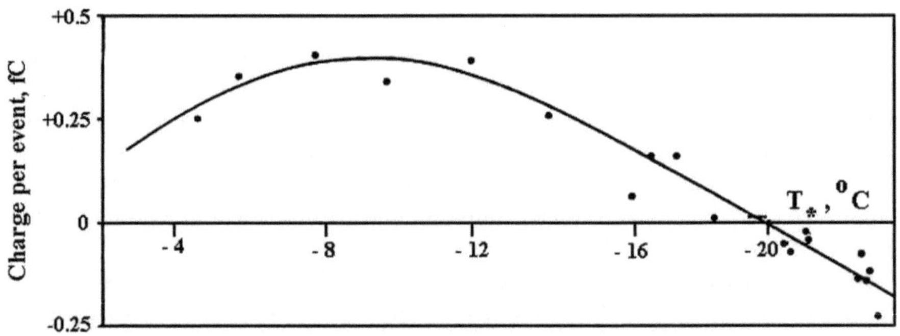

Fig. 3.10 Charge transferred to a riming graupel in one collision with an ice particle [67]

Table 3.6 Minimal temperature T_* of change of the charging sign in collision of a graupel and ice particle, and the water-specific density ρ_w in the atmosphere that provides the minimal temperature of this transition

T_*, K	ρ_w, g/m^3	Reference
264.2	0.8	[115]
259.6	0.9	[116]
260.4	1.6	[117]

where the minimal temperature for the sign change is observed. All this testifies to the complexity of the charging process in aerosol collisions.

3.6 Atmospheric Ionization by Cosmic Rays

Let us consider cosmic rays as hard X-ray radiation with photon energies above keV and fluxes of electrons and nuclei starting from MeV energies. Seemingly, cosmic rays are not connected with atmospheric energies because energies of cosmic rays are large compared with atmospheric particles and the intensity of cosmic rays is relatively small. But in reality, ionization processes under the action of cosmic rays influences some electrical processes in the atmosphere. Moreover, discovery of cosmic rays resulted from the study of atmospheric electricity.

In 1912 V.F. Hess [118] revealed in the course of electroscopic measurements in the atmosphere that the rate of formation of charged particles (and, correspondingly, the rate of discharging of electroscopes) increases with increasing altitude up to the altitude of 3.5 km at sea level. C.T.R. Wilson [119] studied this phenomenon from another standpoint using the Wilson chamber in balloon experiments and showed that ionization due to this effect leads to discharging of the Earth. R.A. Mullican made a large contribution to the investigation of cosmic rays and coined the term "cosmic rays" [120]. It was shown on the basis of Mullican's balloon electroscope experiments that the increase of the ionization rate under the action of cosmic rays with increasing altitude stops at altitudes of approximately 10 km. In addition, the relative altitude distribution of the ionization rate under the action of cosmic rays is practically identical for the same latitude and on the depth if measurements are made in lakes. From this it follows that the source of cosmic rays is located far from the Earth and, because of a large penetrating power, cosmic rays contain a high particle energy.

At the first stage of investigations of cosmic rays, Mullican considered them as photons of high energies (γ-rays) [120] that have a cosmic origin. Subsequently this understanding was expanded and various high-energy elementary particles were included in the term "cosmic rays" [121, 122]. Moreover, study of cosmic rays led to discovering some elementary particles, including positrons, π-mesons, and μ-mesons.

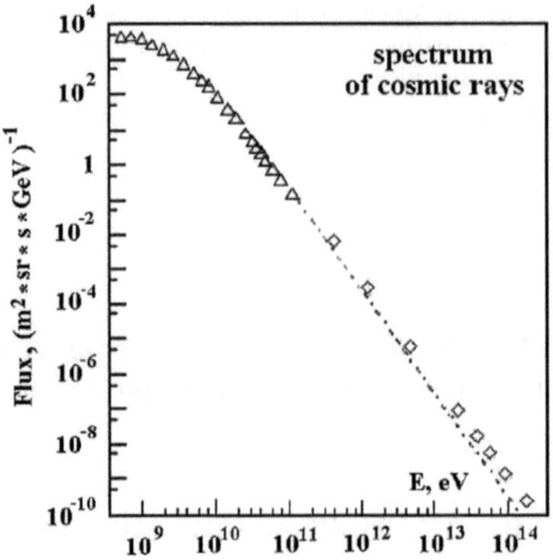

Fig. 3.11 Spectrum of cosmic rays in its low-energy part [123–125]

In considering cosmic rays from the contemporary standpoint [126–132], we are interested in the low-energy part of their spectrum which is given in Fig. 3.11. One can estimate a typical energy of singly charged elementary particles scattered in the troposphere as a result of the Coulomb interaction with nuclei of air molecules on the basis of the relation:

$$\Lambda_N \sigma N_m \sim 1 \tag{3.7}$$

Here $\Lambda_N \approx (6-8)$ km is the troposphere scale, $N_m \sim 10^{19}\,\mathrm{cm}^{-3}$ is the number density of air molecules, and σ is the cross section of scattering of colliding particles. The latter is equal to $\sigma = \pi\rho^2$, where $\rho \sim Ze^2/\varepsilon$, ε is the energy of an incident particle, and $Z \approx 7$ is the charge of nuclei of air molecules. From this it follows that singly charged particles of energies below $\varepsilon \sim 10\,\mathrm{MeV}$ are scattered in the Earth's atmosphere. From another standpoint, charged particles from outside the atmosphere deviate under the action of the Earth's magnetic field and cannot penetrate the atmosphere only excluding trajectories at the Earth's poles. In particular, only singly charged particles with energies $\varepsilon > 14\,\mathrm{GeV}$ may reach equatorial regions of the Earth [133]. The particles of such energies do not scatter on air molecules as a result of their electrostatic interaction with bound electrons.

Along with atmospheric ionization under the action of cosmic rays, the Earth's radioactivity makes a contribution to ionization of air molecules near the Earth's surface although while discovering this phenomenon it was accepted that ground atmosphere ionization follows from atmospheric radioactivity [134, 135] because the rate of the ionization process depends on the meteorological parameters [136]. A typical ionization rate near the Earth's surface is $4-8\,\mathrm{cm}^{-3}\mathrm{s}^{-1}$ [137], although near Earth regions with a heightened radioactivity the rate of ionization reaches $10\,\mathrm{cm}^{-3}\mathrm{s}^{-1}$ [138]. The rate of ion production near the ocean's surface, where ionization only proceeds under the action of cosmic rays, has a value of $2\,\mathrm{cm}^{-3}\mathrm{s}^{-1}$ [139].

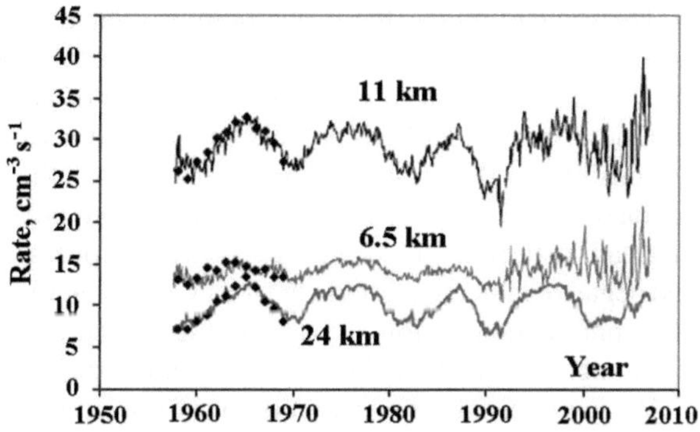

Fig. 3.12 Ionization rate of atmospheric air by cosmic rays at different altitudes [141, 142]

As for regions near the Earth's surface, ionization of air molecules is also caused by γ-rays and α-particles. γ-rays are emitted by soil and rocks and create an average ionization rate of $\sim$4 cm^{-3}s^{-1} at altitudes up to 200−300 m [140]; α-particles result from the decay of Rn_{222} with an average rate of $\sim$3 cm^{-3}s^{-1} at altitudes up to 200−300 m [140]. This shows the contribution of radioactivity to ionization of air molecules.

The peculiarity of air ionization by cosmic rays is that the maximum rate of ion production corresponds to altitudes of 11−15 km (the Plotzer maximum [143]). The maximum ionization rate at these altitudes is 30–40 cm^{-3}s^{-1} [142, 144, 145]. Figure 3.12 contains the rates of air ionization under the action of cosmic rays at different altitudes [141, 142]. The total rate of ionization of air molecules per column of the atmosphere is 4.5×10^7 cm^{-2}s^{-1} [146]. From the ratio of these values one can estimate a typical layer thickness $h \sim 10$ km that gives the main contribution to ionization of atmospheric air. From this it follows that absorption of cosmic rays takes place over the total tropospheric thickness. Next, taking a typical depth $h \sim 10$ km for absorption of cosmic rays and a typical number density of air molecules in this region, $N \sim (10^{18} - 10^{19})$ cm^{-3}, one can obtain a typical cross section of absorption of cosmic rays to be on the order of $(10^{-24}–10^{-25})$ cm^2. This corresponds to the cross section resulting from nuclear interaction between protons and nitrogen nuclei.

One can add to this that ionization of air by fast particles has a cascade character [132], so that a large energy of incident particles is shared between several forming particles which interact more strongly with air molecules. Hence, the cascade processes intensify ionization of atmospheric air molecules. Because fast particles of cosmic rays are fully ionized, for their penetration in the Earth's atmosphere an energy of several GeV is required in order to overcome the action of the Earth's magnetic field. Hence the part of cosmic radiation in the form of fast particles from the Sun is produced in its central part. X-ray radiation of photon energies of several keV freely penetrate the Earth's atmosphere. Note also that variations in the intensity of cosmic rays influence certain atmospheric properties, including the Earth's climate. In particular, the global cloud covering correlates with the intensity of cosmic rays [147].

The stage of atmospheric electricity is related to atmosphere discharging results from ionization under the action of cosmic galactic rays that are emitted partially from the Sun. In particular, Fig. 3.12 gives the rate of atmosphere ionization at different altitudes. As a result of this process, molecular ions are formed; that is, this process is responsible for the presence of molecular ions in the atmosphere. Moving under the action of the Earth's electrical field, forming ions create atmospheric electric currents that leads to the Earth discharging. Thus, this process supports the presence of positive and negative molecular ions in the lower atmosphere whose decay proceeds as a result of recombination. Below we analyze the ionization process in the lower atmosphere assuming this process to be a result of hard radiation emitted from the solar corona.

We now consider ionization of air at low altitudes assuming that it results from the action of hard radiation. Because a small amount of radiation does not change the space distribution of molecules, we have for a typical length parameter according to Fig. (2.5) $\Lambda_N \approx 6\,\text{km}$ at altitudes 11−15 km, where the temperature is $T \approx 200\,\text{K}$ and the number density of air molecules $N \sim 4 \times 10^{18}\,\text{cm}^{-3}$. In this analysis we assume that air ionization in the troposphere results from X-ray radiation with keV photon energy. This radiation releases electrons from the K-shell of air components. In considering K-shell electrons akin to electrons in hydrogen-like ions and assuming the photon energy $\hbar\omega$ to be large compared to the ionization potential J for helium-like atoms (i.e., $\hbar\omega \gg J$), we have for the cross section of photoionization in this case [148, 149]:

$$\sigma_\omega = \frac{\sigma_o}{Z_{ef}^2} \cdot \left(\frac{J}{\hbar\omega}\right)^{7/2} \tag{3.8}$$

We use the asymptotic formula for the cross section of photoionization of hydrogen-like ions (e.g., [148, 149], and Z_{ef} is the effective charge of the nucleus and other electron that acts on a test electron, $\sigma_o = 1.1\,\text{Å}^2$, and we take into account that each K-shell electron may take part in the photoionization process.

Guided by ionization at altitudes 11−15 km, we first consider that it is created due to each of the air components, that is, by nitrogen and oxygen molecules and argon atoms. Their number density at this altitude is cN, where $N \sim 4 \times 10^{18}\,\text{cm}^{-3}$ is the total number density of air molecules, and c is the concentration of a given component, that is, 0.79, 0.20, and 0.01 for nitrogen, oxygen, and argon, respectively. We also use for the ionization potentials for He-like ions of nitrogen, helium, and argon the values 667, 871, and 4121 eV, respectively [111]. Accounting for various air components, one can generalize Equation $\Lambda_N \sigma N \sim 1$ for photon absorption in air to the form

$$\frac{2\Lambda_N \sigma_o N(z_o)}{(\hbar\omega)^{7/2}} \left[\frac{c(N_2)J(N_2)^{7/2}}{Z_{ef}^2(N_2)} + \frac{c(O_2)J(O_2)^{7/2}}{Z_{ef}^2(O_2)}\right] = \ln 2, \tag{3.9}$$

Here the argument indicates to which air component this parameter, $Z_{ef} = Z - 0.75$, is related, as it is obtained for a helium-like ion in the variation method. Solution of this equation gives $\hbar\omega = 3.3\,\mathrm{keV}$ that corresponds to the wavelength of photons $\lambda = 3.8\,\text{Å}$. Because the photon energy $\hbar\omega$ is below the potential ionization of the argon atom, we exclude argon from consideration in (3.9). One can also find the energy of photons that is responsible for absorption of photons at the altitude of $h = 6\,\mathrm{km}$ where we take the number density of air molecules to be $N = 1 \times 10^{19}\,\mathrm{cm}^{-3}$. This corresponds to the photon energy $\hbar\omega = 4.2\,\mathrm{keV}$ and to the wavelength of photons $\lambda = 2.9\,\text{Å}$.

The question concerns the source of these photons. One can expect, in the first place, it is the solar corona [150–152]. The corona temperature during a quiet period is $(1-2) \times 10^6\,\mathrm{K}$, and the number density of electrons is on the order of $10^9\,\mathrm{cm}^{-3}$. This electron temperature and the number density of charged particles may be increased by several orders of magnitude as a result of solar flares [153, 154] during which a mass of the solar substance is ejected outside the Sun and interacts with the solar corona. Along with electrons and protons, the emitted substance contains ions of other elements. Hard X-ray radiation with the wavelength of several Å results from photorecombination and dielectron recombination of electrons and multicharge ions [155, 156], and the ion charge in these processes must exceed $Z = 16-18$ in order for this radiation to reach the troposphere. In this case ionization of the troposphere is realized by photons of energy of a few keV. Inasmuch as the corona is not a regular object, the rate of atmospheric ionization parameters in this channel varies in time. Moreover, the flux of hard X-ray photons may vary by orders of magnitude during solar flares. In addition, this corona spectrum is determined by radiation of the corona plasma, and the spectral intensity of this radiation decreases sharply with increasing photon energy. Hence, X-rays may be considered as only one component of cosmic rays that cause the ionization of air molecules of the troposphere.

References

1. H.S.W. Massey, J.de Physique. **40**, C7 (1979)
2. Ja.L. Alpert, *Propagation of Radiowaves in Atmosphere.* (Moscow, Nauka, 1972)
3. M.C. Kelley, *The Earth's Ionosphere: Plasma Physics and Electrodynamics* (Acad. Press, San Diego, 1989)
4. K. Davies, *Ionospheric Radio* (Peter Peregrinus Ltd, London, 1990)
5. M.C. Kelley, *The Earth's Ionosphere. Plasma Physics and Electrodynamics.* (New York, Academic Press, 1985)
6. B.E. Brunelli, A.A. Laminadze, *Physics of Ionosphere* (Nauka, Moscow, 1988)
7. M.H. Rees, *Physics and Chemistry of Upper Atmosphere* (Cambridge University Press, Cambridge, 1989)
8. R.W. Shunk, A.E. Nagy, *Ionospheres: Physics, Plasma Physics, and Chemistry* (Cambridge University Press, Cambridge, 2000)
9. B.M. Smirnov, *Cluster Processes in Gases and Plasmas* (Wiley, Berlin, 2010)
10. A. Einstein, Ann. Phys. **17**, 549 (1905)
11. A. Einstein, Ann. Phys. **19**, 371 (1906)
12. A. Einstein, Zs.für. Electrochem. **14**, 235 (1908)

13. H.W. Ellis, R.Y. Pai, E.W. McDaniel, E.A. Mason, L.A. Viehland, Atomic Data Nucl. Data Tabl. **17**, 177 (1976)
14. H.W. Ellis, E.W. McDaniel, D.L. Albritton, L.A. Viehland, S.L. Lin, E.A. Mason, Atomic Data Nucl. Data Tabl. **22**, 179 (1978)
15. H.W. Ellis, M.G. Trackston, E.W. McDaniel, E.A. Mason, Atomic Data and Nucl. Data Tabl. **31**, 113 (1984)
16. L.A. Viehland, E.A. Mason, Atom. Data Nucl. Data Tabl. **60**, 37 (1995)
17. B.M. Smirnov, *Nanoclusters and Microparticles in Gases and Vapors* (De Gruyter, Berlin, 2012)
18. B.M. Smirnov, *Reference Data on Atomic Physics and Atomic Processes* (Springer, Heidelberg, 2008)
19. H. Israel, *Atmospheric Electricity* (Keter Press Binding, Jerusalem, 1973)
20. R. Reiter, *Phenomena in Atmospheric and Environmental Electricity* (Elsevier, New York, 1992)
21. R.P. Feynman, R.B. Leighton, M. Sands. *The Feynman Lectures of Physics.* vol.2 (Reading, Addison-Wesley, 1964)
22. J.J. Thomson, Philos. Mag. **47**, 334 (1924)
23. P. Langevin, Ann. Chem. Phys. **8**, 245 (1905)
24. J. Sayers, Proc. Roy. Soc. **A169**, 83 (1938)
25. W. Mächler, Zs. Phys. **104**, 1 (1936)
26. F. Arnold, D. Krankowsky, K.H. Marien, Nature **267**, 30 (1977)
27. E. Arijs, J. Indels, D. Nevejans, Nature **271**, 642 (1978)
28. F. Arnold, H. Bohringer, G. Henschen, Geopys. Res. Lett. **5**, 653 (1978)
29. E.E. Ferguson, Geopys. Res. Lett. **5**, 1035 (1978)
30. F. Arnold, G. Henschen, E.E. Ferguson, Planet. Space Sci. **29**, 185 (1981)
31. E. Arijs, D. Nevejans, J. Indels, Nature **288**, 684 (1980)
32. F. Arnold, G. Henschen, Nature **275**, 521 (1978)
33. F. Arnold, R. Fabian, Nature **283**, 55 (1980)
34. F. Arnold, A.A. Viggiano, H. Schlager, Nature **297**, 371 (1982)
35. G. Henschen, F. Arnold, Geopys. Res. Lett. **8**, 999 (1981)
36. D.R. Bates, Planet. Space Sci. **30**, 1272 (1982)
37. R.C. Whitten (ed.), *The Stratospheric Aerosol Layer.* (Berlin, Springer, 1982)
38. R.S. Narcisi, A.D. Bailey, J. Geophys. Res. **70**, 3787 (1965)
39. R.S. Narcisi, W. Roth, Adv. Electr. Molec. Phys. **29**, 79 (1970)
40. R.S. Narcisi et al., J. Geophys. Res. **43**, 2849 (1974)
41. F. Arnold, D. Krankowsky, Geopys. Res. Lett. **1**, 243 (1974)
42. R.S. Narcisi et al., J. Atmos. Terr. Phys. **33**, 1147 (1971)
43. R.S. Narcisi, in: *Physics and Chemistry of the Upper Atmosphere,* ed. by B.M. McCormac. (Dordrecht, Reidel, 1973), p.171
44. E.E. Ferguson et al., Ann. Geopys. **25**, 819 (1969)
45. E.E. Ferguson et al., Ann. Geopys. **26**, 589 (1970)
46. R. Johnsen, C.M. Huang, M.A. Biondi, J. Chem. Phys. **63**, 3374 (1975)
47. D.M. Hunten, L. Wallace, J. Geophys. Res. **72**, 69 (1967)
48. T.M. Donahie, R.R. Meier, J. Geophys. Res. **72**, 2803 (1967)
49. J.A. Gibson, M.C.W. Sandford, Nature **239**, 509 (1972)
50. G. Megie, J.E. Blamond, Panet Space Sci. **25**, 1039 (1977)
51. B.R. Clemensha, V.W.J.H. Kichhoff, D.M. Simonich, J. Geophys. Res. **83**, 2599 (1978)
52. G. Megie et al., Planet Space Sci. **26**, 27 (1978)
53. G. Megie et al., Planet Space Sci. **26**, 509 (1978)
54. I. Yegingil, H. Ogelman, N. Kiziloglu, Geophys. Res. **85**, 5507 (1980)
55. A.C. Aikin, R.A. Goldberg, J. Geophys. Res. **78**, 734 (1973)
56. P.A. Zbinden et al., Planet Space Sci. **23**, 1621 (1975)
57. S. Wilhelm, S. Eichkorn, D. Wiedner, L. Pirjola, F. Arnold, Atm. Environ. **30**, 1735 (2004)
58. A. Sorokin, F. Arnold, D. Wiedner, Atm. Environ. **38**, 2030 (2006)

59. F. Arnold, in: *Planetary Atmospheric Electricity,* ed. by F.Leblanc et.al. (Heidelberg, Springer, 2008) p.225
60. S.E. Reynolds, M. Brook, M.F. Gourley, J. Met. **14**, 426 (1957)
61. E.J. Workman, S.E. Reynolds, Phys. Rev. **78**, 254 (1950)
62. B. Vonnegut, Bul. Am. Met. Soc. **34**, 378 (1953)
63. W. Findeisen, Meteor. Zeit. **57**, 201 (1940)
64. W. Findeisen, E. Findeisen, Meteor. Zeit. **60**, 5 (1943)
65. J. Latham, B.J. Mason, Proc. Roy. Soc. **260A**, 523 (1961)
66. J. Latham, B.J. Mason, Proc. Roy. Soc. **260A**, 537 (1961)
67. E.R. Jayaratne, C.P.R. Saunders, J. Hallett. Quar. J. Roy. Met. Soc. **109**, 609 (1983)
68. B.J. Mason, *The Physics of Clouds* (Claredon Press, Oxford, 1971)
69. H.R. Byers, *Elements of Cloud Physics* (University of Chicago Press, Chicago, 1965)
70. N.H. Fletcher, *The Physics of Rainclouds* (Cambridge University Press, London, 1969)
71. S. Twomey, *Atmospheric Aerosols* (Elsevier, Amsterdam, 1977)
72. H. Proppacher, J. Klett, *Microphysics of Clouds and Precipitation* (Reidel, London, 1978)
73. R.G. Flleagle, J.A. Businger, *An Introduction to Atmospheric Physics* (Acad. Press, San Diego, 1980)
74. M.L. Salby, *Fundamentals of Atmospheric Physics* (Acad. Press, San Diego, 1996)
75. B.J. Mason, *The Physics of Clouds* (Oxford University Press, Oxford, 2010)
76. H. Stommel, J. Meteor. **4**, 91 (1947)
77. B.R. Taylor, M.B. Baker, J. Atmos. Sci. **48**, 112 (1991)
78. D. Gregory, Quart. J. Roy. Meteor. Sci. **127**, 153 (2001)
79. W.R. Cotton, R.A. Anthes, *Storm and Cloud Dynamics* (Academ. Press, San Diego, 1989)
80. K.A. Emanuel, *Atmospheric Convection* (Oxford Univ. Press, New York, 1991)
81. R.A. Houze, *Cloud Dynamics* (Academ. Press, San Diego, 1993)
82. P. Squires, Tellus **10**, 256 (1958)
83. A.J. Heymsfield, P.N. Johnson, J.E. Dye, J. Atmos. Sci. **35**, 1689 (1978)
84. T. Heus, H.J.J. Jonker, *J. Atmos. Sci.***65**, 1003 (2008)
85. T. Heus et al, *J. Atmos. Sci.***65**, 2581(2008)
86. J. Warner, Tellus **7**, 450 (1955)
87. W.R. Leaitch, G.A. Isaak, Atmosp. Environ. **25**, 601 (1991)
88. http://en.wikipedia.org/wiki/Liquid-water-content
89. T.G. Kyle, W.R. Sand, Science **180**, 1274 (1973)
90. V.F. Petrenko, I.A. Ryzhkin, J. Phys. Chem. **101B**, 6285 (1997)
91. V.F. Petrenko, R.W. Whitworth, *Physics of Ice.* (Oxford, Oxford UniversityPress, 1999)
92. J. Latham, Quart. J. Roy. Meteor. Soc. **89**, 265 (1963)
93. J.M. Caranti, A.J. Illingworth, S.J. Marsh, J. Geophys. Res. **90D**, 6041 (1985)
94. Y. Dong, J. Yallett, J. Geophys. Res. **97**, 20361 (1992)
95. J.G. Dash, B.L. Mason, J.S. Wettlaufer, J. Geophys. Res. **106**, 20395 (2001)
96. J. Nelson, M. Baker, Atmos. Chem. Phys. Discuss. **3**, 41 (2003)
97. J. Latham, Quart. J. Roy. Meteor. Soc. **107**, 277 (1981)
98. E.R. Williams, R. Zhang, J. Rydock, J. Atmosph. Sci. **48**, 2195 (1991)
99. B.L. Mason, J.G. Dash, J. Geophys. Res. **105**, 10185 (2000)
100. P. Berdeklis, R. List, J. Atmos. Sci. **58**, 2751 (2001)
101. J.P. Kuettner, Z. Levin, J. Atmos. Sci. **38**, 2470 (1981)
102. C.R.R. Saunders, J. Appl. Meteor. **32**, 642 (1993)
103. B.J.P. Marshall, J. Latham, C.R.R. Saunders, Quart. J. Roy. Meteor. Soc. **104**, 163 (1978)
104. V.F. Petrenko, S.C. Colbeck, J. Appl. Phys. **77**, 4518 (1995)
105. C.R.R. Saunders et al., Atmos. Res.**58**, 187(2001)
106. G.M. Caranti, A.J. Illingworth, Nature **284**, 44 (1980)
107. J. Hallett, C.R.R. Saunders, J. Atmos. Sci. **36**, 2230 (1979)
108. J.P. Rydock, E.R. Williams, Quart. J. Roy. Meteor. Soc. **117**, 409 (1991)
109. R.L. Ives, J. Franklin Inst. **226**, 691 (1938)

110. I.M. Imyanitov, *Electrization of Flights in Clouds and Precipitation.* (Leningrad, Gidrometeoizdat, 1970; in Russian)
111. D.R. Lide (ed.), *Handbook of Chemistry and Physics*, 86th edn. (London, CRC Press, 2003–2004)
112. W.D. Keith, C.P.R. Saunders, Atmosph. Res. **25**, 445 (1990)
113. C.P.R. Saunders, in *Planetary Atmospheric Electricity*. ed. by F. Leblanc et.al. (Heidelberg, Springer, 2008) p. 335
114. E.E. Avila, M.B. Baker, E.R. Jayaratne, J. Latham, C.P.R. Saunders, Quart. J. Roy. Meteor. Soc. **113**, 1669 (1999)
115. T. Takahashi, J. Atmos. Sci. **35**, 1536 (1978)
116. R.G. Pereyra, E.E. Avila, N.E. Castellano, C.P.R. Saunders, J. Geophys. Res. **105**, 20803 (2000)
117. C.P.R. Saunders, H. Bax-Norman, C. Emersic, E.E. Avila, N.E. Castellano, Quart. J. Roy. Meteor. Soc. **132**, 2653 (2006)
118. V.F. Hess, *Phys. Zs.***113**, 1084 (1912)
119. C.T.R. Wilson, Proc. Roy. Soc. **37A**, 32D (1925)
120. R.A. Mullican, Proc. Nat. Acad. Sci. **12**, 48 (1926)
121. R.A. Mullican, Science **81**, 211 (1935)
122. R.A. Mullican, *Electrons(- and +), Protons, Photons, Neutrons and Cosmic Rays* (Chicago University Press, Chicago, 1935)
123. S. Swordy, D. Mueller, P. Meyer et al., Astrophys. J. **349**, 625 (1990)
124. C. Amster et al., *Phys. Lett.***667B**, 1 (2008)
125. A. Erediato, *Cosmic-rays physics.* (2008)
126. B.B. Rossi, *Cosmic Rays* (McGraw-Hill, New York, 1964)
127. A.M. Hillas, *Cosmic Rays* (Pergamon Press, Oxford, 1972)
128. K. Sakurai, *Physics of solar cosmic rays* (Tokyo University Press, Tokyo, 1974)
129. T.K. Gaisser, *Cosmic Rays and Particle Physics.* (Cambridge University Press, Cambridge, 1990)
130. R.K.F. Grieder, *Cosmic Rays of Earth: Researcher's Reference Manuel and Data Book* (Elsevier, Amsterdam, 2001)
131. L.I. Miroshnichenko, *Solar Cosmic Rays* (Kluwer, Dordrecht, 2004)
132. V.I. Dorman, *Cosmic Rays in the Earth Atmosphere and Underground* (Kluwer, Dordrecht, 2004)
133. G.A. Bazilevskaya et al., in *Planetary Atmospheric Electricity,* ed. by F. Leblanc et al. (Heidelberg, Springer, 2008) p. 149
134. J. Elster, H. Geitel, *Phys. Zs.***2**, 590 (1901)
135. C.T.R. Wilson, Roy. Soc. Proc. **68**, 151 (1901)
136. J. Elster, H. Geitel, *Phys. Zs.***4**, 526 (1903)
137. A. Hirsikko et al., Boreal Envir. Res. **12**, 265 (2007)
138. K. Nagaraja et al., *Radiat. Meas. J. Atm. Sol-Terr. Phys.***68**, 757 (2006)
139. A. Hensen, J.C.H. van der Haage, J. Geophys. Res. **99**, 10693 (1994)
140. R.G. Harrison, H. Tammet, in *Planetary Atmospheric Electricity.* ed. by F.Leblanc et al. (Heidelberg, Springer, 2008) p. 107
141. H.V. Neher, J. Geophys. Res. **72**, 1527 (1967)
142. H.V. Neher, J. Geophys. Res. **76**, 1637 (1971)
143. G. Plotzer, *Zs. Phys.***102**, 23 (1936)
144. M. Nicolet, Planet. Space Sci. **23**, 637 (1975)
145. A. Kryvolutsky et al., Phys. Chem. Earth **27**, 471 (2002)
146. M.A. Ruderman, J.W. Chamberlain, Planet Space Sci. **23**, 247 (1975)
147. H. Svensmark, E. Frils-Christensen, J. Atmos. Terr. Phys. **59**, 1225 (1997)
148. L.D. Landau, E.M. Lifshitz, *Quantum Mechanics* (Pergamon Press, Oxford, 1965)
149. I.I. Sobelman, *Atomic Spectra and Radiative Transitions* (Springer, Berlin, 1979)
150. http://en.wikipedia.org/wiki/corona
151. M.A. Schwanden, *Physics of the Solar Corona* (Praxis Publishing, Chichester UK, 2005)

152. L. Golub, J.M. Pasachoff, *The Solar Corona* (Cambridge University Press, Cambridge, 2010)
153. S.I. Akasofu, S. Chapman, *Solar-terrestrial Physics* (Claredon Press, Oxdord, 1972)
154. Z. Svestska, *Solar Flares* (Reidel Publ.Co, Dordrecht, 1976)
155. R.K. Janev, L.P. Presnyakov, V.P. Shevelko, *Physics of Highly Charged Ions* (Springer, Berlin, 1985)
156. B.M. Smirnov, *Physics of Atoms and Ions* (Springer, New York, 2003)

Chapter 4
Electric Processes in Atmosphere

Abstract The global electric circuit of the Earth's atmosphere may be modeled by a spherical capacitor whose low electrode is the Earth's surface charged negatively and the positively charged upper electrode is located at an altitude of a few kilometers. The atmospheric conductivity increases with an increasing altitude. The air is charging as a result of precipitation of charged aerosols and by lightning which are distributed nonuniformly over the globe. The power of atmospheric electricity is five to six orders of magnitude less than that due to water evaporation and circulation. Cosmic rays are responsible for air ionization with formation of molecular ions in the troposphere. Global powers of some atmospheric processes are compared with those of electrical processes. Three types of ionization waves propagate through the atmosphere under the action of an electric field—an electron avalanche, streamer, and leader—and when a leader creates a conductive channel between a cloud and Earth, a charge is transferred through this channel from a cloud to Earth. Subsequently a transferred charge propagates underground and may cause destruction.

4.1 Peculiarities of Atmospheric Electricity

Electric phenomena in the Earth's atmosphere include processes of formation of electric charges in the atmosphere and propagation of electric currents through the Earth's atmosphere. Atmospheric electric processes are secondary with respect to water circulation [1] and are governed by known physical laws that allow one to study them from the standpoint of general physical positions. Investigations of electrical processes in the Earth's atmosphere and on the Earth's surface last several centuries. In particular, the study by B. Franklin (United States) in the middle of the eighteenth century with a kite launched on a metal wire in a thundercloud showed the identity of atmospheric electricity with that obtained by friction on a revolving circle. According to Franklin, there are two forms of electric currents, positive and negative, and it is possible to remove electric charges from the atmosphere by using metal wires rising from buildings and going into the Earth. Understanding of the electrical nature of thunderstorms allowed the construction of the lightning rod that arose in the United States in 1752 and at approximately the same time in Czechia [2];

B.M. Smirnov, *Microphysics of Atmospheric Phenomena*,
Springer Atmospheric Sciences, DOI 10.1007/978-3-319-30813-5_4

this was the first application of atmospheric electricity. At the same time, investigations by M. V. Lomonosov (theoretically) and G. V. Richter, who was killed by a lightning discharge during measurements of atmospheric electric currents, allowed us to understand some aspects of atmospheric electricity. In particular, Lomonosov assumed that atmospheric electricity results from friction of dust particles and other atmospheric particles with water drops. This corresponds in principle to the contemporary scheme of charge formation in the atmosphere as a result of collisions of snow (graupels) and liquid aerosols with micron-size ice particles that leads to particle charging [3–12]. Experience obtained in the study of electrical atmospheric processes resulted in understanding the nature of these processes and led to the creation of new physical directions, as plasma physics and physics of high energies.

It should be noted that C. T. R. Wilson made a large contribution to the development of the physics of atmospheric electricity at the beginning of the twentieth century. He constructed and modified a condensation chamber for ion detection which is known as the Wilson chamber and was awarded the Nobel prize in 1927. Measurements on the basis of this chamber together with balloon measurements and the analysis of results allowed Wilson to understand and formulate some principles of atmospheric electricity. Namely, Wilson [13, 14] demonstrated the origin of ions in the atmosphere. He first fulfilled measurements for electric currents in the atmosphere and electric field strengths near the Earth's surface [15, 16]. Wilson proved [17, 18] that the Earth is charged negatively, and cosmic rays (X-ray radiation and fast protons and electrons from the Sun penetrating the lower Earth atmosphere) cause discharging of the Earth [19]. Wilson suggested a global circuit model of the atmosphere, and the charge balance follows from Earth charging by lightning and discharging by ion fluxes resulting from the action of cosmic rays [18, 20]. Charged clouds may lead to electrical breakdown in the upper layers of the atmosphere [19, 20] because of low air density. Glowing objects from this breakdown are now called sprites. Next, Wilson showed the formation of fast electrons at high altitudes of the Earth's atmosphere [21]. This atmospheric phenomenon is known as running electrons [22].

Progress in the study of thunderstorm phenomena proceeded in the first half of the twentieth century and resulted from lightning photographed by the Bois chamber, measurements of electric and magnetic fields of lightning, and the spectroscopic analysis of lightning radiation. This all allowed the construction of a physical picture of thunderstorms in the atmosphere. In particular, Fig. 4.1 shows the distribution of electric charges in a thunderstorm cloud [23] that follows from measurements of parameters of thunderstorms in South Africa and was used for demonstrating the charge distribution in a cloud. These data give a physical picture of lightning as a physical phenomenon [24–30]. One can add to this radar diagnostics of clouds, registration of various waves in thunderstorm weather, laboratory modeling of electric processes in the atmosphere, and the theoretical analysis of electric processes in the atmosphere. Methods of registration of thunderstorm processes in the atmosphere and the physical picture of electrical breakdown of the atmosphere resulting from such measurements are given in [31–35].

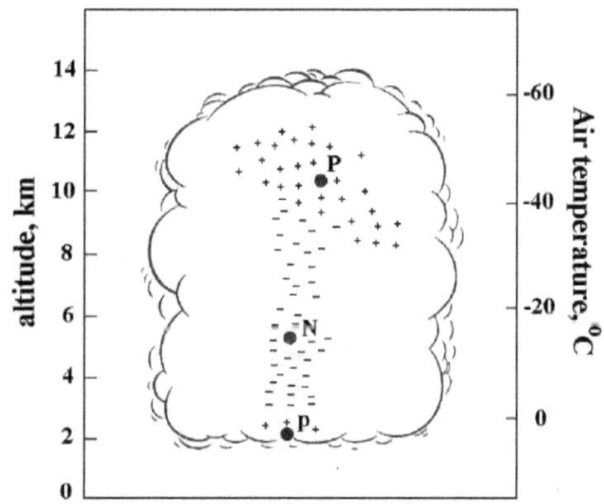

Fig. 4.1 Typical distribution of charges in a thunderstorm cloud [23] constructed on the basis of measurements of cloud electric fields in South Africa. In this case the positive charges are $P = 40\,\mathrm{C}$, $p = 10\,\mathrm{C}$; the negative charge is $N = -40\,\mathrm{C}$

The analysis of experimental data at the first stage of atmospheric research exhibits that electricity is generated inside the atmosphere [36]. As follows from Fig. 4.1, basic electrical processes proceed in an atmospheric layer of a depth of approximately 10 km. Next, a typical electric field strength due to separation of charges in a cloud is below 1 kV/cm, and the voltage for a charged cloud layer is on the order of tens of megavolts. Discharging of clouds results in electrical breakdown between clouds and Earth in the form of lightning and provides the negative Earth charging [18, 20, 37–39]. The negative Earth charge creates an electric field in the atmosphere, so that weak electric currents occur under the action of the electric field leading to the Earth discharging. The properties and dynamics of atmospheric clouds [38, 40–46] are an important element of atmospheric electricity.

Note that we deal with the average values of atmospheric electric currents and other parameters of atmospheric electricity [47–49] which are summed over the globe and are averaged over time. In this consideration we utilize the dipole model of the cumulus cloud [17, 19, 50, 51] in contrast to Fig. 4.1, which is described better by the tripolar model [52–54]. The dipole moment of a charged cloud is from 20 up to $100\,\mathrm{C \cdot km}$ [55] with a distance between charge centers on the order of 1 km.

On the basis of concepts that have existed, at least, from the middle of the twentieth century, one can represent a general scheme of atmospheric electricity. The first process of this scheme consists of the formation of positively and negatively charged aerosols as a result of collisions of neutral water aerosols located in different aggregate states. As an element of clouds, charged aerosols move together with clouds, but a difference in mass of positive and negative aerosols leads to a charge separation as a result of gravitational falling of the aerosols, and the charges of lower and higher cloud edges become different (see Fig. 4.1). Subsequent electrical breakdown of the atmosphere in the form of lightning leads to the Earth charging, mostly negatively, and this creates an atmospheric electric field. The Earth discharging process proceeds due to currents of positive and negative molecular ions formed under the action of cosmic rays, and this takes place in a dry part of the atmosphere that is free from clouds.

Of course, this scheme does not account for all the variety of electrical processes in the atmosphere, but it includes the main channels of atmospheric electricity, and we focus in this book on these stages of atmospheric electricity. These processes include the elementary processes of collisions of atomic particles in the atmosphere (i.e., collisions involving electrons and atmospheric molecules, atoms, and ions), and the subsequent analysis to a great extent is based on elementary processes in the atmosphere. These elementary processes determine other properties of the atmosphere, and we include in consideration also phenomena in the lower part of the ionosphere and its properties at altitudes below 200 km. This allows us to analyze phenomena that adjoin atmospheric electricity, for example, sprites, that is, glowing atmospheric structures at altitudes below 100 km. Because most elementary processes in the atmosphere were studied in detail, this is justified.

The Earth's atmosphere is a conductive matter, as first shown by Coulomb [56]. Wilson [15, 16] demonstrates that an atmospheric current in fair weather is determined by formation of ion pairs. The subsequent measurement of the rate of ion production [57] led to the understanding of the role of cosmic rays in atmosphere ionization [19]. In addition to this, the Earth is charged negatively [17, 18] and therefore an electric field exists in the lower atmosphere. This allows one to consider the Earth an electric machine on the basis of the global circuit model [18, 20, 58], and the parameters of this model follow from averaged results of measurements. One can model the lower atmosphere by a spherical capacitor, and the electric current density between plates of this capacitor is assumed to be independent of the geographic point, so that electrical parameters of this capacitor are determined by average electrical parameters of the Earth's atmosphere and the Earth as a whole. The Earth is charged negatively, and the average electric field strength near the Earth's surface is $E_o = 130\,\mathrm{V/m}$ which corresponds to the Earth's negative charge $Q = E_o R_\oplus^2 = 5.8 \times 10^5\,\mathrm{C}$ [47, 59] ($R_\oplus = 6300\,\mathrm{km}$ is the Earth radius). The Earth's voltage with respect to a surrounding space is equal approximately to $U_o = 240{-}300\,\mathrm{kV}$ [60–62]. Within the framework of a spherical capacitor model, the electric field strength drops linearly with altitude, and the voltage is zero at the upper electrode of this capacitor. From this we obtain for a distance between electrodes $L = 2U_o/E_o \approx 5{-}6\,\mathrm{km}$; that is, electrical processes in the Earth's atmosphere proceed at altitudes of several kilometers.

Let us use an appropriate model for the Earth's electrical system assuming that the Earth's charge is screened by an excess positive charge at atmospheric altitudes up to L, so that the total positive atmospheric charge compensates the Earth's charge. This takes place if positive and negative molecular ions are present in the atmosphere with the difference ΔN of their number densities. Then the Poisson equation for the electric field strength $E(h)$ as a function of the altitude h has the form

$$\frac{dE}{dh} = 4\pi e \Delta N \tag{4.1}$$

Assuming the difference of the number densities of positive and negative ions ΔN to be independent of the altitude, obtain the solution of this equation:

$$E = E_o\left(1 - \frac{h}{L}\right), \quad L = \frac{E_o}{4\pi e \Delta N} \tag{4.2}$$

From this it follows

$$L = \frac{2U_o}{E_o}, \quad \Delta N = \frac{E_o}{4\pi e L} = \frac{E_o^2}{8\pi e U_o}$$

Thus, we have on the basis of the above models $L \approx 5\,\mathrm{km}$ and $\Delta N \approx 3\,\mathrm{cm}^{-3}$. These parameters have a qualitative character because atmospheric parameters vary in time and space. In addition, we note that according to fulfilled measurements the average current density discharged by the Earth is $2.4 \times 10^{-16}\,\mathrm{A/cm^2}$ over land and $3.7 \times 10^{-16}\,\mathrm{A/cm^2}$ over oceans [35, 47]. This corresponds approximately to the total discharged current for the Earth's atmosphere $I = 1700\,\mathrm{A}$ [37] that gives the average current density in the Earth's atmosphere as $i = I/S = 3.3 \times 10^{-16}\,\mathrm{A/cm^2}$, where $S = 4\pi R_\oplus^2 = 5.1 \times 10^{18}\,\mathrm{cm^2}$ is the area of the Earth's surface. From this one can determine the average conductivity of the Earth's atmosphere $\sigma = i/E = 2.5 \times 10^{-14}\,\mathrm{S/m}$, where we take $E = 130\,\mathrm{V/m}$, and the conductivity unit Siemens is $S = 1/\mathrm{Ohm}$ [63].

The conductivity of the lower atmosphere is determined by molecular ions that have nearby values of mobilities approximately $2\,\mathrm{cm^2/(V \cdot s)}$ [64]. This gives a typical number density of ions $N_i = \sigma/2K = 400\,\mathrm{cm}^{-3}$, where the factor 2 accounts for the currents of negative and positive ions under the action of the Earth's electric field of strength $E = 130\,\mathrm{V/m}$ having opposite directions. The above mobility leads to a typical time $\tau = Q/I \approx 6\,\mathrm{min}$ of the Earth's discharging. As can be seen, a mechanism of the Earth's charging is required to support the negative Earth charge. This process results from the electrical breakdown between clouds and the Earth's surface through lightning [37] and also owing to precipitation of negatively charged aerosols. Lightning alone transfers a negative charge to the Earth's surface and provides the Earth's charging [38, 39]. In turn, the creation of a high voltage between clouds and the Earth's surface is necessary for lightning formation [32, 34] that determines the lightning specifics. Although we consider negative charging of the Earth by lightning, the above analysis does not prohibit positive Earth charging as a result of electrical breakdown between clouds and the Earth's surface. According to measurements [65], the ratio of numbers of negative and positive lightning is 2.1 ± 0.5, and the ratio of currents transferred by negative and positive charges is 3.2 ± 1.2 [65].

Continuing to exploit the global electric model of the Earth's atmosphere, we obtain that the atmosphere resistance R_a is equal to

$$R_a = \frac{U_o}{I} = 150\,\Omega \tag{4.3}$$

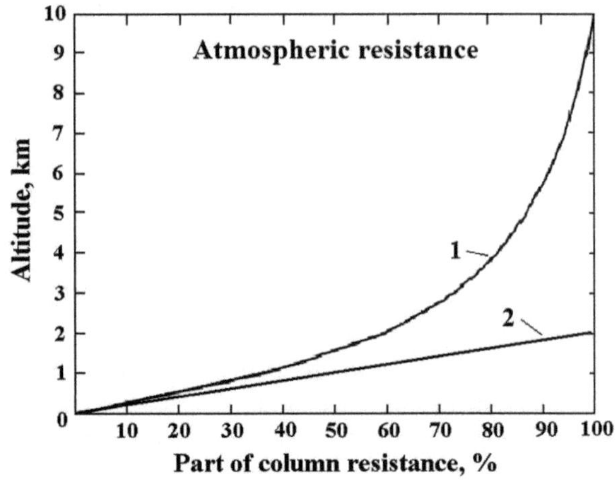

Fig. 4.2 Contribution to the atmospheric resistance from atmospheric layers up to indicated altitudes: *1* according to the analysis [61], *2* the model with $\sigma = const$ at $h < h_o$

Figure 4.2 gives the distribution of the average atmosphere resistance over altitudes [60, 61]. Indeed, the atmosphere resistance is given by

$$R_a = \int_0^\infty \frac{dh}{\sigma(h)}, \tag{4.4}$$

where the total conductivity $\sigma(h)$ is averaged over the Earth's surface. As can be seen, the main contribution to this resistance follows from layers with small conductivity. We give in Fig. 4.2 the above simple approximation where the electric field is concentrated in the lowest altitudes $h \leq h_o$ with the constant conductivity $\sigma(h) = const$. In this approximation on the basis of the ionosphere voltage $U_o = 250\,\text{kV}$ and the electric field strength $E_o = 130\,\text{V/m}$ we have $h_o = U_o/E = 2\,\text{km}$. This approximation is represented in Fig. 4.2 by curve 2.

The global electric circuit model of the atmosphere requires high conductivity of the underground and upper atmosphere layers [66]. This is fulfilled for the lower atmosphere. The conductivity of the underground varies from $10^{-7}\,\text{S/m}$ for rocks up to $10^{-2}\,\text{S/m}$ for clay and limestone soil [66, 67]. The average conductivity of oceans is $3\,\text{S/m}$ [68] and depends on the presence of sodium and potassium salts in the water. The conductivity of the lower ionosphere at the altitude of 80 km is on the order of $10^{-7}\,\text{S/m}$ [60, 61] and exceeds that of the troposphere ($\sim 10^{-14}\,\text{S/m}$). Indeed, the ionosphere conductivity is created by electrons; the electron number density of the ionosphere exceeds that of tropospheric ions and the number density of molecules for scattering charged atomic particles is lower in the ionosphere. Thus, the conditions of the global electric circuit model hold true.

In considering a scheme of the processes of atmospheric electricity, we start from water circulation because atmospheric electricity is its secondary phenomenon. Based on [37–39], we represent in Fig. 4.3 basic processes of water circulation that

Fig. 4.3 Processes of water circulation in the Earth's atmosphere that lead subsequently to circulation of atmospheric electricity

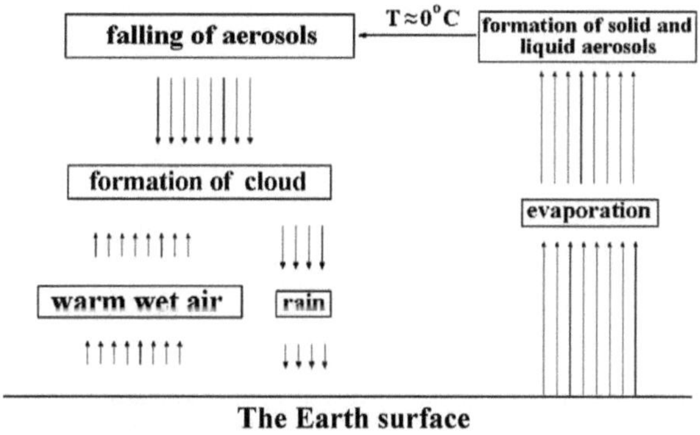

lead to a chain of electric processes in the Earth's atmosphere. The beginning of these processes is evaporation of water from the Earth's surface. Usually evaporated water is found in the form of vapor (free water molecules) near the Earth's surface until the pressure of a water vapor is less than the saturated vapor pressure at the Earth's temperature. But when this vapor propagates up in layers with low temperatures, and its partial condensation takes place starting at certain altitudes where the vapor pressure exceeds the saturated vapor pressure at a current air temperature, vapor excess is transformed in a condensed phase. As a result, aerosols, that is, micron-size water particles, are formed. They fall down and are incorporated into forming clouds and mist. The densest clouds, cumulus clouds, contain charged aerosols, and an aerosol charge hinders association of aerosols as a result of their contact. Streams of wet and warm air from the Earth cause subsequent growth of aerosols as a result of the attachment of water molecules to aerosols and lead to electrical discharge in the form of lightning. Neutral aerosols grow quickly ($\sim$10 min) and fall on the Earth's surface in the form of rain. Precipitation of water to the Earth's surface takes place from other clouds containing neutral aerosols, as well as from mist, fog, and haze. In the latter case, cold atmosphere regions with aerosols are located close to the Earth's surface.

Some processes of water circulation accompany electric processes in the atmosphere which are represented in Fig. 4.4. They include a chain of processes that lead to Earth's charging and discharging. Atmospheric electricity starts from the formation of charged aerosols in the atmosphere as a result of collisions of aerosols located in different aggregate states. Indeed, water may be found in liquid, solid (ice), and amorphous (snow) states. Water in accordance with its chemical structure is a weak electrolyte that contains H_3O^+ and OH^- ions. If two aerosols in different aggregate states touch each other, a contact potential occurs at the interface. This means that one of the aerosols contains an excess of H_3O^+ ions, whereas another aerosol contains an excess of OH^- ions. Hence, after separation of aerosols they will be charged. The same takes place if aerosols in a certain aggregate state collide with a metallic rod. Charging as a result of collisions of aerosols located in different aggregate states was first demonstrated experimentally in 1957 [3].

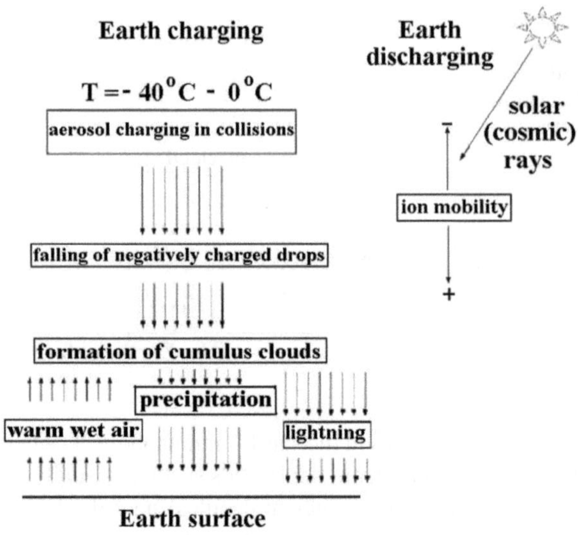

Fig. 4.4 Electric processes in the Earth's atmosphere as a sum of charging and discharging processes. Charging of the Earth includes the process of collision of two water aerosols in different aggregate states that leads to formation of charged aerosols, gravitational falling of charged aerosols that separates atmospheric charges, and electrical breakdown of the atmosphere in the form of lightning. The Earth's discharging results from drift of molecular ions (positive and negative) formed in a dry atmosphere under the action of solar or cosmic rays [69]

If aerosols become charged after collisions, this does not lead to the origin of an electric field because formed aerosols are located in the same space region. An electric field occurs if aerosols fall under the action of gravitational forces, and the falling velocities of positively and negatively charged aerosols are different. As a result, an atmospheric charge is separated and an electric field arises. Falling of charged aerosols corresponds to the passing of an electric current to the Earth and leads to the Earth's charging, mostly negatively, if charged aerosols reach the Earth's surface. But if charged aerosols are found above land, they form a cumulus cloud and the lower edge of this cloud carries a charge. Then the current causes the Earth's charging as a result of electrical breakdown of the atmosphere by means of lightning.

4.2 Lightning

Lightning is an important stage in the chain of electrical processes in the Earth's atmosphere. Detailed experimental study of this phenomenon exhibits the complexity of its nature [31–34] and theoretical analysis of the development and nature of lightning [70–74] allows one to understand a detailed character of separate stages of this phenomenon. Below we do not dwell on these details, but describe lightning in general and analyze some aspects of this problem. This electric breakdown of air proceeds due to a cloud's electrical charge and may be both between a cloud and the Earth's surface and between clouds. A basic stage of lightning is an electric current through a conducted lightning channel that determines the Earth's charging. We give average lightning parameters that are used below. A cloud-earth electric field results from charge separation in the atmosphere due to the drop of charged aerosols, and a high electric charge is accumulated on the lower edge of a cumulus cloud, more often negative. The electric potential of the lower edge of a cumulus cloud is 20−100 MV [75] with respect to the Earth's surface, and lightning passes a distance of several kilometers at a cloud-earth distance of 1–5 km (e.g., Fig. 4.5). Lightning transfers a part of a cloud charge to the Earth. An average charge transferred by middle lightning is approximately 20 C, and each lightning consists of 3–4 flashes [32, 34]; that is, lightning consists of several current pulses, flashes, which are propagated along almost the same channel. Each flash of middle lightning transfers a charge of 4−5 C, and the subsequent phase of a weak current follows after a flash [31–34]. Air is heated and expands as a result of the propagation of an electric current, and the air pressure in the conductive channel is equalized by the pressure of surrounding air. If the rate of air expansion exceeds the sound speed 3.3×10^4 cm/s [76], a shock wave arises. In addition, Fig. 4.6 shows the hierarchy of typical times in lightning development, and these times are given from observational data [31–34].

The first stage of lightning evolution consists in propagation of a stepwise leader [31–34] which creates a conductive channel. A stepwise leader is a weakly glowing phenomenon that propagates along a broken line with a segment length of tens of meters. The first stage of propagation of each signal has a streamer nature, and a

Fig. 4.5 Lightning discharge propagated along one channel [77]

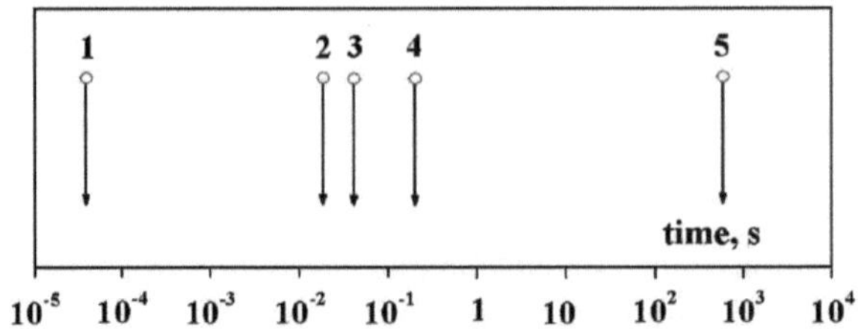

Fig. 4.6 Hierarchy of times for lightning processes: *1* time of the maximum lightning pulse, *2* total time of propagation of stepwise leader, *3* time interval between neighboring flashes, *4* total lightning time, *5* time of discharging of cloud aerosols

plasma case is formed near the streamer head that provides partial air ionization near the wavefront [71–74]. The average velocity of propagation of the stepwise leader is on the order of 10^7 cm/s which corresponds to the electron drift velocity in these electric fields. When a conductive channel is created, an electric current propagates along it and causes glowing of the channel plasma.

The following stage of lightning evolution is the recurrent stroke which proceeds with a velocity up to 5×10^9 cm/s that corresponds to propagation of an electric signal in conductors. The recurrent stroke is short and lasts some μs, and the lightning discharge includes all the stages of channel creation and propagation of the electric current for an individual flash lasts less than 10^{-3} s. However, during this time the conductive channel expands, and an inserted energy is consumed on air heating, dissociation, and partial ionization both in a conductive channel and in adjoining regions.

Note that the lightning electrical energy is high enough and is concentrated in a restricted space region. In order to demonstrate the scale of this energy, let us compare the energy of a lightning flash that carries a charge of 5 C at the voltage 50 MV with an energy of a bullet moving with a velocity 760 m/s. Equalizing the lightning energy and kinetic energy of the bullet, we obtain the gun shell mass to be 3 ton. This estimation demonstrates the danger of lightning for houses and buildings without a lightning protector that allows carrying a lightning charge into the Earth due to an artificial channel. From this it follows also that a carried charge is dangerous for nearby objects.

One can add to this that the processes of water evaporation from the Earth's surface proceed more or less uniformly over the Earth's surface, as well as charging processes in cumulus clouds, whereas the processes involving lightning proceed in small regions of the Earth's surface. Therefore the local energy fluxes at charge transfer from a cloud to the Earth's surface are high enough. Indeed, the energy flux in water evaporation that is consumed subsequently as water condensation is $7 \times 10^{-3}\,\mathrm{W/cm^2}$ on average, and the average energy flux at the Earth's discharging is $1 \times 10^{-10}\,\mathrm{W/cm^2}$. The local energy fluxes in lightning are higher because they relate to a low area of the Earth's surface. Taking according to the above data the rate of lightning discharges to be 100 flashes per second [75, 78, 79], and energy consumed on an area 1 m^2 during a specific time on the order of 1×10^{-4} s, and assuming the development of lightning to proceed under the action of a voltage 50 MV, we obtain for the local power flux $1 \times 10^9\,\mathrm{W/cm^2}$. This flux characterizes power propagation, rather than power deposition, but this exhibits a high power density in comparison with an initial process of water transport.

In order to demonstrate the scale of the thunderstorm phenomenon, let us use a simple model of a cumulus cloud in the form of a charged disk with a charge $Q = 20\,\mathrm{C}$ that is transferred by middle lightning. Requiring the electric field strength near the disk surface E to be below the breakdown electric field strength $E_c = 30\,\mathrm{kV/cm}$ for dry air, we obtain that the charge density at the disk is below $\sigma = E/(2\pi) = 5 \times 10^{-9}\,\mathrm{C/cm^2}$ which corresponds to a disk radius above $R = \sqrt{Q/(\pi\sigma)} = 400\,\mathrm{m}$. This demonstrates the threshold character of thunderstorm phenomena, wherein a cloud size must exceed a certain threshold value. Indeed, thunderstorm phenomena are developed in volcanic dust if a cloud size exceeds 500 m, and the size of a usual cloud must exceed 3 km for thunderstorms in atmospheric air.

On the basis of this consideration one can construct a simple model of lightning by representing it as an electrotechnical scheme whose elements are resistance, capacity, and inductance. In spite of the roughness of this approach based on the analogy between processes of atmospheric electricity and electrotechnical devices, this approach may be useful for understanding some aspects of electrical processes in gases (e.g., [71–73, 80]. Below we apply this approach to propagation of the electric current along the conductive channel of lightning. This model is represented in Fig. 4.7. Here a cloud is modeled by a disc with elementary charges that are fastened to certain points of these electrodes, the Earth's surface is modeled by a finite plane of high conductivity, and a metal rod may join these electrodes at some time in order to provide charge transfer between electrodes. The disk conductivity in transversal directions is assumed to be large, thus this scheme models a transfer of an electric charge between electrodes.

Let us estimate the accuracy of operation with averaged parameters over the globe. Approximately 100 lightning flashes are observed on average over the globe according to observed data [75, 78, 79]. More precise observations of flashes are fulfilled on the basis of the NASA satellite [81], according to which 44 lightning flashes are observed per second all over the globe. Other measurements of the frequency of lightning flashes give 70 flashes per second [82] and 45 flashes per second [83].

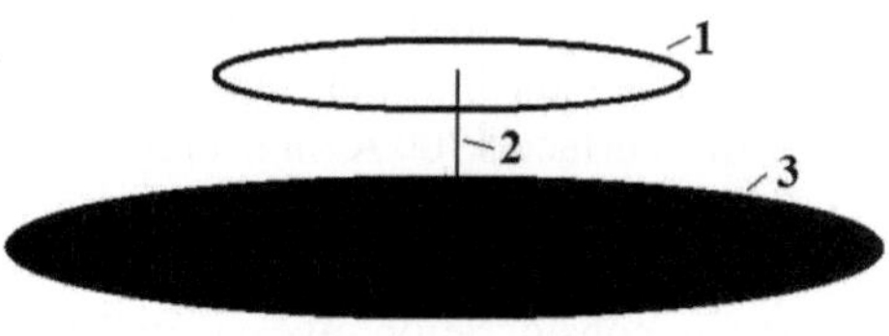

Fig. 4.7 Simple lightning model: *1* charged cloud, *2* metal rod for transferring a charge from a charged cloud to the Earth, *3* Earth's surface

Fig. 4.8 Typical lightning of Catatumbo [84]

A degree of coincidence of the above values testifies to their accuracy. Next, the most common observations of thunderstorm phenomena relate to continental and coastal areas, whereas inside oceans and in the polar regions they are not practically observed. Approximately 70 % of thunderstorms are observed in a tropical land in accordance with the concept that thunderstorms are observed in warm seasons.

A champion for rates of flashes is a village Kifuki in the Democratic Republic of the Congo. This village is located at an altitude of 975 m above sea level and the intensity of lightning flares equals $160/(\mathrm{km}^2 \cdot \mathrm{year})$ whereas an average of this value over the globe is $3/(\mathrm{km}^2 \cdot \mathrm{year})$. Another example of a region with a higher probability of thunderstorms is the entry of the Catatumbo River and Maracaibo Lake in Venezuela where thunderstorms are often called the Catatumbo phenomenon (Fig. 4.8) [84–86]. In this region thunderstorms are observed at night and last approximately 10 h for 150 days per year. The total number of lightning flashes is 1.2×10^6 per year. The Catatumbo region is surrounded on three sides by ridges of the Andes and Cordilleras 3700 m in height above sea level which causes a specific character of air fluxes. In addition, methane emerges from surrounding marshes, and its presence in the atmosphere simplifies lightning generation. Lightning arises often between clouds and reaches 10 Km in length. Assuming that each flash transfers a charge of 5 C on average [31], one can obtain that the contribution of the Catatumbo region to the lightning current is 0.2 A compared to the total electric current to the Earth's surface 1700 A on average. Thus, thunderstorms are distributed nonuniformly over the globe surface. Because we deal below with averaged parameters for atmospheric electrical processes in the analysis of atmospheric electricity, the subsequent consideration gives a qualitative picture of this phenomenon.

4.3 Ionization Wave in Air

One can see an analogy of lightning with laboratory gaseous discharge that is, by definition, a pass of an electric current under the action of an electric field. Being guided by this analogy, we analyze briefly some forms of gas discharges in atmospheric air and extract those that are close to lightning by their nature. The simplest form of gas discharge as a self-consistent system is based on the Townsend ionization equilibrium [87–92] which corresponds to a uniform partially ionized gas. This ionization equilibrium is realized under restricted gas sizes at a gas pressure below the atmospheric one. Figure 4.9 gives a simplified distribution of the electrical potential if the gas discharge is realized in a cylinder tube of small diameter compared to its length. This figure exhibits that a gas discharge plasma inside the tube is a self-consistent system. Indeed, one can extract in Fig. 4.9 two principal regions of gas discharge, the cathode region, where reproduction of electrons takes place, and the positive column, where the electric field strength is relatively small and supports the electric current in the region of the positive column. Note that a gas discharge plasma chooses this character of the distribution of the electric potential in the tube such that this corresponds to the minimum total discharge voltage. This distribution of the electric potential over the tube length is realized both for glow discharge, where reproduction of electrons results from generation of secondary electrons as a result of the cathode bombardment by ions, and also in the arc, where electrons near the cathode are formed by electron thermoemission.

At high pressures, ionization equilibrium in a gas discharge has the form of an ionization wave that propagates between electrodes. The principles of this form of gas discharge were elaborated by Raether [93–102], Loeb [91, 103–107], and Mick [108–112], starting from the 1930s of the twentieth century. The concept of the streamer as an ionization wave consists in redistribution of the electric potential in a space between two electrodes, so that the electric field strength increases near the front of the ionization wave and exceeds the breakdown one there. This leads to intense gas ionization near the front of the ionization wave. Photoionization of a gas is an important element of this wave, thus a photon emitted near the front of the ionization wave is absorbed at some distance from it, and formed electrons ionize the gas in this region. This increases the velocity of propagation of the ionization wave. It should be noted that according to the ionization equilibrium the electric field strength exceeds the threshold for gas breakdown in some region. This is realized also

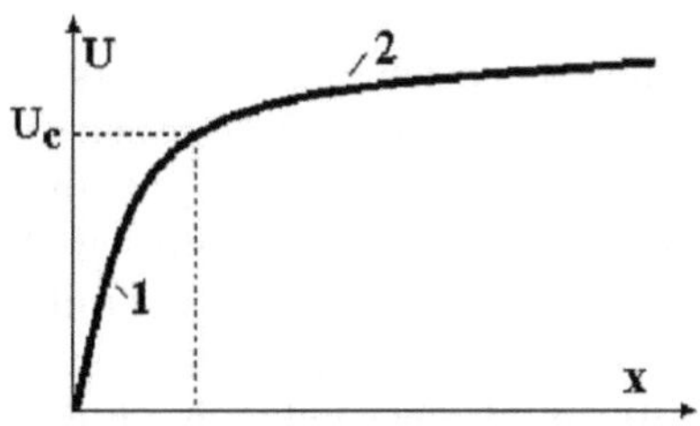

Fig. 4.9 Voltage distribution U along a discharge tube for the Townsend discharge form (x is a distance from the cathode): *1* cathode region, *2* positive column; U_c is the cathode region voltage

in the Townsend regime, and the choice between the Townsend regime and regime of the ionization wave of gas discharge depends on the degree of the overvoltage [113, 114].

The concept of origin and generation of the ionization wave under the action of an external electric field is the object of several books [72–74, 93, 108, 109, 115–120]. The character of processes of origin and propagation of the ionization wave is based on general principles of nonlinear phenomena in plasmas [121–127]), according to which plasma dynamics is determined by self-consistent fields resulting from the Coulomb interaction between electrons and ions that leads to collective plasma processes. Under an external field these collective processes cause specific redistribution of plasma components. The ionization wave and the streamer as a type of an ionization wave satisfy the general principles of plasma dynamics that now lead to amplification of the electric field near the wavefront.

Within the framework of the lightning model under consideration (Fig. 4.7) one can roughly divide the lightning process into two stages. The first stage corresponds to the connection of electrodes by a thin conductive wire, and the second stage is an electric current passing through this wire. In reality, the connection of a wire with electrodes is similar to the formation of a conductive channel as a result of propagation of an ionization wave. There are three forms of ionization waves in air under the action of high fields that may be considered as three stages of development of the ionization wave in air. The first type is the electron avalanche, a group of electrons formed from a seed electron by their multiplication in a high electric field. When a number of electrons in an avalanche becomes enough large and reaches roughly $\sim 10^8$ electrons [100] drifting in an electric field, a forming plasma creates a strong electric field, and then an avalanche is transformed into a streamer as a result of redistribution of plasma electrons. In contrast to electron avalanches, where an external electric field strength must exceed the threshold for electrical breakdown, streamers may propagate at average fields that are less than the breakdown electric field. On the tail of a streamer the electric field drops as we move from its head, and electrons relax there to thermal energies. Then in electronegative gases such as air, electrons may be transformed in negative ions behind the streamer front. This leads to a decrease of the plasma conductivity; that is, streamers cannot propagate over large distances. But if negative ions are destroyed behind the ionization wave, this ionization wave may propagate over large distances. As we estimate below, this takes place if the temperature of the conductive channel exceeds 1000 K. This heating leads to formation of the leader, the ionization wave with a heated conductive channel.

A streamer form of gas discharge may be realized in atmospheric air [71, 72, 80, 93, 108, 109, 115, 117, 119, 120], but under stationary conditions this discharge is unstable and transfers in an arc [80, 120]. Barrier discharge is realized also at atmospheric pressure and its plasma consists in separate conductive filaments [128–130]. This discharge is burnt between two electrodes and a dielectric insertion on one electrode (or this electrode is covered by a dielectric film). As a result, the barrier discharge consists of many microdischarges—filamentary discharges [131–133]—with the following parameters of individual microdischarge [128–130]. A filament radius is on the order of 0.1 mm, the lifetime of an individual microdischarge is of a few nanoseconds, and microdischarge transfers a charge of $(10^{-10} - 10^{-9})$ C. The average energy of electrons for a plasma of barrier discharge is 1−10 eV, and in the barrier discharge of the power of 100 W approximately 10^{11} conductive channels are present simultaneously. Comparing parameters of the conductive lightning channel at the stage of its formation with barrier discharge of the filamentary structure, one can conclude that in spite of their visual analogy, these objects are different according to their nature. Indeed, microdischarges of a barrier discharge are supported by motionless force centers on the dielectric surface, and individual microdischarges are fastened to these centers, whereas such centers are absent in air. Hence filamentary discharge is unstable in air; these filaments are joined in one channel. One other gas discharge at atmospheric pressure, corona discharge [74, 134, 135], may be connected with thunderstorm phenomena in the atmosphere because high electric fields can create a corona near objects of low transverse sizes, as needles, rods, or wires. Near these objects the electric field is intensified, and electric currents are present only in the vicinity of these objects.

Let us consider the nature of ionization waves in more detail. If the electric field strength in a given space region exceeds the threshold for electron multiplication and a seed electron is located in this region, the exponential growth of a number of electrons proceeds in time until the field of this electron avalanche may be compared with an external field. One can estimate the size of an electron avalanche on the basis of the Meek criterion [93, 105, 108, 109, 116, 119], according to which the maximum size of the electron avalanche is compared to the Debye–Hückel radius of the avalanche plasma, and then an avalanche is transformed in a streamer. The current size of an electron avalanche is $r \sim \sqrt{D_e t}$, where D_e is the diffusion coefficient of electrons in air, and t is the time from the beginning of multiplication of a seed electron. We take as the maximum size of an electron avalanche the Debye–Hü ckel radius r_D of a forming plasma

$$r_D = \sqrt{\frac{T_e}{8\pi N_e e^2}}, \tag{4.5}$$

where T_e is the electron temperature and N_e is the number density of electrons. From the equation $r_D = r$ and $N_e = 3n/4\pi r^3$, where n is the number of electrons in an electron avalanche, we have

$$n \approx \frac{T_e r}{6e^2} \tag{4.6}$$

It is convenient to introduce a distance $r_o = \sqrt{6D_e t_o}$ at which an electron displaces during time t_o for one multiplication act. Here $D_e \sim 10^3\,\mathrm{cm}^2/\mathrm{s}$ is the diffusion coefficient of electrons in atmospheric air. In this consideration we assume that a distance r_o significantly exceeds the mean free path $\lambda \sim 0.1\,\mu$m of electrons in atmospheric air. Taking a number of avalanche electrons $n = \exp(k)$ at a time when an avalanche is transformed in a streamer, one can obtain from Formula (4.5) a number of multiplications k from a seed electron up to a streamer as

$$k = \ln \frac{T_e r_o \sqrt{k}}{6e^2} \tag{4.7}$$

For definiteness, we take for the specific electric field strength near the head of a streamer $E/N = 400\,\text{Td}$ that exceeds the threshold electric field strength fourfold for electrical breakdown of atmospheric air. At this specific electric field strength the specific first Townsend coefficient in nitrogen is $\alpha/N \approx 3 \times 10^{-17}\,\text{cm}^2$ [136]. Taking the electron drift velocity at this electric field strength to be $2 \times 10^7\,\text{cm/s}$ [137], one can find $t_o \approx 0.05\,\text{ns}$. From this we have $r_o = \sqrt{6D_e t_o} \approx 5\,\mu\text{m}$ which significantly exceeds the mean free path of electrons in atmospheric air that justifies the above assumption about the diffusion character of electron motion. Then Formula (4.7) gives for a number of electron generations $k \approx 16$. Because of the exponential dependence of a number of electrons on time, the number of electron generations k depends weakly on used parameters. This gives to the number density of electrons in the streamer head in the course of transition avalanche-streamer $N_e \sim 10^{13}\,\text{cm}^{-3}$. Note that according to the analysis [80, 120], the transition from an electron avalanche to a streamer at the reduced electric field strength in atmospheric air $E = 30\,\text{kV/cm}$ proceeds after $k = 17$ acts of electron multiplications. In addition, we give above the number of electrons in an electron avalanche when it is transformed in a streamer, that according to [100] is equal to $n \sim 10^8$. From this we find for this case $k = 18$. Summing these results, we find an average value $k = 17 \pm 1$. This gives the number of electrons doubling up to ripening of an electron avalanche

$$\kappa = \frac{k}{\ln 2} = 25 \pm 1 \tag{4.8}$$

Thus when development of an electron avalanche is finished, it forms the head of a streamer. Photons are emitted from the streamer head and, creating a seed electron ahead of the streamer front in a region with a high electric field strength, the streamer front displaces in a new region. As can be seen, photoionization of air before the front of the ionization wave is important for wave propagation. Let us determine a typical velocity v_{st} of streamer propagation. Because the cross section of photoionization in air in the course of streamer propagation is approximately $1 \times 10^{-18}\,\text{cm}^2$ [138], the mean free path of the photon that ionizes air before the streamer front is $\lambda_p \sim 2\,\text{mm}$. The size of an electron avalanche is $r = r_o\sqrt{k} \approx 20\,\mu\text{m}$, where $r_o \sim 5\,\mu\text{m}$ is the way in which an electron propagates during one act of multiplication. Thus the ionization wave moves with the electron drift velocity $v_d \sim 10^7\,\text{cm/s}$, and a signal propagates on a distance $\lambda_p \sim 2\,\text{mm}$ with the light speed. From this we find the streamer velocity as

$$v_{st} = \lambda_p \left(\frac{r}{v_d} + \frac{\lambda_p}{c} \right)^{-1} \tag{4.9}$$

which gives $v_{st} \sim 1 \times 10^9\,\text{cm/s}$. One can add to this that when a seed electron is formed at some distance from the streamer front, new electrons may propagate to the streamer front or outside it depending on the electric field direction. Correspondingly, streamers may be positive or negative in accordance with the direction of propagation of a new avalanche. Below we consider negative streamers.

Let us ascertain the role of formation of atomic negative ions O^- in streamer propagation. Evidently, a streamer propagates at electric fields near its front if the rate of air ionization in these fields exceeds the rate of formation of negative ions. Evidently, at the threshold electric field strength $E_c = 30\,\text{kV/cm}$ and atmospheric pressure at the reduced electric field strength $E/N = 100\,\text{Td}$ the rates of ionization and electron attachment are identical. If the specific electric field strength increases, the first Townsend coefficient increases sharply, whereas the corresponding value for dissociative electron attachment varies weakly. In particular, the reduced first Townsend coefficient in nitrogen is equal to $\alpha/N = 7 \times 10^{-20}\,\text{cm}^2$ at the specific electric field strength $E/N = 100\,\text{Td}$ [139]. In the nitrogen case we also have $\alpha/N = 4 \times 10^{-18}\,\text{cm}^2$ at the specific electric field strength $E/N = 200\,\text{Td}$ [140], $\alpha/N = 3 \times 10^{-17}\,\text{cm}^2$ at $E/N = 400\,\text{Td}$ [136], and $\alpha/N = 7 \times 10^{-16}\,\text{cm}^2$ at $E/N = 800\,\text{Td}$ [136]. In the same units the reduced coefficient of electron dissociative attachment to oxygen molecules in air does not exceed the value $\eta/N = 1 \times 10^{-18}\,\text{cm}^2$, because the maximum cross section for electron attachment corresponds approximately to this value [141–144]. Evidently, at the threshold of electron multiplication the rates of molecule ionization and electron attachment are identical, and at higher electric field strength the ionization process dominates.

The theoretical analysis of streamer propagation is based on the simplest equations—the balance equations for the number densities of electrons and ions, and the Poisson equation—taking into account redistribution of electrons and ions in the vicinity of the streamer front, and ions may be considered motionless because of a small ion mobility [145–147]. Although this problem would seem to be simple, its solution may be different depending on the connection used between plasma parameters, and mostly due to nonlinearity and nonstationarity of this plasma. The numerical analysis of the streamer problem [148] exhibits a narrow front of this ionization wave. In Fig. 4.10 we give parameters of the positive streamer in atmospheric air [149] as the dependence on a distance z from the wavefront for the electron number density N_e, the electric field strength E, and the rate of photoionization S_{ex} at the wave axis for atmospheric air under certain conditions [148]. Apart from details of this evaluation, we note that according to Fig. 4.10, the excitations and ionization rates are characterized by large gradients, and the ionization processes proceed in a narrow space region. From this it follows that the analytical solution of this problem is based on small gradients and hence is characterized by low accuracy. In addition, this ionization wave is close to the soliton (solitary wave) [150], where the electric field is concentrated in a narrow space region near the wavefront. One can conclude from this that it requires careful attention to results of analytical calculations that are represented in several books [71, 72, 80, 93, 108, 109, 115, 117, 119, 120]. On the other hand, this fact increases the role of numerical methods in streamer analysis [151]. In addition to this, according to the character of development, a streamer has a branch structure because in the course of its propagation some arms grow behind the wavefront and then these arms are branched. This structure is typical for fractal objects and analytical methods are not available for their analysis.

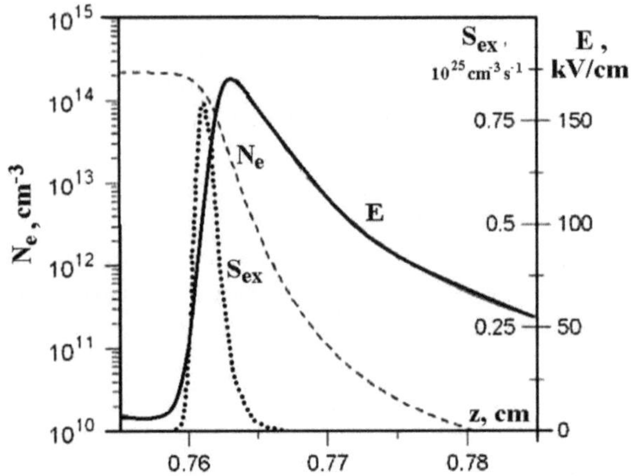

Fig. 4.10 Distribution of streamer parameters in air at the streamer axis before the streamer front [149]: x is a distance from the streamer front, E is the electric field strength, N_e is the electron number density, S_{ex} is the rate of excitation of nitrogen molecules in the state $N_2(C^3\Pi_u$ as a result of collisions of electrons with nitrogen molecules

Let us consider experimental methods of study of ionization waves that primarily are connected using the Wilson chamber [94–102]. In this case electrodes are placed inside the Wilson chamber, and a high pulse voltage is transferred to electrodes. The gas pressure in the chamber is decreased leading to a decrease of the gas temperature. Usually water or alcohols are impurities of a gas-filled chamber, and the saturated vapor pressure for these impurities is lower than the impurity pressure in the chamber. Then an impurity excess is condensed in the chamber with formation of liquid droplets, and electrons and ions are nucleation centers. Therefore this method of using various versions [93, 119] allows one to obtain the space distribution of electrons and ions at a given time, that is, at a certain stage of development of the ionization wave. The results of such investigations [93, 94, 105, 108, 116] allow checking the validity of various versions of electron avalanche evolution. From the standpoint of this analysis it is important that this method allows the measurement of the electron drift velocity in gases at electric field strengths which exceed the breakdown. Figure 4.11 gives the dependence of the electron drift velocity in air on the electric field strength in a range of electric field strengths above the breakdown electric field strength measured by the described method. As can be seen, the measured electron drift velocity in air in a range of electric field strengths above the breakdown slightly exceeds the value 10^7 cm/s that we use in estimations.

Important information about streamers and sprites follows from photography of these objects because contemporary devices allow one to obtain the resolution of 1 ns [156–161]. There is no point in subsequent improvement of this resolution because streamer radiation is created by radiative transitions

$$N_2(C^3\Pi_u \rightarrow B^3\Pi_g) \tag{4.10}$$

with the radiative lifetime approximately 40 ns and by

$$N_2^+(B^2\Sigma_g \rightarrow X^2\Sigma_g) \tag{4.11}$$

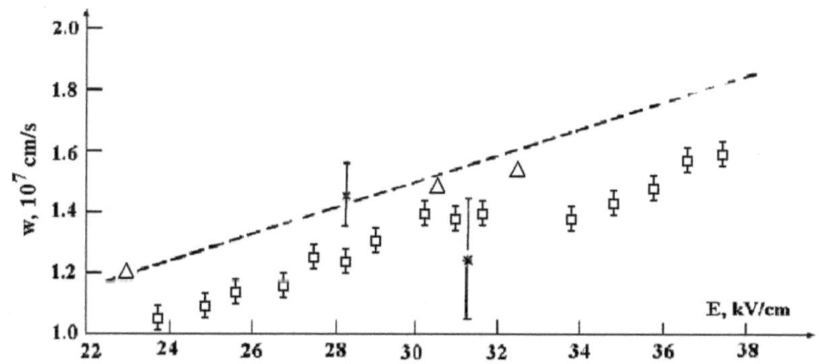

Fig. 4.11 Reduced to the atmospheric pressure experimental values of the electron drift velocity in air in the limit of a low electron number density. *Squares* [152], *triangles* [153], *crosses* [154], *dotted line* [155]

with the radiative lifetime about 60 ns [162, 163]. Photography of streamers and sprites exhibits their branched structure [161, 164] which is typical of fractal structures. This raises doubts about application of analytical methods for the streamer analysis where streamer parameters are continuous functions of space and time. The analysis of streamer properties and its evolution based only on analytical methods is represented in books [72–74, 80, 93, 108, 109, 115–120].

The spectroscopy of a streamer plasma is reduced mostly to the analysis of the above radiative transitions (4.10) and (4.11) and are characterized by the wavelengths 337.1 and 391.4 nm, if they take place between the ground vibrational states. Measurements of the intensity of radiation for corresponding electron-vibrational transitions in ionized air located in an external electric field [165–169] give information about the excitation of molecules in collisions with electrons and allow the determination of the energy distribution function of electrons. The stepwise character should be noted of excitation of nitrogen molecules in states $b^1\Pi_u$, $b'\,^1\Sigma_u^+$, $c_4'\,^1\Sigma_u^+$ whose radiation leads to photoionization of oxygen molecules in accordance with the main channel of ionization in a streamer propagated in atmospheric air [170]. In this case excited nitrogen molecules are formed effectively in electron-molecule collisions [171, 172] and may be destroyed in collisions with nitrogen and oxygen molecules [162, 163]. Subsequent collisions of excited molecules with electrons also lead to formation of excited states of nitrogen molecules that are responsible for propagation of the ionization wave in air in accordance with [170, 173].

We now prove one more streamer peculiarity that corresponds to a narrow front of a streamer as an ionization wave. A typical electron number density at the streamer front is $N_e \sim (10^{13} - 10^{14})\,\mathrm{cm}^{-3}$ (see, e.g., Fig. 4.11). The excess electron number per unit area of the wavefront σ follows from the Poisson equation

$$\sigma = \int (N_e - N_i)dx = \frac{E_{\max}}{4\pi e} \tag{4.12}$$

Assuming the maximum electric field strength near the wavefront E_{max} is close to the breakdown $E_c = 30\,kV/cm$, one can obtain for the charge excess in the ionization wave per unit area

$$\sigma \sim 2 \times 10^{10} \frac{e}{cm^2}, \tag{4.13}$$

or $\sigma \sim 3 \times 10^{-9}\,C/cm^2$. For definiteness, we consider a negative streamer with an excess negative charge at the streamer front. If the entire excess charge is concentrated at the ionization front, its depth δ is equal to

$$\delta = \frac{\sigma}{\Delta N_e} = 6\,\mu m \tag{4.14}$$

Thus, we obtain the streamer width to be relatively small in accordance with the above assumptions. From this it follows that the ionization wave is not able to transfer a charge 5C [34] in the leader stage. The largest part of an indicated charge is transferred after creation of the conductive channel when it overlaps a space between a cloud and the Earth.

In conclusion of the streamer analysis we note that streamer propagation proceeds through formation of seed electrons due to the photoionization process and then forming electrons move in an electric field of the streamer head. In the positive streamer electrons move against the direction of the electric field, whereas in the negative streamer they move in the electric field direction. From this it follows that the positive streamer requires photoionization with generation of seed electrons ahead of its front [174]. A negative streamer may develop without seed electrons; that is, ionization is possible owing to other mechanisms [175, 176]. This character of streamer propagation leads to a lower branching for positive streamers in comparison with negative ones [138]. Streamer discharge is realized at high pressure and long distances between electrodes. A general character of development of streamer discharge is such that photons emitted near the streamer head create seed electrons that are multiplied under the action of a high electric field and form an electron avalanche, a cloud of electrons that subsequently becomes a head of the propagating streamer.

In reality lightning starts from charged cumulus clouds that create an electric field with a strength of approximately two orders of magnitude lower than that causing electrical breakdown in dry air. Therefore an occasional seed electron in air cannot lead to electrical breakdown and a large number of electrons ($\sim 10^8$ [100]) are required for streamer existence. Then a heightened electric field strength near the streamer head provides air ionization which is necessary for propagation of this ionization wave. But various known mechanisms do not allow the transformation of a seed electron in a plasma [70]. Evidently, experiments [177, 178] allow one to explain the above transformation.

The experiments [177, 178] were fulfilled in the "thunderstorm" system which included a series of detectors of γ radiation, X-ray radiation, and radio waves in the Tien Shan mountains at altitudes from 3340 m up to 4000 m above sea level. A radio

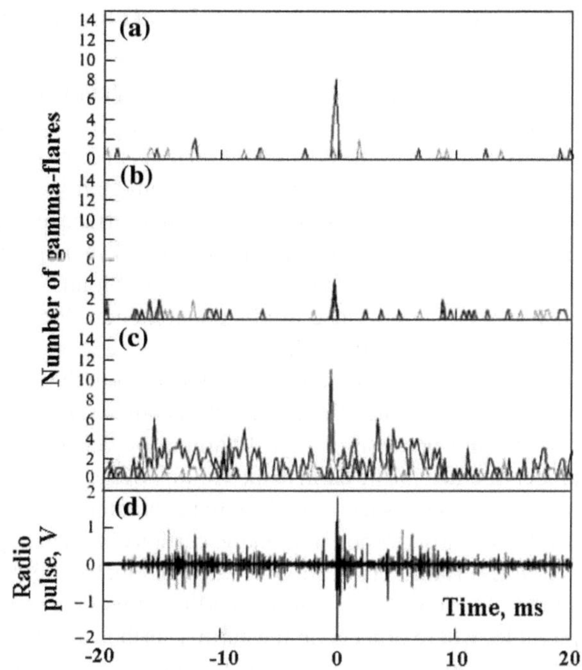

Fig. 4.12 Correlation between γ-flares detected at three stations (**a**, **b**, **c**) and radio signal from lightning (**d**) [178]

wave detector recorded emission of short radio waves in the range of 0.1–30 MHz, and this emission characterized lightning radiation. Several devices located in different places allowed one to detect γ radiation with photon energies in the range 100−500 keV. These experiments showed the correlation of γ radiation and radiation of radio waves. One can conclude from this that lightning is caused by bright atmospheric showers. An experimental result given in Fig. 4.12 confirms this conclusion which shows the correlation between flashes of γ radiation and radio radiation from lightning. Several detectors of γ radiation were used in this experiment, so that a detector "a" of Fig. 4.12 was located at a station at a distance from station "c" of 1100 m, and altitudes of these stations differed by 600 m. One can conclude from this experiment that bright showers propagated through the atmosphere create γ radiation. According to [178], a shower formed from an initial particle of energy of 10^{15} eV gives 10^6 electrons, and an energy of 10^{18} eV leads to formation of 10^{10} electrons. During thunderstorm weather, these electrons are transformed in an avalanche that causes the development of lightning. Other aspects of the ionization wave are considered in the following together with corresponding elementary processes in excited air.

4.4 Propagation of Strong Electric Current Through Soil

In analyzing lightning as an atmospheric phenomenon, we are guided by a simple model given in Fig. 4.7, and this model includes two stages of this phenomenon, namely, creation of the conductive channel and passing of an electric current through this channel. This model assumes the conductivity of electrodes to be relatively large, and a cloud charge transfers to the earth's surface and spreads over the earth quickly. In reality this assumption is partially fulfilled, and lightning consists of several current pulses. This means the redistribution of the cloud charge and spreading of the earth's charge are of importance in this phenomenon. We consider below the character of charge propagation in earth.

To prevent buildings and some objects from the action of lightning, lightning rods have been used since the eighteenth century. They take away the lightning electric current, and because of high lightning power, this current can act on the nearest objects located on the earth in the vicinity of the lightning conductive channel. For this reason, it is necessary to analyze the character of propagation of the electric current inside the earth. Depending on the earth's conductivity, a certain charge is collected near the surface. This charge creates an electric field near the surface and locks up a pass of a new charge into the earth. If this field is high enough, air breakdown is possible, where objects on the earth's surface are the second electrode. Though this field is significantly weaker than the lightning field, it may lead to damages. In order to escape electric breakdown at earth, it is necessary to carry the lightning charge in earth. In the following we analyze this process within the framework of the geometry of Fig. 4.13.

In this geometry, a lightning rod is finished by a metal sphere of a radius R_o that is located underground. We first consider the regime where the earth conductivity is high enough, and the lightning electric current I is carried inside the earth. An equilibrium is established in this case, and a charge Q is located on the metal sphere that corresponds to the electric field on the sphere surface $E_o = Q/R_o^2$, and the electric field strength from the charged sphere at a distance R from it is $E = Q/R^2$. This charge causes an underground current whose density i is equal according to Ohm's law to

$$i = \Sigma_e E = \frac{Q\Sigma_e}{R^2}, \tag{4.15}$$

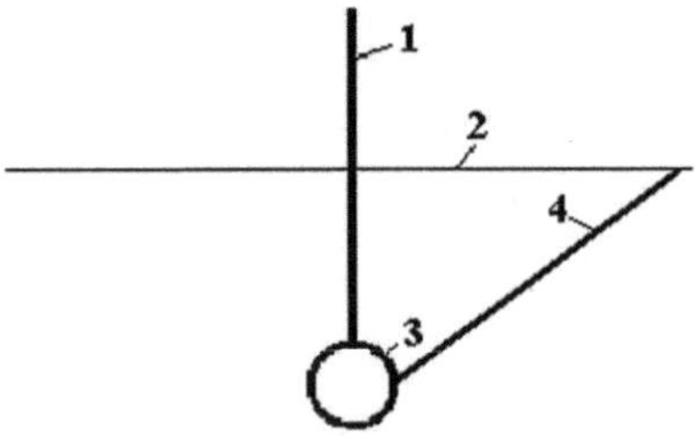

Fig. 4.13 Scheme of propagation of an electric signal through a lightning rod into earth: *1* wire for current propagation, *2* earth boundary, *3* metal sphere inside the Earth, *4* direction of current propagation to the Earth's surface

where Σ_E is the earth conductivity, and we assume that the metal sphere is buried deeply. From this we obtain for the total underground current $I = 4\pi Q\Sigma_e$, and because it is equal to the lightning current I_l, we have the sphere charge Q and the electric field strength E_o at the sphere surface given by

$$Q = \frac{I_l}{4\pi\Sigma_e}, \; E_o = \frac{I_l}{4\pi\Sigma_e R_o^2} \tag{4.16}$$

The land conductivity ranges from $\sim 10^{-8}$ S/m for marble up to $\sim 10^{-2}$ S/m for wet clay or limestone [66]; the ocean conductivity is about 4 S/m [68]. These values are large compared with atmospheric conductivity, that is, 10^{-14} S/m. If the soil consists of dry sand which is a mixture of micron-size $Si\,O_2$-particles, its conductivity is small. But if the sand is wet, it becomes a weak electrolyte, especially if a salt is dissolved in the water. In this case a current passes through the water, and the conductivity of such a sand may be higher than that of a clay. In order to estimate a scale of parameters for charge transport through soil, let us take, for definiteness, soil conductivity to be $\sim 10^{-2}$ S/m and an underground sphere radius to be 1 m. Formulas (4.16) give $Q = 0.03\,\mathrm{C}$, $E_o = 3 \times 10^6\,\mathrm{V/cm}$. One can see that the sphere charge Q is small compared to the charge (4–5) C that is transferred in one lightning flash, whereas the equilibrium electric field strength near the sphere exceeds by two orders of magnitude the breakdown electric field strength in atmospheric air ($E_c = 30\,\mathrm{kV/cm}$), although the breakdown electric field strength of the soil is several orders of magnitude higher than that for atmospheric air. Hence, depending on the soil conductivity, we have two regimes of propagation of the lightning electric current through a lightning rod and the soil. In the first case the lightning current passes inside the earth and propagates there over large distances, whereas in the second case due to low soil conductivity a metal sphere collects the electric charge of a lightning flash, and soil electrical breakdown takes place.

Let us consider some peculiarities of current propagation underground. Therefore, if there is an air gap between the metal sphere and soil, the electrical breakdown takes place in this gap. Next, if the metal sphere is buried at a depth below 100 m in this estimate, at the exit of the electric field on the earth's surface the electric field strength may exceed the breakdown; that is, electrical breakdown and damages are possible outside. We note that in these cases we assume a soil to be uniform. If metal conductors such as wires and tubes are found underground, a certain part of the electric current will go along the conductors. Favorable conditions prevent damage if the metal sphere of the lightning rod has a connection with some metal tubes, and then the lightning electric current will be carried by these metal tubes.

References

1. R. Braham, J. Meteorol. **9**, 227 (1952)
2. http://en.wikipedia.org/wiki/Lightning-rod
3. S.E. Reynolds, M. Brook, M.F. Gourley, J. Meteorol. **14**, 426 (1957)
4. E.R. Javaratne, C.P.R. Saunders, J. Hallett, Q. J. R. Meteorol. Soc. **109**, 609 (1983)
5. J. Latham, Q. J. R. Meteorol. Soc. **89**, 265 (1963)
6. C.P.R. Saunders, H. Bax-Norman, E.E. Avila, N.E. Castellano, Q. J. R. Meteorol. Soc. **130**, 1395 (2004)
7. T. Takahashi, J. Atmos. Sci. **35**, 1536 (1978)
8. C.P.R. Saunders, in *Planetary Atmospheric Electricity*, ed. by F. Leblanc, et al. (Springer, Heidelberg, 2008), p. 335
9. J. Latham, Q. J. R. Meteorol. Soc. **107**, 277 (1981)
10. C.P.R. Saunders, H. Bax-Norman, C. Emersic, E.E. Avila, N.E. Castellano, Q. J. R. Meteorol. Soc. **132**, 2653 (2006)
11. J.M. Caranti, A.J. Illingworth, S.J. Marsh, J. Geophys. Res. **90D**, 6041 (1985)
12. C.P.R. Saunders, S.L. Peck, J. Geophys. Res. **103**, 13949 (1998)
13. C.T.R. Wilson, Nature **62**, 149 (1900)
14. C.T.R. Wilson, Nature **68**, 102 (1903)
15. C.T.R. Wilson, Proc. Camb. Philos. Soc. **13**, 363 (1906)
16. C.T.R. Wilson, Proc. R. Soc. **80A**, 537 (1908)
17. C.T.R. Wilson, Proc. R. Soc. **92A**, 555 (1916)
18. C.T.R. Wilson, Philos. Trans. R. Soc. **221A**, 73 (1921)
19. C.T.R. Wilson, Proc. R. Soc. **37A**, 32D (1925)
20. C.T.R. Wilson, J. Frankl. Inst. **208**, 1 (1929)
21. C.T.R. Wilson, Proc. Camb. Philos. Soc. **22**, 534 (1925)
22. A.V. Gurevich, K.P. Zybin, Usp. Fiz. Nauk **171**, 1177 (2001)
23. D. Malan, Ann. Geophys. **8**, 385 (1952)
24. B.F.J. Schonland, *Atmospheric Electricity* (Methuen, London, 1932)
25. J.A. Chalmers, *Atmospheric Electricity* (Claredon Press, Oxford, 1949)
26. B.F.J. Schonland, *Atmospheric Electricity* (Methuen, London, 1953)
27. B.F.J. Schonland, *The Lightning Discharge*, Handbuch der Physik (Springer, Berlin, 1956), p. 576
28. H. Israël, *Atmospheric Electricity*, vol. 1, Fundamentals : Conductivity (Academische Verlagsgeselschaft, Leipzig, 1957)
29. H. Israël, *Atmospheric Electricity*, vol. 2, Fields, Charges, Currents (Academische Verlagsgeselschaft, Leipzig, 1961)
30. D. Malan, *Physics of Lightning* (The English Univ. Press Ltd., London, 1963)
31. M.A. Uman, *Lightning* (McGrow Hill, New York, 1969)
32. M.A. Uman, *About Lightning* (Dover, New York, 1986)
33. M.A. Uman, *The Lightning Discharge* (Academ. Press, New York, 1987)
34. V.A. Rakov, M.A. Uman, *Lightning, Physics and Effects* (Cambr. Univ. Press, Cambrigge, 2003)
35. R. Reiter, *Phenomena in Atmospheric and Environmental Electricity* (Elsevier, New York, 1992)
36. G.C. Simpson, Mon. Weather Rev. **34**, 16 (1906)
37. R.P. Feynman, R.B. Leighton, M. Sands, *The Feynman Lectures of Physics*, vol. 2 (Addison-Wesley, Reading, 1964)
38. B.J. Mason, *The Physics of Clouds* (Claredon Press, Oxford, 1971)
39. R.G. Fowler, in *Applied Atomic Collision Physics*, vol. 5, ed. by H.S.W. Massey, E.W. McDaniel, B. Bederson (Acad.Press, New York, 1982), p. 35
40. H.R. Byers, *Elements of Cloud Physics* (University of Chicago Press, Chicago, 1965)
41. N.H. Fletcher, *The Physics of Rainclouds* (Cambridge Univ. Press, London, 1969)

42. S. Twomey, *Atmospheric Aerosols* (Elsevier, Amsterdam, 1977)
43. H. Proppacher, J. Klett, *Microphysics of Clouds and Precipitation* (Reidel, London, 1978)
44. R.G. Flleagle, J.A. Businger, *An Introduction to Atmospheric Physics* (Acad. Press, San Diego, 1980)
45. M.L. Salby, *Fundamentals of Atmospheric Physics* (Acad. Press, San Diego, 1996)
46. B.J. Mason, *The Physics of Clouds* (Oxford Univ. Press, Oxford, 2010)
47. H. Israel, *Atmospheric Electricity* (Keter Press Binding, Jerusalem, 1973)
48. J.A. Chalmers, *Atmospheric Electricity* (Pergamon Press, Oxford, 1967)
49. R.H. Golde (ed.), *Lightning* (Acad. Press, London, 1977)
50. G.C. Simpson, Phil. Trans. R. Soc. **A209**, 379 (1909)
51. G.C. Simpson, Proc. R. Soc. **A114**, 376 (1927)
52. M. Stolzenburg, W.D. Rust, B.F. Smull, T.C. Marshall, J. Geophys. Res. **103**, 14059 (1998)
53. M. Stolzenburg, W.D. Rust, T.C. Marshall, J. Geophys. Res. **103**, 14078 (1998)
54. M. Stolzenburg, W.D. Rust, T.C. Marshall, J. Geophys. Res. **103**, 14097 (1998)
55. M. Brook, T. Ogawa, The cloud discharge, in *Lightning*, ed. by R.H. Golde (Acad. Press, London, 1977), p. 191
56. C.A. Coulomb, *Collection de Memoires Relatifs a la Physique. Tome 1* (Gauthier-Villars, Paris, 1784), p. 113
57. V.F. Hess, Phys. Zs. **113**, 1084 (1912)
58. G.C. Simpson, Phil. Trans. R. Soc. **A205**, 61 (1906)
59. C.B. Moore, B. Vonnegut, Point-discharge currents, in *Lightning*, ed. by R.H. Golde (Acad. Press, London, 1977), p. 51
60. M. Rycroft et al., Space Sci. Rev. **137**, 83 (2008)
61. M.J. Rycroft et al., in *Planetary Atmospheric Electricity*, ed. by F. Leblanc, et al. (Springer, Heidelberg, 2008), p. 83
62. E. Williams, Atmos. Res. **91**, 140 (2009)
63. http://en.wikipedia.org/wiki/Siemens
64. B.M. Smirnov, *Cluster Processes in Gases and Plasmas* (Wiley, Berlin, 2010)
65. J. Latham, I.M. Stromberg, The thunder cloud, in *Lightning*, ed. by R.H. Golde (Acad. Press, London, 1977), p. 99
66. W. Lowrie, *Fundamentals of Geophysics* (Cambr. Univ. Press, Cambridge, 2007)
67. M. Uyeshima, Surv. Geophys. **28**, 199 (2007)
68. N. Olson, A. Kuvshinov, Earth Planet Sci. **56**, 525 (2004)
69. B.M. Smirnov, Phys. Uspekhi **184**, 1153 (2014)
70. J.R. Dwyer, M. Uman, Phys. Rep. **534**, 147 (2014)
71. E.M. Bazelyan, Y.P. Raizer, Usp. Fiz. Nauk. **170**, 753 (2000). (Phys. Uspekhi **43**, 753 (2000))
72. E.M. Bazelyan, YuP Raizer, *Lightning Physics and Lightning Protection* (IOP Publ, Bristol, 2000)
73. E.M. Bazelyan, Y.P. Raizer, *Physics of Lightning and Lightning Protection* (Fizmatlit, Moscow, 2001). (in Russian)
74. Y.P. Raizer, *Physics of Gas Discharge* (Intellect, Dolgoprudnyi, 2009). (in Russian)
75. K. Berger, The earth flash, in *Lightning*, ed. by R.H. Golde (Acad. Press, London, 1977), p. 119
76. W.H.J. Childs, *Physical Constants* (Chapman and Hall, London, 1972). (Wiley, New York, 1979)
77. http://photo-element.ru/ps/photo-of-lightning/photo-of-lightning.html
78. C.E.P. Brooks, Geophys. Mem. (Lond.) **24**, 147 (1925)
79. R.E. Orville, D.W. Spencer, Mon. Weather Rev. **107**, 934 (1979)
80. E.M. Bazelyan, Yu.P Raizer, *Spark Discharge* (CRC Press, Roca Baton, 1997)
81. http://en.wikipedia.org/wiki/Distribution-of-lightning
82. D.A. Gurnett, P. Zarka, R. Manning et al., Nature **409**, 313 (2001)
83. H.J. Christian, R.J. Blakeslee, D.J. Boccippio et al., J. Geophys. Res. **108**, 4005 (2003)
84. http://maxcls.ya.ru/replies.xml?item-no=369
85. http://en.wikipedia.org/wiki/Catatumbo-lightning

86. http://www.guardian.co.uk/world/2010/mar/05/venezuela-lightning-el-nino
87. J.S. Townsend, Electrifician **50**, 971 (1903)
88. J.J. Thomson, *Conduction of Electricity Through Gases* (Cambr. Univ. Press, Cambridge, 1904)
89. J.S. Townsend, *Electricity in Gases* (Claredon Press, Oxford, 1915)
90. J.S. Townsend, *Motion of Electrons in Gases* (Claredon Press, Oxford, 1925)
91. L.B. Loeb, *Fundamental Processes of Electrical Discharges in Gases* (Wiley, New York, 1939)
92. J.S. Townsend, *Electrons in Gases* (Hutchinson'ssci. Techn. Publ, London, 1947)
93. H. Raether, *Electron Avalanches and Breakdown in Gases* (Butterworths, London, 1964)
94. E. Flegler, H. Raether, Zs. Tech. Phys. **16**, 435 (1935)
95. H. Raether, Phys. Zs. **15**, 560 (1936)
96. H. Raether, Zs. Tech. Phys. **18**, 564 (1937)
97. H. Raether, Zs. Phys. **107**, 91 (1937)
98. H. Raether, Zs. Phys. **110**, 611 (1938)
99. H. Raether, Zs. Phys. **112**, 464 (1939)
100. H. Raether, Arch. Electrotech. **34**, 49 (1940)
101. H. Raether, Zs. Phys. **117**, 375 (1941)
102. H. Raether, Ergebn. ex. Naturwissenshaft. **22**, 73 (1949)
103. L.B. Loeb, J.M. Meek, J. Appl. Phys. **11**, 438 (1940)
104. L.B. Loeb, J.M. Meek, J. Appl. Phys. **11**, 459 (1940)
105. L.B. Loeb, J.M. Meek, *Mechanism of Electric Spark* (Standf. Univ. Press, Stanford, 1941)
106. L.B. Loeb, Phys. Rev. **81**, 287 (1951)
107. L.B. Loeb, Phys. Rev. **94**, 227 (1954)
108. J.M. Meek, J.D. Crags, *Electrical Breakdown in Gases* (Claredon Press, Oxford, 1953)
109. J.M. Meek, J.D. Crags, *Electrical Breakdown in Gases* (Wiley, New York, 1998)
110. J.M. Meek, Proc. Phys. Soc. **52**, 547 (1940)
111. J.M. Meek, Proc. Phys. Soc. **52**, 822 (1940)
112. J.M. Meek, Phys. Rev. **57**, 722 (1940)
113. H. Tholl, Zs. Phys. **172**, 536 (1963)
114. H. Tholl, Zs. Naturforsch **19a**, 346 (1964)
115. F. Llewellyn-Jones, *Ionization and Breakdown in Gases* (Methuen, London, 1957)
116. L.B. Loeb, *Basic Processes of Gaseous Electronics* (Univ. Californ. Press, Berkeley, 1961)
117. F. Llewellyn-Jones, *Ionization, Avalanches and Breakdown* (Methuen, London, 1967)
118. E. Nasser, *Fundamentals of Gaseous Ionization and Plasma Electronics* (Wiley-Interscience, New York, 1971)
119. E.D. Lozanskii, O.B. Firsov, *Theory of Spark* (Atomizdat, Moscow, 1975). (in Russian)
120. E.M. Bazelyan, Y.P. Raizer, *Spark Discharge* (CRC Press, Roca Baton, 1997)
121. B.B. Kadomtzev, *Plasma Turbulence* (Acad. Press, New York, 1965)
122. R.Z. Sagdeev, A.A. Galeev, *Nonlinear Plasma Theory* (Benjamin, New York, 1969)
123. V.N. Tsytovich, *Nonlinear Effects in Plasmas* (Plenum Press, New York, 1970)
124. G. Ecker, *Theory of Fully Ionized Plasmas* (Academic Press, New York, 1972)
125. B.B. Kadomtzev, Cooperative Phenomena in Plasma. Rev. Plasma Phys. ed. by V.D. Shafranov **22**, 1 (2001)
126. V.E. Fortov, I.T. Iakubov, *The Physics of Non-ideal Plasma* (World Scientific, Singapore, 2000)
127. V.E. Fortov, I.T. Iakubov, A.G. Khrapak, *Physics of Strongly Coupled Plasma* (Claredon Press, Oxford, 2006)
128. U. Kogelschatz, Pure Appl. Chem. **62**, 1667 (1990)
129. H.-E. Wagner et al., Vacuum **71**, 417 (2003)
130. A. Chirokov, A. Gutsol, A. Fridman, Pure Appl. Chem. **77**, 487 (2005)
131. B. Eliasson, M. Hirth, U. Kogelschatz, J. Phys. **20D**, 1421 (1987)
132. U. Kogelschatz, B. Eliasson, W. Egli, Pure Appl. Chem. **71**, 1819 (1999)
133. K.G. Kostov et al., Braz. J. Phys. **39**, 322 (2009)

134. L.B. Loeb, *Electrical Coronas* (Cambr. Univ. Press, London, 1965)
135. R.S. Sigmond, Corona discharge, in *Electrical Breakdown in Gases*, ed. by J.M. Meek, J.D. Craggs (Wiley, New York, 1978), p. 319
136. J. Jones, Br. J. Appl. Phys. (J.Phys.) **1D**, 769 (1968)
137. J. Dutton, J. Chem. Phys. Ref. Data **4**, 577 (1975)
138. N. Liu, V.P. Pasko, J. Geophys. Res. **109**, A04301 (2004)
139. B.W. Ward, Nature **208**, 994 (1965)
140. A.E.D. Heylen, Nature **183**, 1545 (1959)
141. W.W. Lozier, Phys. Rev. **46**, 268 (1934)
142. H.D. Hagstrum, J.T. Tate, Phys. Rev. **59**, 354 (1941)
143. D. Rapp, D.D. Briglia, J. Chem. Phys. **43**, 1480 (1965)
144. F.D. Findley, C.J. Fortin, D.R. Snelling, Chem. Phys. Lett. **3**, 204 (1969)
145. A.A. Kulikovsky, J. Phys. **27D**, 2556 (1994)
146. A.A. Kulikovsky, J. Phys. **30D**, 441 (1997)
147. A.A. Kulikovsky, Phys. Rev. **57E**, 7066 (1998)
148. NYu. Babaeva, G.V. Naidis, J. Phys. **29D**, 2423 (1996)
149. G.V. Naidis, Phys. Rev. **79E**, 057401 (2009)
150. A.V. Shelobolin, J. Phys. **40D**, 6669 (2007)
151. X. Lu, G.V. Naidis, M. Laroussi, K. Ostrikov, Phys. Rep. **540**, 123 (2014)
152. K.R. Allen, K. Philips, Proc. R. Soc. **274A**, 163 (1963)
153. J.S. Townsend, H.T. Tizard, Proc. R. Soc. **A88**, 336 (1913)
154. W. Riemann, Zs. Phys. **122**, 216 (1944)
155. R.A. Nielsen, N.E. Bradbury, Phys. Rev. **51**, 69 (1937)
156. E.M. Veldhuizen, P.C.M. Kemps, W.R. Rutgers, IEEE Trans. Plasma Sci. **30**, 162 (2002)
157. E.M. Veldhuizen, W.R. Rutgers, J. Phys. **35D**, 2169 (2002)
158. W.J. Yi, P.F. Williams, J. Phys. **35D**, 205 (2002)
159. T.M.P. Briels, E.M. Veldhuizen, U. Ebert, IEEE Trans. Plasma Sci. **33**, 264 (2005)
160. S.V. Pancheshnyi, M. Nudovna, A.Y. Starikovskii, Phys. Rev. **71E**, 016407 (2005)
161. U. Ebert, Plasma Sources Plasma Technol. **15**, S118 (2006)
162. S.V. Pancheshnyi, S.M. Starikovskaia, A.Y. Starikovskii, Chem. Phys. Lett. **294**, 523 (1998)
163. S.V. Pancheshnyi, S.M. Starikovskaia, A.Y. Starikovskii, Chem. Phys. **262**, 349 (2000)
164. U. Ebert, D.D. Sentman, J. Phys. **41D**, 230301 (2008)
165. S.V. Pancheshnyi, S.V. Sobakin, S.M. Starikovskaia, A.Y. Starikovskii, Plasma Phys. Rep. **26**, 1054 (2000)
166. P. Paris, M. Aints, M. Laan, F. Valk, J. Phys. **37D**, 1179 (2004)
167. Y. Kim, W.S. Kang, J.M. Park et al., IEEE Trans. Plasma Sci. **32**, 18 (2004)
168. P. Paris, M. Aints, F. Valk et al., J. Phys. **38D**, 3894 (2005)
169. Y.V. Shcherbakov, R.S. Sigmond, J. Phys. **40D**, 474 (2007)
170. M.B. Zheleznyak, A.K. Mnatsakanyan, S.V. Sizykh, High Temp. **20**, 357 (1982)
171. D.C. Cartwright, S. Trajmar, A. Chutjian, W. Williams, Phys. Rev. **16A**, 1041 (1977)
172. J.T. Fons, R.S. Schappe, C.C. Lin, Phys. Rev. **53A**, 2239 (1996)
173. A.X. Zheleznyak, G.V. Naidis, S.V. Sizykh, High Temp. **20**, 423 (1982)
174. S.K. Dhall, P.F. Williams, J. Appl. Phys. **62**, 4696 (1987)
175. P.A. Vitello, B.M. Penetrante, J.B. Bardsley, Phys. Rev. **49E**, 5574 (1994)
176. A. Rocco, U. Ebert, W. Hundsdorfer, Phys. Rev. **66E**, 035102 (2002)
177. A.V. Gurevich et al., Phys. Lett. A **325**, 389 (2004)
178. A.V. Gurevich, A.N. Karashtin, V.A. Ryabov et al., Usp. Phys. Nauk. **179**, 779 (2009)

Part II
Elementary and Transport Atmospheric Processes

"A detailed understanding of the ionospheres of the Earth and planets requires knowledge of the rates of many ionic reactions as well as of electron recombination coefficients and photoionization cross sections" (H.S.W. Massey)

Chapter 5
Electron Processes in Atmospheric Air

Abstract Elementary processes in atmospheric air are responsible for atmospheric properties at various altitudes, especially for ionosphere properties. Electron processes in atmospheric air are analyzed and include electron attachment to the oxygen molecule and dissociative recombination with molecular ions of air. An equilibrium between electrons and negative or positive ions in hot air is considered. Some aspects of leader propagation in atmospheric air are analyzed.

5.1 Three-Body Electron Attachment to Oxygen Molecule

In the analysis of the upper atmosphere, starting from an altitude of 70–80 km and finishing by a E_2-layer of the ionosphere at altitudes approximately 200 km, we consider this atmosphere as nonequilibrium air excited by solar radiation. Hence the properties of this system are determined by elementary processes in this system involving electrons, ions, and neutral atomic particles (atoms and molecules) in the ground and excited states, as well as by transport processes. A general position in the study of this system is contained in the above epigraph by Massey in 1979 [1]. The study of processes that take place in the Earth's ionosphere developed in this way, and on the basis of elementary processes and transport phenomena in this system one can compose parameters of the weakly ionized plasma of the ionosphere and its evolution. Guided by this, below we analyze in detail elementary processes involving atomic particles and microparticles that are present in the atmosphere, as well as transport processes with their participation.

Electrons as the main component of a plasma determine the plasma conductivity and other plasma parameters. But electrons are absent usually in a quiet low atmosphere because of the attachment of electrons to oxygen molecules. For thermal electrons this process proceeds in three-body collisions with oxygen and nitrogen molecules

$$e + O_2 + O_2(N_2) \to O_2^- + O_2(N_2) \tag{5.1}$$

B.M. Smirnov, *Microphysics of Atmospheric Phenomena*,
Springer Atmospheric Sciences, DOI 10.1007/978-3-319-30813-5_5

A typical time of electron attachment to oxygen molecules in atmospheric air is $\tau \approx 10^{-8}$ s at room temperature. Electrons accelerated by an electric field may attach to oxygen molecules as a result of the pair collision process

$$e + O_2 \rightarrow O^- + O, \tag{5.2}$$

The cross section of this process as a function of the electron energy has a maximum at the electron energy about 7 eV [2], and at the electron energy $\varepsilon = 6.5$ eV the rate constant of the process (5.2) is $k = 2 \times 10^{-10}$cm^3/s [3]. Hence electrons are absent in the atmosphere, and negative ions partake in ion-molecular reactions with radicals (e.g., [2, 4]), and finally this leads to formation of the most stable negative ions. It should be noted that the basic sort of positive and negative ions in a quiet atmosphere may be different depending on admixtures in the atmosphere. But stable ions are characterized by a large electron binding energy, and such negative ions are NO_2^- with the electron binding energy 2.3 eV, NO_3^- with the binding energy 3.7 eV, and CO_3^- with the binding energy 2.8 eV [5]. For comparison, the electron binding energy in the negative ion O_2^- is 0.43 eV, and the electron binding energy in the negative ion O^- is equal approximately to 1.5 eV. Two last types of negative ions are formed in air at the first process stage, and subsequently they are converted in stable negative ions as a result of corresponding ion-molecular reactions.

Electrons are formed in an air plasma as a result of molecule ionization in collisions with electrons accelerated by an electric field. First an energy from an electric field is transferred to electrons, and molecules obtain this energy in collisions with electrons. In addition, electrons are lost in an air plasma by attachment to oxygen molecules. At thermal energies this process proceeds in three-body collisions according to the scheme

$$e + O_2(^3\Sigma_g^-) + O_2(N_2) \rightarrow O_2^-(^2\Pi_u) + O_2(N_2), \tag{5.3}$$

and at room temperature the rate constant of the three-body process is $K(N_2) = 1 \times 10^{-31}$ cm^6/s [2, 6–9], if the nitrogen molecule is a third particle, and $K(O_2) = (2.4 \pm 0.4) \times 10^{-30}$ cm^6/s, where the oxygen molecule is a third particle. The latter value is the statistical average of measurements [6, 8–15] and corresponds to the temperature $T = 300$ K. A high attachment rate for the oxygen molecules as a third particle is explained by the resonant character of this process, where an electron is captured with formation of an autodetachment state of the negative oxygen molecular ion O_2^-, and then this state is stabilized as a result of the resonant charge exchange of the negative ion in collision with the oxygen molecule [16]. As a result, the electron binding energy is consumed partially by molecule vibrational excitation. The process (5.3) leads to the following typical time of electron attachment of a thermal electron to air molecules at atmospheric pressure

$$\tau_{at} = \frac{1}{c(O_2)N^2[c(N_2)K(N_2) + c(O_2)K(O_2)]}, \tag{5.4}$$

where N is the density of air molecules, and $c(N_2)$ and $c(O_2)$ are the concentrations of nitrogen and oxygen molecules in air. From this we have for a typical attachment time for an electron in atmospheric air at room temperature:

$$\tau_{at} = 13\,\text{ns} \tag{5.5}$$

Thus, a thermalized electron in air attaches fast to the oxygen molecule in cold air. Correspondingly, the air conductivity drops quickly after thermalization of an electron resulting from ionization of nitrogen or oxygen molecules.

Figure 5.1 gives the temperature dependence for the process (5.3) with the oxygen molecule as the third body. The Maxwell distribution function for electrons and molecules is realized in this case with the same temperature that is indicated. As a result of participation of electrons with different electrons the resonant character of this process is lost. We now consider the nature of the process of electron attachment to the oxygen molecule. This process proceeds in two stages:

$$e + O_2 \to (O_2^-)^{**}, \quad (O_2^-)^{**} + O_2 \to O_2^- + O_2, \tag{5.6}$$

where the first stage is formation of an autodetachment state $(O_2^-)^{**}$, and in the subsequent collision of a negative molecular ion with an oxygen molecule a negative ion transfers in a stable state. The rate constant of electron attachment to the oxygen molecule with formation of an autodetachment state owing to simultaneous electron collisions with two molecules is larger than that without formation of an autodetachment state, and the rate constant of electron attachment in this case is inversely proportional to the lifetime of an autodetachment state.

In the case of electron attachment to the oxygen molecule, the mechanism is realized that corresponds to formation of an autodetachment state of the oxygen molecular ion as an intermediate state. This mechanism is often called the Bloch–Bradbary mechanism [16] after the researchers who first studied this process. To explain this mechanism, we give in Fig. 5.2 the potential curves of an oxygen molecule and its negative ion that is cut from the series of potential curves for the oxygen molecule [17]. In this process, an electron is captured in an autodetachment state of

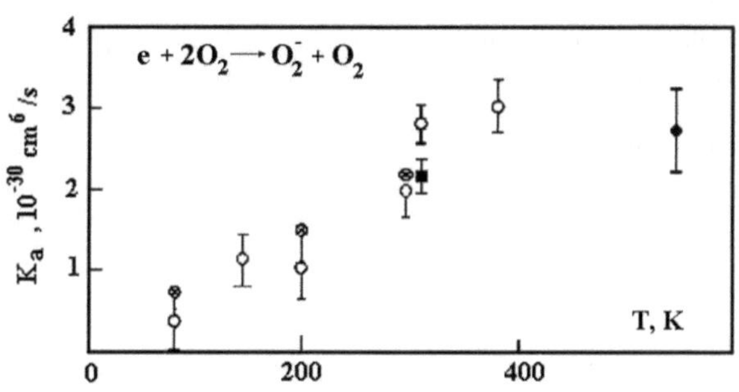

Fig. 5.1 Temperature dependence for the rate constant of three-body electron attachment to an oxygen molecule with an oxygen molecule as the third body: cross inside circle—[7], open circle—[8], closed circle—[11], closed square—[9]

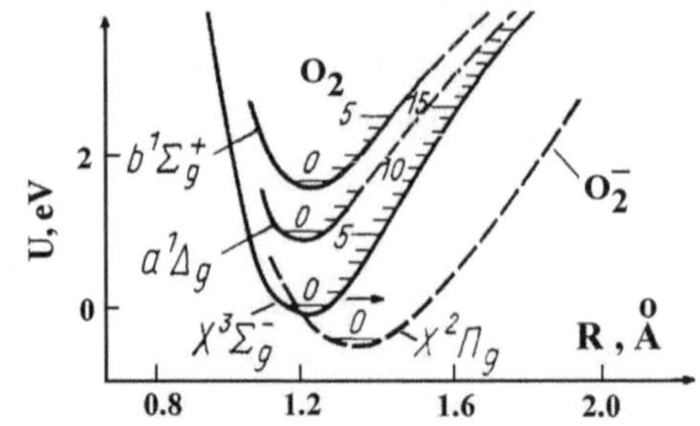

Fig. 5.2 The lowest potential curves (electron terms) for the oxygen molecule and negative molecular ion that may be responsible for electron attachment to an oxygen molecule

a negative molecular ion, and the transition in a stable state of the negative molecular ion results in the process of charge exchange:

$$O_2^-(v_1) + O_2(v_2) \rightarrow O_2^-(v_1') + O_2(v_2') \tag{5.7}$$

As a result of this process, vibrationally excited negative ion transfers in a lower state, and the negative ion becomes stable if the energy of vibration excitation of the negative ion is below the electron binding energy in the negative ion.

If the width of the autodetachment level Γ is less than the transition energy between neighboring vibrational levels $\hbar\omega$, the excitation cross section of an autodetachment state of the negative ion is characterized by extrema for which the electron energy coincides with the excitation energy of a corresponding vibration level of the autodetachment state. But if the electron energy distribution function is smooth, the resonant structure of the rate constant is lost, as follows from Fig. 5.3, where the temperature of oxygen is given, and an electron is accelerated in an electric field. Figure 5.4 contains the dependence of this three-body rate constant on the electron energy in the case of a monoenergetic beam of electrons.

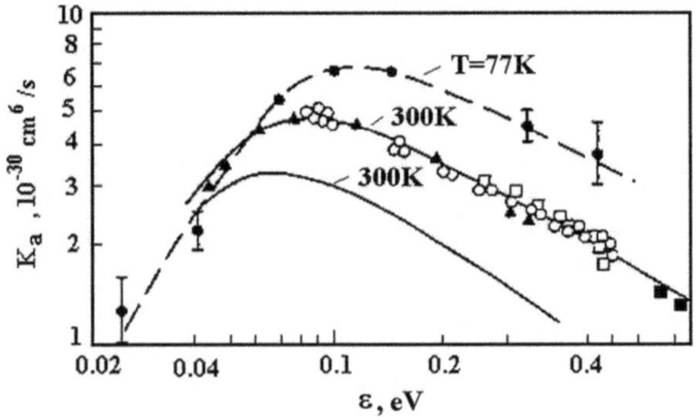

Fig. 5.3 Dependence on the average electron energy for the rate constant of three-body electron attachment to an oxygen molecule located in an electric field. Signs: experiments [8, 11], solid curve: theory [18]

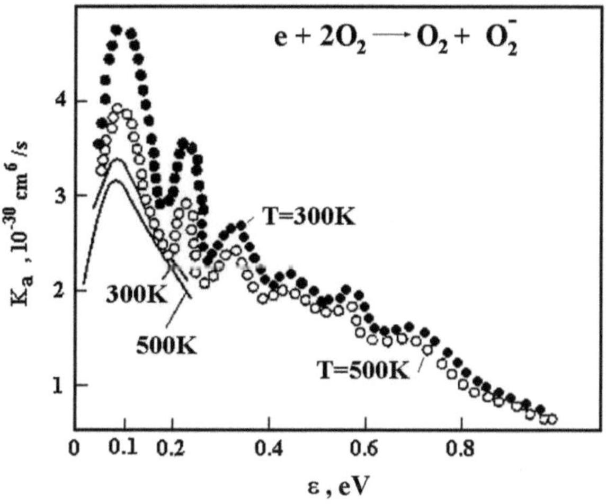

Fig. 5.4 Rate constant of three-body electron attachment to an oxygen molecule for a monoenergetic beam of electrons injected in oxygen. Experiment: [15], theory [18] under the assumption that the width of an autodetachment level is independent of the distance between nuclei

5.2 Dissociative Electron Attachment to Oxygen Molecule

Electron attachment to a molecule is a result of electron transition from a free to a bound state, and an excess electron energy must be removed from the electron-molecule system. In the case of the three-body electron attachment process (5.1), this energy is transferred to a third particle. The electron attachment process is possible in pair electron-molecule collisions, where an energy excess is transferred to the kinetic energy of atoms, and then dissociation of the molecule proceeds according to scheme

$$e + O_2 \to O^- + O \qquad (5.8)$$

In this case an electron is captured in a repulsive electron term (see Fig. 5.5), so that an atom and negative ion fly away. When the distance between them exceeds the distance where the electron terms of the molecule and negative molecular ion intersect, the atomic negative ion becomes stable.

Based on the character of the electron attachment process, one can connect the cross section of this process with parameters of the autodetachment state. The Breit–Wigner formula [20–22] gives for the cross section σ_{cap} of electron capture in an autodetachment electron term

$$\sigma_{cap} = \frac{\pi \hbar^2}{2m_e \varepsilon} \int \frac{dR \Gamma^2(R) \left|\varphi_o(R)\right|^2}{[\varepsilon - E(R)]^2 + \Gamma^2(R)/4} \qquad (5.9)$$

Here R is the distance between molecule nuclei or the reaction coordinates, $\Gamma(R)$ is the width of the autodetachment state at a given distance between nuclei, $E(R)$ is the energy difference between molecular and autodetachment terms, ε is the electron

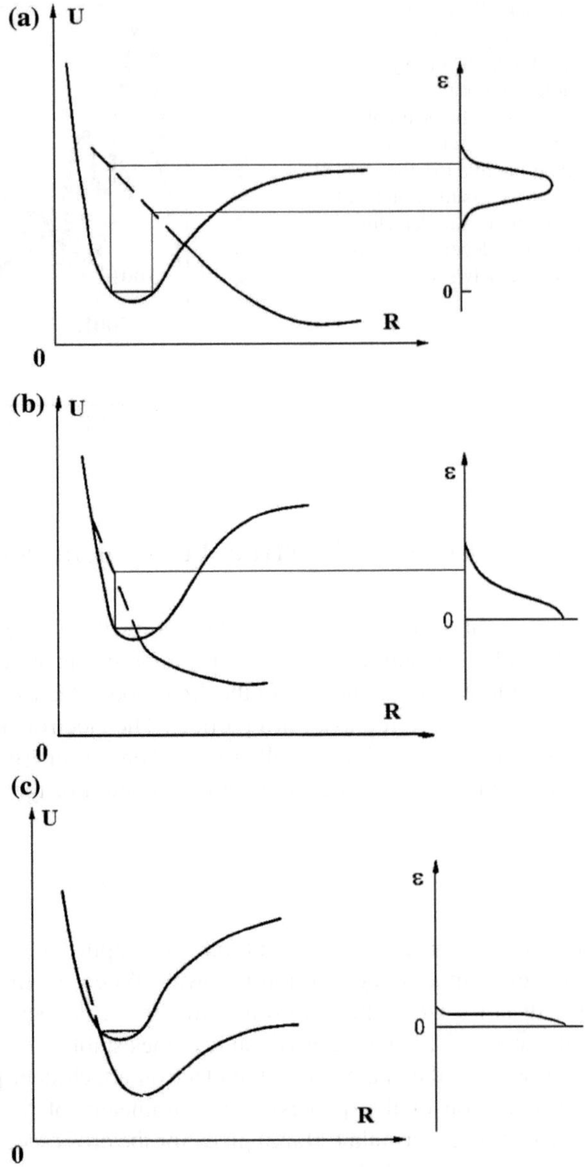

Fig. 5.5 Possible positions of the potential curves for the molecule (an electron binding state) and negative ion (a repulsive curve) for the electron attachment process (*left*) and the electron spectrum (*right*) as a result of this process [19]

energy, $\varphi_o(R)$ is the nuclear wave function for the molecule, and we account for the Born–Oppenheimer approximation according to which in the course of electron capture the distance between nuclei does not vary. This approximation holds true if typical electron times for this process significantly exceed a typical time of nuclear motion. In the course of motion of nuclei along the autodetachment term, decay of this state is possible with the return of the system in the molecular term. Hence, the cross section of formation of a stable negative ion is given by

$$\sigma_{at} = \frac{\pi\hbar^2}{2m_e\varepsilon}\int dR \frac{\Gamma^2(R)(|\varphi_o(R)|^2}{[\varepsilon - E(R)]^2 + \Gamma^2(R)/4} \exp\left(-\int\limits_R^{R_c} \frac{\Gamma(R')dR'}{\hbar v_R}\right), \qquad (5.10)$$

where $v_R = dR/dt$ is the velocity of nuclear motion along the reaction coordinate. This expression has a resonant structure, and the integral is conserved near the resonance $\varepsilon = E(R_\varepsilon)$, and R_ε is a reaction coordinate where the resonance takes place. Expanding $E(R)$ in a resonance range and assuming $\Gamma(R)$ to be independent of the coordinate R, one can obtain

$$\sigma_{at} = \frac{\pi^2\hbar^2}{2m_e\varepsilon}\frac{\Gamma(R_\varepsilon)|\varphi_o(R_\varepsilon)|^2}{|dE/dR|} \exp\left(-\int\limits_R^{R_c} \frac{\Gamma(R')dR'}{\hbar v_R}\right), \qquad (5.11)$$

where the derivation $|dE/dR|$ is taken at a distance R_ε. This formula exhibits that the electron attachment cross section is determined by the autodetachment level width at the resonance.

The integral relation for the electron attachment cross section is useful. Integrating formula (5.10) over electron energies, one can obtain

$$\int\limits_0^\infty \sigma_{at}(\varepsilon)d\varepsilon = \frac{\pi^2\hbar^2}{m_e}\left\langle \Gamma(R)\exp\left(-\int\limits_R^{R_c} \frac{\Gamma(R')dR'}{\hbar v_R}\right)\right\rangle \qquad (5.12)$$

It is convenient to rewrite this relation in the form:

$$\langle \Gamma\exp(-\zeta)\rangle = \frac{m_e\varepsilon_{\max}}{\pi^2\hbar^2}\int \sigma_{at}(\varepsilon)d\varepsilon, \ \zeta = \left(\int\limits_R^{R_c} \frac{\Gamma(R')dR'}{\hbar v_R}\right), \qquad (5.13)$$

where $\varepsilon_{\max}$ is the electron energy that corresponds to the maximum of the cross section of electron attachment, $\exp(-\zeta)$ is the probability of surviving of an autodetachment state during nuclear flying away, and the triangle parentheses mean an average over distances of electron capture. This formula allows one to estimate parameters of the autodetachment state. In particular, for the process (5.8) we have near the resonance at the electron energy $\varepsilon_{\max} = 6.7\,\text{eV}$ for the right-hand side of (5.13)

$\langle \Gamma \exp(-\zeta) \rangle = 3.2\,\text{meV}$ on the basis of measurements [23] and $\langle \Gamma \exp(-\zeta) \rangle = 2.9\,\text{meV}$ from the experiment [24]. Roughly, the cross section of dissociative electron attachment to the oxygen molecule in the ground state and the molecular temperature $T = 300\,\text{K}$ has the resonant structure with the maximum at the electron energy approximately 7 eV, and the width of this resonance about 3 eV. The maximum cross section in this case is about $2 \times 10^{-18}\,\text{cm}^2$. Parameters of electron attachment to an oxygen molecule are given in Table 5.1 [25].

The electron attachment cross section allows the determination of the rate constant of this process in a plasma. Assuming the Maxwell distribution function of electrons with the electron temperature T_e and a low width of the autodetachment level $\Gamma \ll T_e$, we obtain after an average over electron energies, using Formula (5.13) for the electron attachment cross section:

$$k_{at} = \frac{2}{\sqrt{\pi} T_e^{3/2}} \int_0^\infty \sqrt{\frac{2\varepsilon}{m_e}} \sigma_{at}(\varepsilon) \sqrt{\varepsilon} \exp\left(-\frac{\varepsilon}{T_e}\right) d\varepsilon$$
$$= \left(\frac{2\pi}{m_e T_e}\right)^{3/2} \hbar^2 \left\langle \Gamma(R) \exp\left(-\frac{E(R)}{T_e} - \zeta\right) \right\rangle, \qquad (5.14)$$

where the angular brackets mean averaging over positions of nuclei and $E(R)$ is the excitation energy of the autodetachment state at a given distance between nuclei. In the case $E'_R \Delta R \gg T_e$ the main contribution to the integral gives a small range of distances between nuclei as compared to ΔR. Then, assuming a weak dependence $\Gamma(R)$ for excitation energies $E(R) \sim T_e$, from Formula (5.14) for the rate constant of electron attachment we obtain

$$k_{at} = \hbar^2 \left(\frac{2\pi}{m_e}\right)^{3/2} \frac{\Gamma \varphi^2(R)}{\sqrt{T_e} E'_R} \exp(-\zeta) \qquad (5.15)$$

It is necessary to keep in mind that the above formulas correspond to certain quantum numbers of a free electron that are determined by the quantum numbers of the molecule and an autodetachment state of the negative ion. The transition of the electron system from the molecular to autodetachment state of a given symmetry proceeds

Table 5.1 Cross section of electron attachment to the oxygen molecule σ_{at} and partial cross sections of this process σ_g and σ_u via the autodetachment states $^2\Pi_g$ and $^2\Pi_u$, correspondingly [25], when the captured electron is found in the states π_g and π_u

ε, eV	σ_{at}	σ_g	σ_u
5.75	9.1	1.1	7.9
6.70	13.7	2.2	11.4
7.80	5.7	1.4	4.3
8.40	2.4	0.70	1.7

The cross sections are given in $10^{-19}\,\text{cm}^2$

only for a certain orbital momentum or other quantum numbers of the captured electron.

The process (5.8) proceeds via the autodetachment states $^2\Pi_g$ and $^2\Pi_u$. Electron capture in these autodetachment states leads to different angular dependence in the distribution of formed negative ions, and simultaneous measurement of the electron attachment cross section as a function of the energy of an incident electron and the scattering angle of forming ions allows one to separate processes with different autodetachment states. Table 5.2 gives the results of such a separation for channels of the process at various electron energies ε [25]. From this it follows that the main contribution to the process (5.8) follows from the autodetachment state $^2\Pi_g$ of the negative oxygen molecular ion.

Another aspect of electron attachment to the oxygen molecule follows from the mass-spectrometric analysis of the products of the electron attachment process in the pulse regime. The possibilities of this method are demonstrated by Fig. 5.6 [26, 27] which contains time-of-flight (TOF) spectra for the process (5.8). In this method, the kinetic energy of ions is determined by measurement of the delay time for the ion pass to the mass spectrometer of negative oxygen ions formed in this process. Figure 5.6 exhibits one single resonance for the process (5.8) in the energy range between 4 and 9 eV and according to the time-of-flight analysis, a doublet character of the cross section of electron attachment follows from measurements at high electron energies. The ion translational energy as a function of the electron energy across the resonance is plotted in Fig. 5.6. This gives, as expected, a straight line with a slope of 0.5 because the translational energy is divided between O and O^-. Extrapolation to zero kinetic energy yields the thermodynamic threshold for the dissociative attachment reaction. This mass-spectrometric study thus gives no indication of the contribution of a π_g state in a range between 4 and 9 eV as follows from the angular distribution measurements [26, 27]. In fact, by measuring electron attachment to oxygen clusters it appears that the π_g state is only accessible at low-energy electrons and finally results in the relaxed ground-state negative ion (see below). The threshold of the process (5.8) corresponds to the initial electron energy $\varepsilon = 3.56\,\text{eV}$, because the dissociation energy of the oxygen molecule is $D = 5.12\,\text{eV}$ and the electron affinity of the oxygen atom equals $EA = 1.46\,\text{eV}$.

Table 5.2 Quantum numbers of transition states for electron capture in the autodetachment states of the oxygen molecules and the angular distribution of captured electrons [28] (l is the angular moment of an attaching electron, m is its projection on the molecular axis, and θ is the angle between the molecular axis and the incident electron)

Molecule	Autodetachment state	Electron parity	l	m	Angular distribution
$O_2(^3\Sigma_g^-)$	$O_2^-(^2\Pi_u)$	-1	1	1	$\sin^2\theta$
$O_2(^1\Sigma_g^+)$	$O_2^-(^2\Pi_u)$	-1	1	1	$\sin^2\theta$
$O_2(A^3\Sigma_u^+)$	$O_2^-(^2\Pi_u)$	1	2	1	$\sin^2 2\theta$

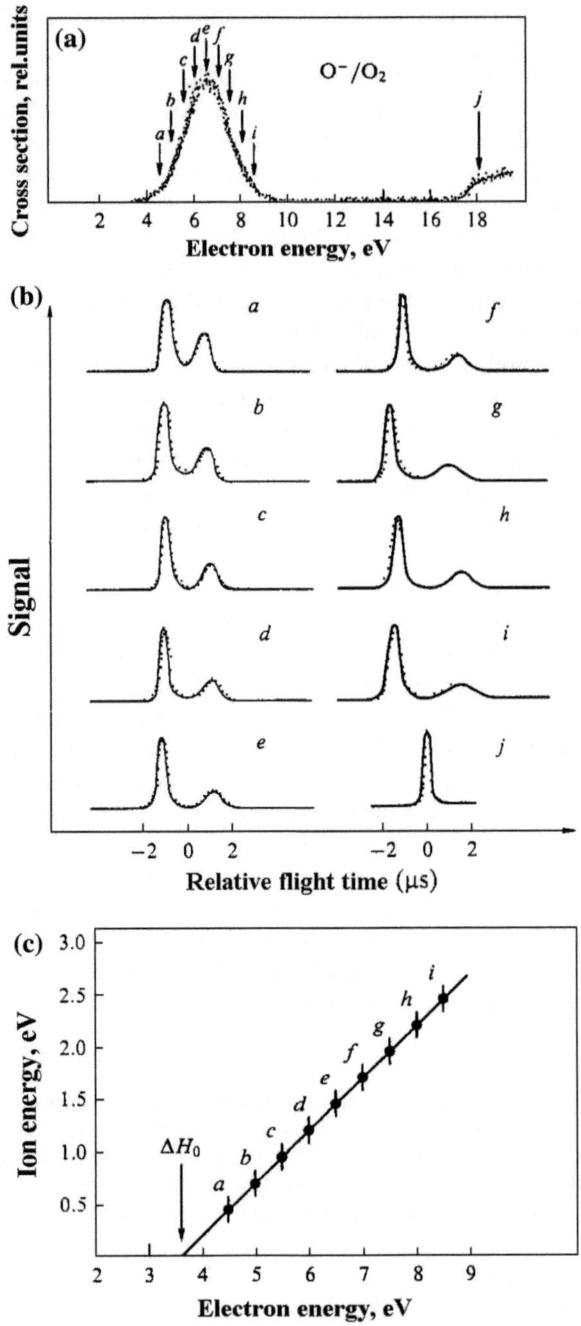

Fig. 5.6 Connection of the electron energy in the process of electron attachment to the oxygen molecule and the energy distribution function of ions determined on the basis of the TOF (time-of-flight) method [26, 27]. **a** Electron energy dependence for the rate of electron attachment for the oxygen molecule with formation of the negative atomic ions; arrows show at which electron energies the energy distribution function of ions is measured which is given at **b**; **c** the dependence of the ion energy for the distribution function maximum on the energy of an incident electron

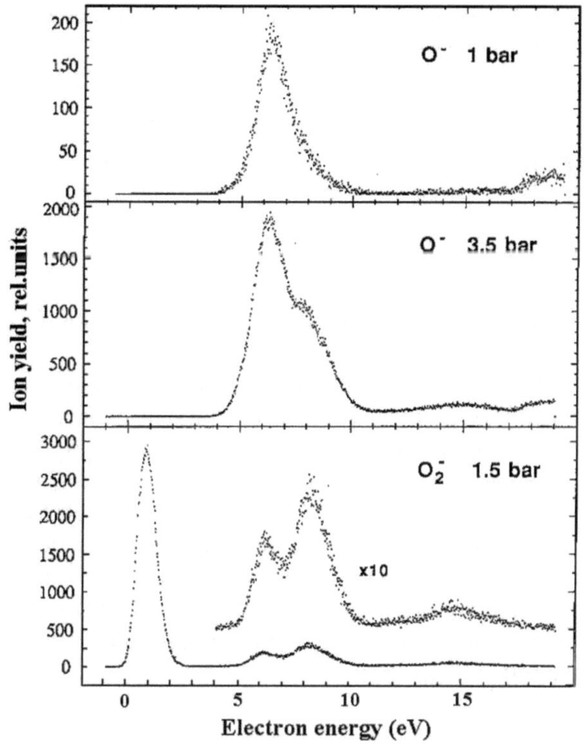

Fig. 5.7 Ion signals of indicated types as a function of the energy of an incident electron at a given oxygen pressure [26, 27]

Along with the energy distribution of negative ions, this method allows the determination of the ion type. Figure 5.7 represents the results of such measurements for the process (5.8). According to Fig. 5.7, molecular negative oxygen ions O_2^- are formed mostly at low energies of incident electrons, whereas the maximum of the distribution function for atomic oxygen negative ions O^- corresponds to the electron energy of several electronvolts. The symmetry of electron states of the oxygen molecule and negative molecular ion is important for the process (5.8). Let us analyze their symmetry starting from an oxygen atom with the electron shell $O(2p^4)$ where valence electrons are assumed to be identical and to be located in an effective spherically symmetric electric field of the atomic core. Then within the model of molecular orbits, the ground state of the oxygen molecule is characterized by the following distribution of electrons over molecular orbitals $O_2(\sigma_g^2\pi_u^4\pi_g^2)^3\Sigma_g^-$, so that according to the usual notations we denote the angular momentum projection onto the molecular axis by σ, π, subscripts g, u mean the symmetry of the electron wave functions as a result of reflection with respect to the plane that is perpendicular to the molecular axis and bisects it, and the superscripts denote the number of electrons in a given state. As a demonstration of the above analysis, Table 5.2 gives the connection of quantum numbers of the initial and intermediate electron states of molecular

Table 5.3 Parameters of low autodetachment states of the ground electron term of the oxygen negative molecular ion $O_2^-(X^2\Pi_g)$ [29–33]; v is the vibrational number of the state, E_a is the excitation energy from the ground state of the oxygen molecule of the autodetachment level, and Γ is the width of this level

v	4	5	6	7	8	9	10
E_a, meV	90	215	338	458	577	694	809
Γ, meV	0.004	0.036	0.12	0.26	0.46	0.74	1.1

particles of oxygen and the angular distribution of captured electrons for oxygen molecules [28].

The ground state of the oxygen negative molecular ion is $O_2^-(\sigma_g^2\pi_u^4\pi_g^3)^2\Pi_g$. Vibrationally excited states of this electron state of the oxygen negative molecular ion with the vibrational quantum numbers $v \geq 4$ correspond to the autodetachment states formed as a result of collisions of a slow electron and oxygen molecule. Parameters of these autodetachment states are given in Table 5.3.

But collisions via this autodetachment state do not lead to formation of atomic negative ions because of a small resonant electron energy compared to the dissociation energy of the oxygen molecule. The broad resonance in the cross section of electron attachment to the oxygen molecule in the range of 5–8 eV corresponds to the following process

$$e + O_2(\sigma_g^2\pi_u^4\pi_g^2)^3\Sigma_g^- \to \left[O_2^-(\sigma_g^2\pi_u^3\pi_g^4)^2\Pi_u\right]^{**} \to O^-(^2P) + O(^3P), \qquad (5.16)$$

$$e + O_2(\sigma_g^2\pi_u^4\pi_g^2)^1\Delta_g \to \left[O_2(\sigma_g^2\pi_u^3\pi_g^4)^2\Pi_u\right]^{**} \to O^-(^2P) + O(^3P) \qquad (5.17)$$

Autodetachment states, as intermediate states of the electron attachment processes, are formed by the joining of an electron to the oxygen molecule. Because in other cases reconstruction of the molecular structure proceeds by a change of electron distribution over orbitals, they are characterized by lower cross sections or rates of the electron attachment process.

As follows from Fig. 5.5, vibrational excitation of molecules widens the range of distances between nuclei at which electron capture takes place. Therefore electron capture for vibrationally excited molecules becomes possible at lower energies than those for the molecule in the ground state, and the resonance in the cross section of electron attachment to an excited molecule is broadened. In addition, excitation of the molecule leads to an increase in the cross section of electron attachment because the probability of survival of the autodetachment state decreases. These peculiarities of the attachment process are demonstrated in Fig. 5.8 where the cross section of electron attachment to the oxygen molecule as a function of the electron energy is given for different temperatures [34]. As shown, an increase of the temperature of

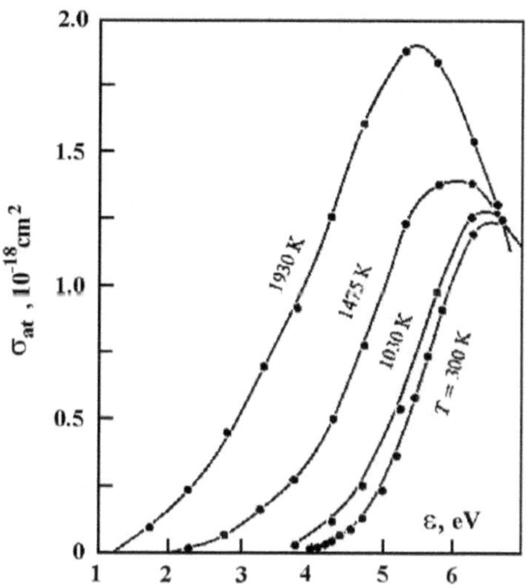

Fig. 5.8 Cross section of dissociative electron attachment to the oxygen molecule as a function of the electron energy at different oxygen temperatures [34]

oxygen molecules leads to a shift of the resonance in the electron attachment to low energies, to its broadening, and to an increase of the maximum cross section of electron attachment [24, 28, 35].

In the case of a small probability of survival of the autodetachment state in which an electron is captured, the electron attachment cross section increases with the vibrational state number of the molecule. In particular, in the case of the oxygen molecule, when the maximum cross section of the electron attachment process for the molecular ground state is 1.2×10^{-18} cm^2, the maximum cross section of electron attachment for the first vibrational state is estimated as 4×10^{-18} cm^2. Note that the cross section of electron capture into the autodetachment state exceeds this value significantly. In addition, at a temperature of 1930 K only 21 % of molecules are found in the ground vibrational state, and the above temperature dependence of the electron attachment cross section confirms a small probability of survival of the autodetachment state in the course of its evolution.

Alongside vibrational excitation, electron excitation of molecules influences the efficiency of the electron attachment process. Measurements of electron attachment to the metastable oxygen molecule $O_2(^1\Delta_g)$ show that the main resonance is similar to the process (5.16) and proceeds according to the scheme (5.8). Note that the configurations of the electron shell are identical for the ground $^3\Sigma_g^-$ and metastable $^1\Delta_g$ states within the framework of the model of molecular orbitals. The maximum of the cross section for process (5.8) is shifted with respect to the cross section maximum of processes (5.16) by about 1 eV (the excitation energy of $O_2(^1\Delta_g)$) to smaller electron energies, and the maximum cross section is several-fold more than that in the case of processes (5.16) [36–38]. In particular, according to [38], the maximum cross section of electron attachment is 6.1×10^{-18} cm^2 at an energy 5.7 eV, and the resonance width is 2.7 eV. From this, on the basis of Formula (5.13), we have $\langle \Gamma \exp(-\zeta) \rangle = 14\,\text{meV}$. This exceeds the same value for the ground state of the oxygen molecule by several times.

The electron dissociative attachment process to the oxygen molecule in air is effective in cold atmospheric air located in a strong electric field. Then an electron acquires energy from an electric field, and if the electron energy reaches the value of the resonant energy for excitation of autodetachment levels, the processes (5.16) and (5.17) are realized. Because the threshold of the process of electron dissociative attachment is about 3.5 eV, this process is realized in strong fields, so that it is observed at specific electric field strengths above 10 Td. The process of electron dissociative attachment in strong electric fields is reviewed in [39, 40]. According to this analysis, there is a discrepancy between different measurements. Let us introduce the coefficient η of dissociative attachment of an electron to an oxygen molecule in analogy with the first Townsend coefficient, so that $1/\eta$ is the mean free path of an electron with respect to its dissociative attachment to an oxygen molecule, and the specific coefficient of electron attachment η/N (N is the number density of oxygen molecules) is an effective cross section of electron attachment. In spite of the discrepancy of measured values of the specific coefficient of electron attachment, one can determine roughly that the maximum of η/N corresponds to the specific electric field strength $E/N = 30{-}50\,\text{Td}$ and the maximum of η/N is $(2{-}3) \times 10^{-18}\,\text{cm}^2$.

5.3 Dissociative Recombination of Electrons and Molecular Ions

The dissociative recombination process proceeds according to the scheme

$$e + AB \rightarrow A^* + B \qquad (5.18)$$

and the behavior of electron terms that take part in this process is analogous to electron attachment to molecules given in Fig. 5.5. For its nature, the process of dissociative recombination is analogous to the process of dissociative electron attachment to a molecular ion that in the case of oxygen is given by (5.8). In both cases, the formation process of a bound electron state proceeds through a bound state with an energy above the boundary of the continuous spectrum. This intermediate state is an autodetachment or autoionization state, and decay of this state leads to dissociation of the molecule or molecular ion. In this context it is necessary to note the finite number of electron-atom bound states and infinite electron-ion number states. Therefore there is a finite number of autodetaching states in electron-atom interaction and an infinite number of autoionizing states in electron-ion interaction. For this reason, the values of the recombination coefficient for different molecular ions cannot differ by several orders of magnitude, as takes place for the electron attachment process.

It is convenient to apply a simple model [41, 42] to the dissociative recombination process assuming an incident electron to be classical and introducing the region of a strong electron-ion interaction that is found inside the sphere of a radius R_o in a space of distances between an electron and molecular ion. Then the cross section of dissociative recombination is given by

$$\sigma_{rec} = \pi \rho^2,$$

where ρ_o is the impact parameter of electron-ion collisions that leads to a distance of closest approach of R_o. Let us use the relation between collision parameters (e.g., [43])

$$-\frac{e^2}{R_o \varepsilon} = 1 - \frac{\rho_o^2}{R_o^2},$$

where e is an electron and ion charge and ε is the electron energy. On the basis of the above formulas we obtain in the limit $e^2/R_o \gg \varepsilon$

$$\sigma_{rec} = \pi \frac{e^2 R_o}{\varepsilon}, \tag{5.19}$$

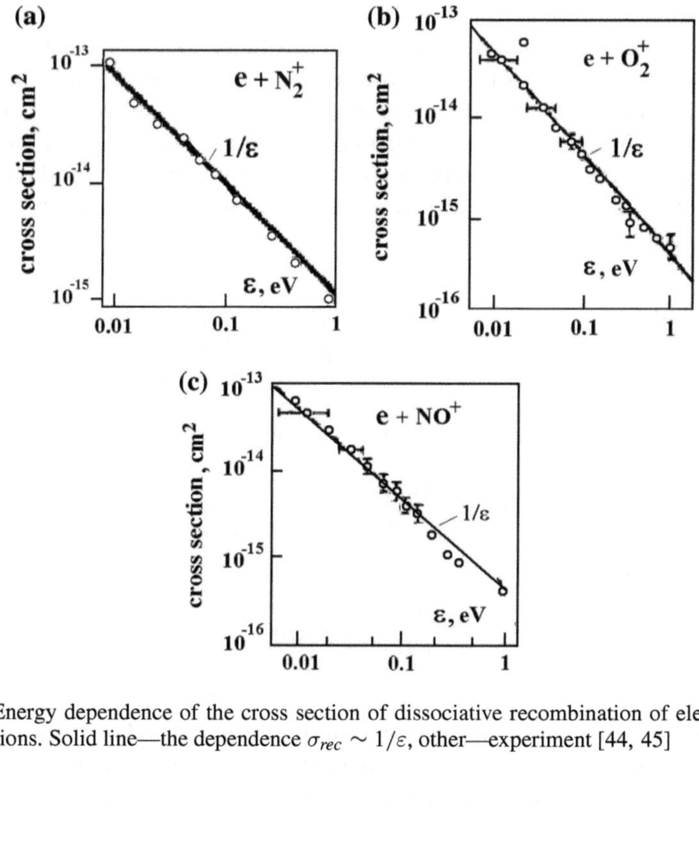

Fig. 5.9 Energy dependence of the cross section of dissociative recombination of electrons and molecular ions. Solid line—the dependence $\sigma_{rec} \sim 1/\varepsilon$, other—experiment [44, 45]

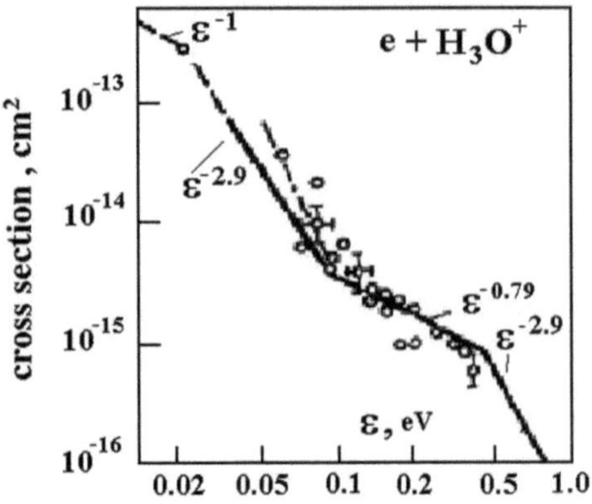

Fig. 5.10 Energy dependence for the cross section of dissociative recombination for molecular ion H_3O^+. Open circles—experiments [46, 47], lines—approximations

Figures 5.9 and 5.10 contain the energy dependencies for the measured cross sections of dissociative recombination involving molecular ions N_2^+, O_2^+, NO^+, and H_3O^+ at room temperature. One can see that the energy dependence (5.19) for the cross section of dissociative recombination is fulfilled at low energies $\varepsilon \leq \varepsilon_o$. Table 5.4 gives the values of parameters R_o and ε_o for the cross section of dissociative recombination that follow from the above experiment. As shown, the parameter R_o restored from measurements is small compared to the size of a molecular ion, whereas the concept of this model is based on the inverse relation between these parameters.

As follows from Table 5.4, the above model of a region of strong interaction holds true if a size of this region is large compared with the size of a molecular ion. One can analyze the process of dissociative recombination from another standpoint, taking into account that the dependence $\sigma_{rec} \sim 1/\varepsilon$ also results from the Breit–Wigner formula given in the form (5.11)

$$\sigma_{dr} = \frac{\pi^2 \Phi \hbar^2}{2m_e \varepsilon} \exp(-\zeta), \ \Phi = \frac{\Gamma(R_\varepsilon)|\varphi_o(R_\varepsilon)|^2}{|dE/dR|} \exp\left(-\int_{R_\varepsilon}^{R_c} \frac{\Gamma(R')dR'}{\hbar v_R}\right), \quad (5.20)$$

Let us represent the parameter Φ of this formula as

$$\Phi = \frac{\Gamma(R_\varepsilon)|\varphi_o(R_\varepsilon)|^2}{|dE/dR|} = \frac{\Gamma}{\Delta E}, \quad (5.21)$$

where

$$\Delta E = \frac{|dE/dR|}{|\varphi_o(R_\varepsilon)|^2} \quad (5.22)$$

is the difference of energies of electron terms for the initial and final transition states related to a transition range. Note that in contrast to the process of electron attachment of electrons to a molecule, where the width of the autodetachment state at the

Table 5.4 Parameters of dissociative recombination of electrons and molecular ions [48]

Ion	R_o, Å	ε_o, eV	$\sigma_{rec}\varepsilon$, $10^{-16}\text{cm}^2 \cdot \text{eV}$	Φ	α, $10^{-7}\text{cm}^3/\text{s}$
N_2^+	0.2	0.1	10	0.17	2
O_2^+	0.1	1	3.0	0.05	2
NO^+	0.14	1	3.3	0.06	4
H_3O^+	1.4	0.03	56	0.9	12

The parameter R_o is taken from Formula (5.19), ε_o is the upper limit where the dependence (5.19) holds true according to measurements, and α is the coefficient of dissociative recombination

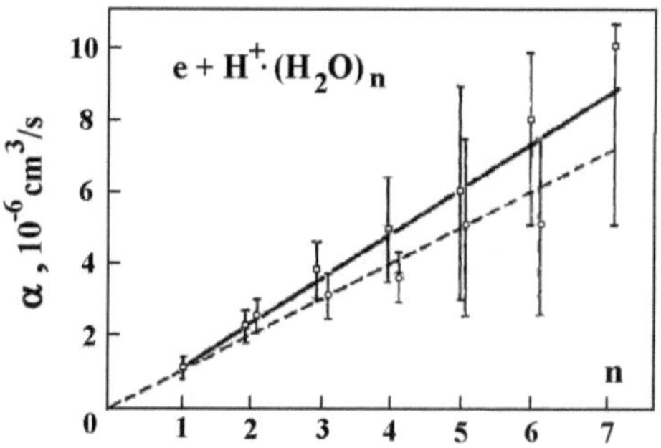

Fig. 5.11 Coefficient of dissociative recombination of electrons and complex ions $H^+ \cdot (H_2O)_n$. Squares—experiment [68], open circles—measurements [67], solid line—Botcher formula [70] $\alpha_n = \alpha_1(1.3n - 0.3)$, chain line—Formula $\alpha_n = \alpha_1 n$

threshold is zero $\Gamma(R_c) = 0$, for the case of the autoionization state $\Gamma(R) = const$ in the vicinity of the threshold $R \approx R_c$. Therefore the parameter $\Phi(R)$ does not vary strongly in the transition range, and one can accept a model assumption $\Phi(R) = const$. Then assuming the probability of survival of the autoionization state to be one $\zeta \ll 1$, the dependence (5.19) for the cross section of dissociative recombination $\sigma_{rec} \sim 1/\varepsilon$ can be obtained. This dependence is confirmed more or less by experimental data in some range of electron energies (see Figs. 5.9 and 5.10).

Table 5.4 also contains parameters of the coefficients of dissociative recombination for the indicated molecular ions at room temperature. These values are obtained by averaging the experimental data [45, 49–57] for recombination of the molecular ion N_2^+, data [45, 51, 53, 58–62] for O_2^+, [44, 45, 63–67] for NO^+, and [66, 68, 69] for H_3O^+. Evidently, in the range where $\sigma_{rec} \sim 1/\varepsilon$, the temperature dependence of the coefficient of dissociative recombination has the form $\alpha \sim T_e^{-1/2}$, if the electron and vibrational temperatures coincide.

In addition we note that the rate constant of dissociative recombination involving complex ions increases with the increasing size of the recombining ion. As a demonstration, Fig. 5.11 contains the dependence of the dissociative recombination coefficient of electrons and complex ions $H^+ \cdot (H_2O)_n$ on the number of water molecules in a complex ion if this process proceeds in a gas at room temperature.

5.4 Electron Equilibrium in Hot Air

The motion of electrons of small energies in air is mostly determined by their elastic collisions with nitrogen molecules. The diffusion cross section of this process in thermal collisions is approximately $\sigma^* \approx 2 \times 10^{-16}\,\mathrm{cm}^2$ [71, 72] that corresponds to the mean free path of thermal electrons in nitrogen approximately $0.02\,\mu\mathrm{m}$ at atmospheric pressure. Because we deal with larger distances of motion, the diffusion character of electron motion in air is important. The diffusion cross section of electron scattering on a nitrogen molecule increases with increasing electron energy, and at the electron energy in a range 1–8 eV it is equal approximately to $\sigma^* \approx (1-2) \times 10^{-15}\,\mathrm{cm}^2$ [72, 73]. Thus, the diffusion character of electron motion in atmospheric air takes place at distances that exceed micron-size scales.

Inasmuch as air is an electronegative gas, negative ions are formed in it effectively, and at thermal electron energies this process proceeds in three-body collisions according to scheme (5.1). But the electron affinity of the oxygen molecules $EA(O_2)$ (or the electron binding energy in a negative molecular ion) is not large $EA(O_2) = 0.45\,\mathrm{eV}$ [2, 5]. Hence, air heating leads to destruction of negative ions. Assuming the thermodynamic equilibrium for electrons and negative ions according to the equation

$$e + O_2(^3\Sigma_g^-) \leftrightarrow O_2^-(^2\Pi_u) \tag{5.23}$$

we have that the relation between the number density of free electrons N_e, the number density of molecular negative ions $N(O_2^-)$, and the number density of oxygen molecules $N(O_2)$, is determined by the Saha formula [74, 75]

$$\frac{N_e N(O_2)}{N(O_2^-)} = \frac{g_e g(O_2)}{g(O_2^-)} \left(\frac{m_e T}{2\pi\hbar^2}\right)^{3/2} \exp\left(-\frac{EA}{T}\right), \tag{5.24}$$

Here m_e is the electron mass, T is the air temperature, the electron statistical weight is $g_e = 2$, the statistical weight of the electron state $O_2(^3\Sigma_g^-)$ of the oxygen molecule is $g(O_2) = 3$, and the statistical weight of the negative ion $O_2^-(^2\Pi_u)$ is equal to $g(O_2^-) = 4$. Above we assume for simplicity the rotational and vibrational parameters for the oxygen molecule and its negative ion to be identical. One can rewrite (5.24) in the form

$$\frac{N_e}{N(O_2^-)} = 78T^{5/2} \exp\left(-\frac{5.2}{T}\right), \tag{5.25}$$

where the temperature T is expressed in 10^3 K, and we use in this equation the value of electron attachment for the oxygen molecule $EA = 0.45\,\mathrm{eV}$ [2, 5].

The ratio (5.25) of the number densities of electrons and oxygen negative ions is given in Fig. 5.12 as a function of the air temperature both for atmospheric pressure and at the pressure 0.3 atm related to the altitude of 6 km. Note that in the presence of electric fields this ratio increases. Thus electrons dominate in the equilibrium (5.23) at temperatures above 1000 K. In principle, at temperatures $T \sim 1000\,\mathrm{K}$ negative

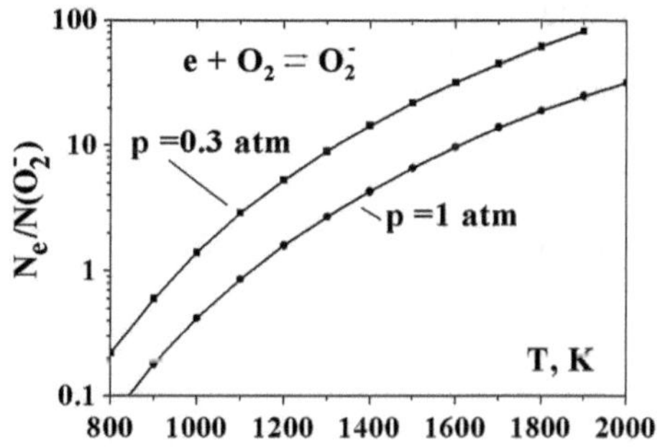

Fig. 5.12 Ratio of the number densities of electrons and oxygen negative molecular ions at thermodynamic equilibrium (5.23) according to Formula (5.25)

ions with large electron binding energies, such as ions O_3^-, NO_2^-, NO_3^- may exist in atmospheric air. The process of formation of such negative ions proceeds with partaking of radicals and small admixtures, so that these ions are formed over a long period of time.

In the same manner we consider an equilibrium between electrons and positive ions in air. This takes place at higher temperatures where air is dissociated. Indeed, one half of nitrogen molecules in atmospheric air becomes dissociated at the temperature 6000 K, and for oxygen molecules this temperature is approximately 4000 K. Analyzing ionization of air, we deal with higher temperatures, and because the thermodynamic equilibrium in equilibrium air is supported between different degrees of freedom, an ionization equilibrium takes place in dissociated air and proceeds according to the scheme

$$e + N^+(^3P) \leftrightarrow N(^4S), \ e + O^+(^4S) \leftrightarrow O(^3P), \tag{5.26}$$

Again we use the Saha formula [74, 75], which in accordance with the equilibrium (5.26), has the following form for nitrogen and oxygen

$$\frac{N_e[N^+]}{[N]} = \frac{g_e g(N^+)}{g(N)} \left(\frac{m_e T}{2\pi\hbar^2}\right)^{3/2} \exp\left(-\frac{J_N}{T}\right),$$
$$\frac{N_e[O^+]}{[O]} = \frac{g_e g(O^+)}{g(O)} \left(\frac{m_e T}{2\pi\hbar^2}\right)^{3/2} \exp\left(-\frac{J_O}{T}\right) \tag{5.27}$$

Here $[N]$, $[O]$, $[N^+]$, and $[O^+]$ are the number densities of nitrogen and oxygen atoms, and their atomic ions correspondingly, the statistical weights are $g(N) = 4, g(O) = 9, g(N^+) = 9, g(O^+) = 4$ for indicated atoms and ions, and $J_N = 14.534\,\text{eV}$, $J_O = 13.618\,\text{eV}$ are the ionization potentials for these atoms. It is convenient to reduce (5.27) to one equation

$$\frac{N_e^2}{N_a} = \left[\frac{9}{2}c_N \exp\left(-\frac{J_N}{T}\right) + \frac{8}{9}c_O \exp\left(-\frac{J_O}{T}\right)\right] \left(\frac{m_e T}{2\pi\hbar^2}\right)^{3/2}, \tag{5.28}$$

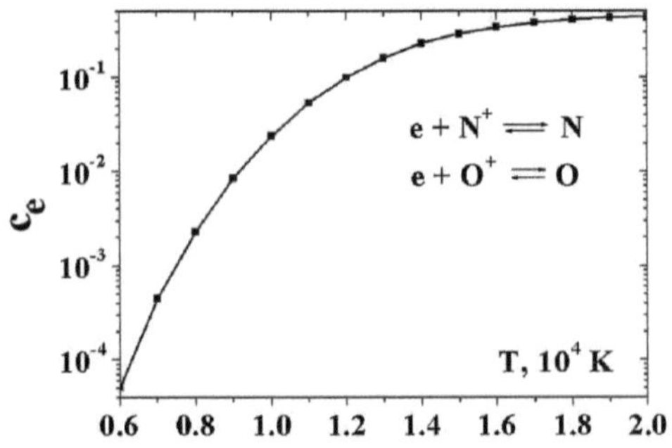

Fig. 5.13 Electron concentration in dissociated equilibrium air at atmospheric pressure as a function of the air temperature in accordance with Formula (5.29)

where $N_a = [N] + [O]$ is the total number density of atoms in dissociated air, and c_N and c_O are concentrations of nitrogen and oxygen atoms in dissociated air; for simplicity, we neglect the contribution of argon in this equilibrium. From this one can represent the electron concentration in hot air as

$$c_e = \frac{N_e}{N_a + 2N_e} \tag{5.29}$$

Figure 5.13 gives the temperature dependence for the electron concentration in hot equilibrium air in accordance with Formulas (5.28) and (5.29). As shown, in the conductive lightning channel where the air temperature according to spectroscopic measurements is 20,000–30,000 K [76–79], atmospheric air is ionized if equilibrium in the lightning channel is established.

In considering an equilibrium of electrons and negative ions in air in a strong electric field, we note that at specific electric field strengths above $E/N = 60\,\mathrm{Td}$ the rate of ionization of oxygen and nitrogen molecules becomes comparable with that of electron attachment to oxygen molecules. The specific electric field strength of electrical breakdown of air is $E_c/N = 100\,\mathrm{Td}$ which corresponds to the electric field strength $E_c = 30\,\mathrm{kV/cm}$ at atmospheric pressure. At this electric field strength the rates of electron attachment and atom ionization by electron impact are equal; that is, $\eta = \alpha$. It is essential that at such electric field strengths the first Townsend coefficient α increases with an increasing electric field strength, whereas the electron attachment coefficient η to an oxygen molecule drops with increasing electric field strength. Therefore, electrical breakdown of atmospheric air takes place at $E > E_c$, and the rate of electron multiplication is characterized by the quantity $\alpha - \eta$.

5.5 Leader Propagation as Preionization Process in Electrical Breakdown

Returning to the lightning model given in Fig. 4.7, one can see that it is necessary to have a conductive channel for charge transport between electrodes, and within the framework of the model of Fig. 4.7 its role is fulfilled by a metal rod. In reality a precursor of electrical breakdown of air for charge transfer from a cloud to Earth is an ionization wave, a leader. In contrast to a streamer, a leader conserves the conduction in a lightning channel behind the wavefront, and negative ions are not formed there because of the high air temperature in the channel. In order to destroy negative ions in a conductive channel, it is necessary to heat air in this channel to the temperature $\Delta T \approx 1000\,\mathrm{K}$, as follows from Fig. 5.12. Note that other temperatures are given in some books and reviews, for example, the temperature $\Delta T = 5000\,\mathrm{K}$ [80, 81] without grounds. One can see disagreement with the data of Fig. 5.12. Possibly the temperature $\Delta T = 5000\,\mathrm{K}$ [80, 81] characterizes a steady-state regime, rather than the threshold for leader existence. Moreover, the conductivity of a mature leader channel is $10^4\,\mathrm{S/m}$ according to [79] which corresponds to the temperature of an equilibrium air of approximately 7000 K.

We now consider air heating during streamer propagation. The rate of air heating after streamer propagation through this channel is given by the balance equation:

$$C_p \frac{dT}{dt} = eEw_eN_e \tag{5.30}$$

Here C_p is the air capacity per unit volume that varies from $1.2 \times 10^{-3}\,\mathrm{J/(cm^3 \cdot K)}$ at the temperature $T = 300\,\mathrm{K}$ up to $2.3 \times 10^{-4}\,\mathrm{J/(cm^3 \cdot K)}$ at the temperature $T = 2000\,\mathrm{K}$ [82]. The electron drift velocity is $w_e = 1 \times 10^7\,\mathrm{cm/s}$ at the breakdown electric field strength and nearby electric field strengths. This gives for the specific rate of air heating:

$$\frac{dT}{N_e dt} = \frac{eEw_e}{C_p} \tag{5.31}$$

Figure 5.14 gives the specific rate of air heating as a result of propagation of the ionization wave. As follows from Fig. 5.14, at the reduced electric field strength $E_c/N = 100\,\mathrm{Td}$ and typical number density of electrons in an avalanche $N_e \sim 10^{14}\,\mathrm{cm}^{-3}$ a time of heating up to temperatures of destruction of negative ions $T > 1200\,\mathrm{K}$ proceeds through times $\tau \sim 10^{-7}\,\mathrm{s}$. This is available for the lightning leader because the ionization wave goes away $\sim$1 cm for this time in atmospheric air.

In this context a leader in air is formed from a streamer if a streamer current heats the channel of this ionization wave such that negative ions are destroyed and the negative charge of the wave channel comes from electrons. As shown, there is a sequence in the development of ionization waves under high electric fields. An electron avalanche is formed from a seed electron, and when a number of electrons in the avalanche become large and the electric field of a plasma is compared with an external electric field, the avalanche is transformed into a streamer. If heating of air in the streamer channel becomes significant and therefore negative ions are destroyed there because of high temperature, a streamer is transformed into a leader. This sequence of transitions between different types of ionization waves is represented in Fig. 5.15.

It should be noted that in the analysis of the heat balance we ignore the heat transport in the transverse direction. Let us estimate this effect. Being guided by a radius of the ionization wave in a range $r = 0.1-1\,\text{cm}$, an increase of this radius during a time τ is $\Delta r \sim 2\chi\tau$, where χ is the thermal diffusivity coefficient. The latter is equal at room temperature to $\chi = 0.2\,\text{cm}^2/\text{s}$, and at the temperature $T = 2000\,\text{K}$ it equals $\chi = 6\,\text{cm}^2/\text{s}$. As can be seen, during a time $\tau \sim 10^{-7}\,\text{s}$ expansion of the conductive channel is negligible, whereas during the lightning lifetime this may be essential.

The specific properties of a lightning leader that propagates from a cloud to earth is the stepwise character of its propagation. Namely, its manner consists of some segments of a length on the order of 50 m, so that the leader stops for some time after each segment, and the average velocity of its propagation is on the order of $10^7\,\text{cm/s}$, whereas inside each segment the leader velocity is, at least, one order of magnitude higher. One can explain why the lightning leader cannot move with a high drift velocity. Indeed, the average electric field strength between a cloud and Earth is taken to be $\overline{E} \approx 200\text{V/cm}$, which is small compared to the breakdown electric field strength in atmospheric air $E_c = 30\,\text{kV/cm}$. Hence it is necessary to redistribute electrons and ions in an ionization wave such that the electric field strength can exceed the breakdown near the head of this wave. In order to support a high electric field strength near the front of an ionization wave it is necessary that this field become small behind the front. But this change is determined by the motion of electrons and therefore if the velocity of the ionization wave exceeds the drift velocity of electrons, the wave is saturated and is stopped.

Figure 5.16 gives the space distribution of the electric field strength in the ionization wave leader. As the leader moves and its propagation velocity exceeds the electron drift velocity, after a certain time the electric potential of the wave cannot provide the electric field strength above the breakdown one, and the wave stops (line 3 of Fig. 5.16). After redistribution of electrons (lines 1, 2 of Fig. 5.16) the wave motion will be resumed. The total area for all curves 1, 2, and 3 are identical and are equal to the same voltage of the ionization wave. Note that redistribution of

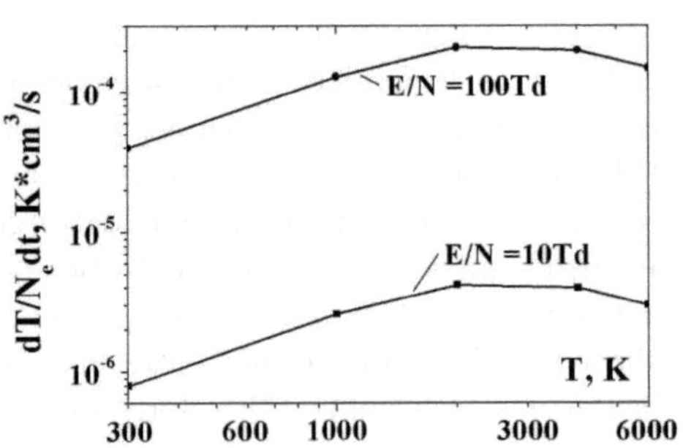

Fig. 5.14 Specific rate of air heating under the action of electrons moved in an electric field according to Formula (5.31). The reduced electric field strength of $E/N = 100\,\text{Td}$ at room temperature and atmospheric pressure correspond to the breakdown electric field strength $E = 30\,\text{kV/cm}$

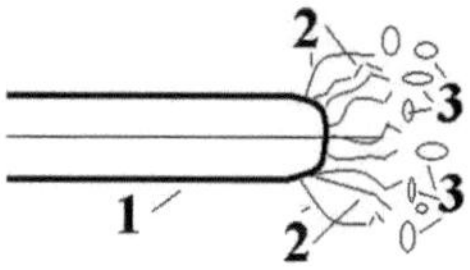

Fig. 5.15 Sequence of different ionization waves in the course of leader propagation. *1*—leader, *2*—streamers, *3*—electron avalanches

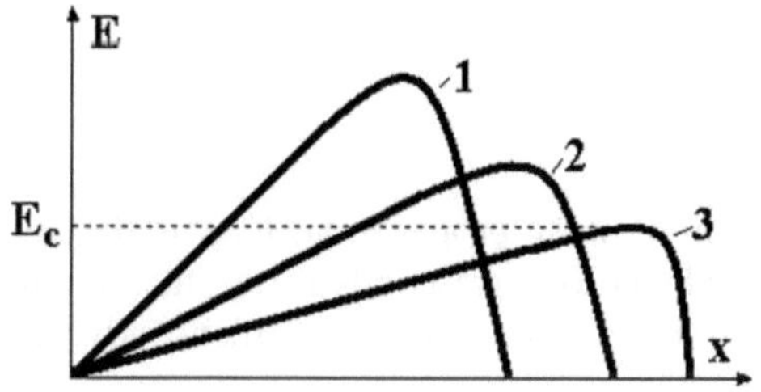

Fig. 5.16 Character of distribution of the electric field strength in a leader. The electric field strength E_c refers to the breakdown threshold. (*1, 2*) distributions of the electric field strength during leader propagation; (*3*) distribution of the electric field strength at leader stop

the electric field in atmospheric air during a thunderstorm is strong. All the electric potential is spent on a restricted length which in $E_c/\overline{E}$ times is less than the cloud-Earth distance. Of course, this does not explain the stopping of a leader, but it exhibits only that a large average velocity of wave propagation is impossible. One can expect that because the drift velocity of electrons is the relaxation time for this process, the ionization wave velocity is restricted by the electron drift velocity. Evidently, stops of a leader are connected with some instabilities in the leader plasma. It is surprising that long-term laboratory studies for leaders in long tubes do not allow understanding this effect in detail. We note also that inasmuch as the electric field is concentrated in a small region compared to a distance between electrodes, this ionization wave may be considered as a solitary wave [83]. This starts from the streamer stage.

We now show that the leader as any ionization wave transfers a small charge or does not transfer it. The role of the leader is only to create a conductive channel. Let us construct the distribution of electrons and ions in a leader such that the electric field distribution corresponds to the curves 1 or 2 of Fig. 5.16. It is important that the electric field strength far from the wavefront is zero. The difference ΔN of the number densities of electrons and ions in this case is given in Fig. 5.17 and is zero at the maximum of the electric field strength. From the Poisson equation we have (we do not follow the sign)

$$\int\limits_{-\infty}^{0} \Delta N dz = \int\limits_{0}^{\infty} \Delta N dz = E_{\max}, \tag{5.32}$$

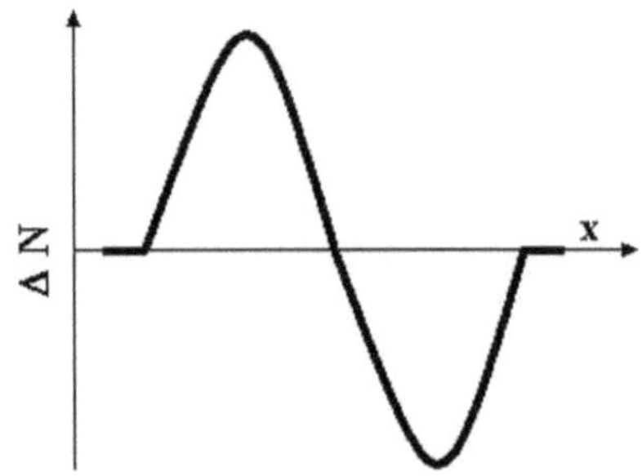

Fig. 5.17 Distribution of the electron number density excess ($\Delta N = N_e - N_i$) in the leader. Coordinate 0 corresponds to the maximum electric field strength

where z is the direction of wave propagation, $E_{\max}$ is the maximum electric field strength, and the coordinate 0 corresponds to the maximum of the electric field strength. From this we find a charge located in the negative and positive regions of coordinates

$$Q = \pi r^2 \int_{-\infty}^{0} \Delta N dz = \frac{E_{\max} r^2}{4e}, \tag{5.33}$$

where r is the radius of a channel along which the leader propagates. If we take $r = 1\,\text{cm}$, Formula (5.33) gives $Q \approx 10^{-9}\,\text{C}$.

Note that the analysis of this book is based on simple models and estimations, on observational data, rate constants of processes, and physical laws. This allows us to remove contradictory information. In particular, the above leader analysis shows that the leader creates a conductive channel, but does not transfer a charge. In contrast, it is accepted according to observational data [76–79] that the leader of middle lightning transfers a charge $Q = 5\,\text{C}$. This statement is represented in books [84–86] and the review [87] gives the linear charge density $q = 3 \times 10^{-3}\text{C/m}$ [87] that, evidently, is the result of the ratio of the transferred charge q to the cloud-Earth distance. In order to show the groundlessness of the above value Q, we first calculate a leader radius R, so that the electric field strength E_b on its boundary according to the Gauss theorem is

$$E_b = \frac{2q}{R} \tag{5.34}$$

Let us take that the air temperature inside the leader exceeds the temperature $T_b \approx 1200\,\text{K}$ at which negative ions are destroyed, and the breakdown electric field strength in atmospheric air at this temperature is $E_b \approx 10^4\,\text{V/cm}$. Then according to Formula (5.34) the leader radius is $R \approx 50\,\text{m}$ that contradicts observational data.

This large radius of the conductive lightning channel results from a large linear charge density. Let us analyze the above value of the specific charge q from another standpoint. If we take a typical area of the cross section of the lightning conductive channel $S \sim 100\,\text{cm}^2$ (a channel radius is approximately 5 cm), one can obtain an excessive number density of the negative charge in the conductive channel $\Delta N_e \sim$

$2 \times 10^{14}\,\text{cm}^{-3}$. Then from the Poisson equation we have for the electric potential U_l on a distance l from the boundary where the electric potential is zero

$$U_l = 2\pi \Delta N_e l^2, \tag{5.35}$$

and it gives $U_l/l^2 = 2 \times 10^8\,\text{V/cm}^2$. From this for the cloud electric potential $U_c \sim 50\,\text{MeV}$ one can obtain $l < 1\,\text{cm}$. One can expect that these contradictions result from an incorrect interpretation of observational data according to which a charge transferred by the lightning leader is $Q = 5\,\text{C}$ for middle lightning.

References

1. H.S.W. Massey, J. de Physique **40**, C7 (1979)
2. H.S.W. Massey, *Negative Ions* (Cambridge University Press, Cambridge, 1976)
3. B.M. Smirnov, *Fundamentals of Ionized Gases* (Wiley, Weinheim, 2012)
4. B.M. Smirnov, *Cluster Ions and Van der Waals Molecules* (Gordon and Breach, Philadelphia, 1992)
5. B.M. Smirnov, *Negative Ions* (McGraw Hill, New York, 1982)
6. L.M. Chanin, A.V. Phelps, M.A. Biondi, Phys. Rev. Lett. **2**, 344 (1959)
7. G.S. Hurst, T.E. Bortner, Phys. Rev. **114**, 116 (1959)
8. L.M. Chanin, A.V. Phelps, M.A. Biondi, Phys. Rev. **128**, 219 (1962)
9. M.H. Hirsch, P.N. Eisner, J.A. Slevin, Phys. Rev. **178**, 175 (1969)
10. R. Hackam, J.J. Lennon, Proc. Phys. Soc. **86**, 123 (1965)
11. J.L. Pack, A.V. Phelps, J. Chem. Phys. **45**, 4316 (1966)
12. V.B. Brodski, S.E. Zagik, Sov. Phys. Tech. Phys. **11**, 498 (1965)
13. J.M. Warman, K.M. Bansal, R.W. Fessenden, Chem. Phys. Lett. **12**, 211 (1971)
14. F.K. Truby, Phys. Rev. **6A**, 671 (1972)
15. D.L. McCorkle, L.G. Christophorou, V.E. Anderson, J. Phys. **5B**, 1211 (1972)
16. F. Bloch, N.E. Bradbery, Phys. Rev. **48**, 689 (1935)
17. F.R. Gilmore, J. Quant. Spectrosc. Radiat. Transf. **5**, 369 (1965)
18. A.A. Radzig, B.M. Smirnov, High Temp. **10**, 29 (1972)
19. E. Illenberger, B.M. Smirnov, Phys. Uspekhi **41**, 651 (1998)
20. G. Breit, E. Wigner, Phys. Rev. **49**, 519 (1936)
21. J.M. Blatt, V.E. Weisskopf, *Theoretical Nuclear Physics* (Wiley, New York, 1952)
22. L.D. Landau, E.M. Lifshitz, *Quantum Mechanics* (Pergamon Press, Oxford, 1965)
23. L.G. Christophorou et al., J. Chem. Phys. **43**, 4273 (1965)
24. L.G. Christophorou, D.L. McCorkle, J.G. Carter, J. Chem. Phys. **54**, 253 (1971)
25. R.L. Brunt, L.J. Kiefer, Phys. Rev. A **2**, 1900 (1970)
26. T. Oster, A. Kühn, E. Illenberger, Int. J. Mass Spectrom. Ion Process. **89**, 1 (1989)
27. E. Illenberger, J. Momigny, *Gaseous Molecular Ions. An Introduction to Elementary Processes Induced by Ionization* (Steinkopf Verlag, Darmstadt, 1992)
28. T.F. O'Malley, H.S. Taylor, Phys. Rev. **176**, 207 (1968)
29. F. Linder, H. Schmidt, Zs. Naturforsch **26a**, 1617 (1971)
30. J.E. Land, W. Raith, Phys. Rev. Lett. **30**, 193 (1973)
31. F. Koike, T. Watanabe, J. Phys. Soc. Jpn. **34**, 1022 (1973)
32. G.J. Schulz, Rev. Mod. Phys. **45**, 423 (1975)
33. G.J. Schulz, in *Electron-Molecule Scattering*, ed. by S.C. Brown (Wiley, New York, 1979), p. 43
34. W.R. Henderson, W.L. Fite, R.T. Brackmann, Phys. Rev. **183**, 157 (1969)

35. D. Spence, G.J. Schulz, Phys. Rev. **188**, 280 (1969)
36. P.D. Burrow, J. Chem. Phys. **59**, 4922 (1973)
37. D.C. Belic, R.I. Hall, J. Phys. **14B**, 365 (1981)
38. T. Jaffke et al., Chem. Phys. Lett. **193**, 62 (1992)
39. J. Dutton, J. Chem. Phys. Ref. Data **4**, 577 (1975)
40. AKh Mnatsakanyan, G.V. Naidis, in *Chemistry of Plasma*, vol. 14, ed. by B.M. Smirnov (Energoatomizdat, Moscow, 1987), p. 227 (in Russian)
41. B.M. Smirnov, *Introduction to Plasma Physics* (Mir, Moscow, 1977)
42. A.V. Eletskii, B.M. Smirnov, Sov. Phys. Uspekhi **25**, 13 (1982)
43. L.D. Landau, E.M. Lifshitz, *Mechanics* (Pergamon Press, London, 1980)
44. C.S. Walls, G.H. Dunn, J. Geophys. Res. **79**, 1911 (1974)
45. P.M. Mul, J.W. Gowan, J. Phys. **12B**, 1591 (1979)
46. R.A. Heppner et al., Phys. Rev. **13A**, 1000 (1976)
47. J.W. McGowan et al., Phys. Rev. Lett. **42**, 373 (1979)
48. B.M. Smirnov, *Physics of Ionized Gases* (Wiley, New York, 2001)
49. A.C. Faire, K.S.W. Champion, Phys. Rev. **113**, 1 (1959)
50. W.H. Kasner, W.A. Rogers, M.A. Biondi, Phys. Rev. Lett. **7**, 321 (1961)
51. C.B. Kretchmer, H.L. Petersen, J. Appl. Phys. **34**, 3209 (1963)
52. W.H. Kasner, M.A. Biondi, Phys. Rev. **137A**, 137 (1965)
53. W.H. Kasner, M.A. Biondi, Phys. Rev. **174**, 139 (1968)
54. F.J. Mehr, M.A. Biondi, Phys. Rev. **181**, 264 (1969)
55. A.J. Cunningham, R.M. Hobson, J. Phys. **5B**, 2320 (1972)
56. A.J. Cunningham, R.M. Hobson, J. Phys. **5B**, 2328 (1972)
57. E.C. Zipf, Geophys. Res. Lett. **7**, 645 (1980)
58. M.H. Mentzoni, J. Appl. Phys. **36**, 57 (1965)
59. M. Hirsh et al., Bull. Am. Phys. Soc. **11**, 495 (1966)
60. R.C. Gunton, Bull. Am. Phys. Soc. **12**, 208 (1967)
61. D. Smith, C.V. Goodal, Planet. Space Sci. **16**, 1177 (1968)
62. D. Smith et al., J. Phys. **3B**, 34 (1970)
63. W.B. Kunkel, A.L. Gardner, J. Chem. Phys. **37**, 1785 (1962)
64. R.C. Gunton, T.M. Shaw, Phys. Rev. **140**, 44 (1965)
65. R.A. Young, G. St. John, Phys. Rev. **152**, 25 (1967)
66. C.S. Weller, M.A. Biondi, Phys. Rev. **172**, 198 (1968)
67. C.M. Huang, M.A. Biondi, B. Johnsen, Phys. Rev. **14A**, 984 (1976)
68. M.T. Leu, M.A. Biondi, R. Johnson, Phys. Rev. **7A**, 292 (1973)
69. W. Lindinger, Phys. Rev. **7A**, 328 (1973)
70. C. Bottcher, J. Phys. **11B**, 3887 (1978)
71. L.G.H. Huxley, Aust. J. Phys. **10**, 240 (1957)
72. L.S. Frost, A.V. Phelps, Phys. Rev. **127**, 1621 (1962)
73. R.W. Crompton, D.J. Sutton, Proc. Roy. Soc. **215A**, 467 (1952)
74. M.H. Saha, Proc. R. Soc. **99A**, 135 (1921)
75. L.D. Landau, E.M. Lifshitz, *Statistical Physics*, vol. 1 (Pergamon Press, Oxford, 1980). §6
76. M.A. Uman, *Lightning* (McGraw Hill, New York, 1969)
77. M.A. Uman, *About Lightning* (Dover, New York, 1986)
78. M.A. Uman, *The Lightning Discharge* (Academic Press, New York, 1987)
79. V.A. Rakov, M.A. Uman, *Lightning, Physics and Effects* (Cambridge University Press, Cambridge, 2003)
80. E.M. Bazelyan, YuP Raizer, *Spark Discharge* (CRC Press, Roca Baton, 1997)
81. E.M. Bazelyan, Y.P. Raizer, *Spark Discharge* (CRC Press, Roca Baton, 1997)
82. N.B. Vargaftik, *Tables of Thermophysical Properties of Liquids and Gases*, 2nd edn. (Halsted Press, New York, 1975)
83. A.V. Shelobolin, J. Phys. **40D**, 6669 (2007)
84. E.M. Bazelyan, Yu.P. Raizer, Usp. Fiz. Nauk **170**, 753 (2000) (Phys. Uspekhi **43**, 753 (2000))
85. E.M. Bazelyan, Yu.P. Raizer, *Physics of Lightning and Lightning Protection* (Moscow, Fizmatlit, 2001) (in Russian)
86. Yu.P. Raizer, *Physics of Gas Discharge* (Dolgoprudnyi, Intellect, 2009) (in Russian)
87. J.R. Dwyer, M. Uman, Phys. Reports **534**, 147 (2014)

Chapter 6
Ion Processes in Atmospheric Air

Abstract Elementary processes involving molecular ions in atmospheric air are considered and include charge exchange processes, ion-molecular reactions, associative ionization, and formation of cluster ions. Molecular ions in the troposphere are formed under the action of cosmic rays, move due to air convection and are destroyed in ion-ion recombination collisions in a dry atmosphere or attach to aerosols in clouds. From detection of atmospheric ions and this analysis it follows that cosmic rays penetrate in all atmospheric layers. The number densities of ions located at different layers are determined by rates of corresponding elementary processes. An equilibrium between molecular ions and charged aerosols is analyzed in a cloud that is neutral on average.

6.1 Evolution of Molecular Ions in Troposphere

Molecular ions determine the conductivity of a low atmosphere at altitudes below 80 km (see Fig. 3.1) because electrons are absent there. The average conductivity of the atmosphere near its surface is 2.5×10^{-14} S/m (see Sect. 4.2) that corresponds to the average number density $N_i = 400\,\mathrm{cm}^{-3}$ of positive and negative atmospheric ions at low altitudes (up to about 3 km). These ions are formed in the atmosphere as a result of ionization of air molecules under the action of cosmic rays and decay in the recombination process and as a result of ion attachment to aerosols in accordance with the scheme of Fig. 3.4. This means that ions make a small contribution to the atmosphere conductivity.

One can estimate a typical ion density in a dry quiet atmosphere on the basis of the current densities in the atmosphere which are on average $2.4 \times 10^{-16}\,\mathrm{A/cm^2}$ over land and $3.7 \times 10^{-16}\,\mathrm{A/cm^2}$ over an ocean [1] or $1500\,\mathrm{e/(cm^2 \cdot s)}$ and $2300\,\mathrm{e/(cm^2 \cdot s)}$, respectively. Taking a typical ion mobility $2\,\mathrm{cm^2/(V \cdot s)}$ according to Table 3.3 data and the average electric field strength $E = 1.3\,\mathrm{V/cm}$, we obtain the ion drift velocity $w = 2.6\,\mathrm{cm/s}$ in this field, and also the ion number density $N_i = 300\,\mathrm{cm}^{-3}$ over land and over an ocean, taking into account that an atmospheric

B.M. Smirnov, *Microphysics of Atmospheric Phenomena*,
Springer Atmospheric Sciences, DOI 10.1007/978-3-319-30813-5_6

current is created of both positive and negative ions. Note that this relates to molecular ions, whereas charged particles may be located in the atmosphere and they do not make any contribution to an atmospheric current.

In accordance with Fig. 3.5, we take the recombination coefficient of positive and negative ions in the atmosphere as $1 \times 10^{-6}\,\mathrm{cm}^3/\mathrm{s}$ at altitudes of several kilometers under consideration. Accounting for a weak pressure dependence for the recombination coefficient of positive and negative ions, we find the number density of ions in a region of the maximum ionization of air is approximately $6 \times 10^3\,\mathrm{cm}^{-3}$, which exceeds the ion number density in lower atmospheric layers. Correspondingly, the time of establishment of ion equilibrium is $t \sim 1/(\alpha N_i) \approx 3\,\mathrm{min}$.

Based on the above ion number density in the troposphere $N_i = 300{-}400\,\mathrm{cm}^{-3}$, we find for a typical recombination time $t \sim (\alpha N_i)^{-1} \sim 5\mathrm{min}$. For this time ions pass $\sim$8 m under the action of the Earth's electric field. This means that ions cannot penetrate low atmospheric layers from altitudes above 10 km and must be formed only at a given altitude. Let us check the possibility of ion transport from high altitudes as a result of convection. Taking into account ion transport as a result of their diffusion and ion recombination in this way, we have the following equation for the ion number density $N_i(z)$ at a distance z from the ion source

$$D_i \frac{d^2 N_i}{dz^2} - \alpha N_i^2 = 0, \tag{6.1}$$

where D_i is the diffusion coefficient for ions in air. For simplicity we take the diffusion coefficient to be independent of the altitude $D_i(z) = const$. Then multiplying the equation by dN_i/dz, we obtain

$$\frac{1}{2}\left(\frac{dN_i}{dz}\right)^2 - \frac{\alpha}{3D_i} N_i^3 = 0 \tag{6.2}$$

In addition to this equation, we assume that the ion number density N_i and its derivation dN_i/dz are zero at an infinite removal from the ion source, that is, at large z. Solution of (6.2) at the indicated boundary condition gives

$$\frac{1}{\sqrt{N_i}} = \frac{1}{\sqrt{N_o}} + \sqrt{\frac{\alpha}{6D_i}} z = \frac{1}{\sqrt{N_o}}\left(1 + \frac{z}{z_o}\right), \; z_o = \sqrt{\frac{6D_i}{\alpha N_o}} \tag{6.3}$$

Here $N_o = 6 \times 10^3\mathrm{cm}^{-3}$ is the maximum number density of ions or that at altitude of maximal ionization, the recombination coefficient is $\alpha \approx 1 \times 10^{-6}\mathrm{cm}^3/\mathrm{s}$, as follows from Fig. 4.3, and $D_L \sim 2 \times 10^4\mathrm{cm}^2/\mathrm{s}$ according to Formula (2.2.6), if we assume that diffusion is determined by air convection. This gives $z_o \sim 40\,\mathrm{m}$. According to Formula (6.3) we obtain for the number density of ions far from the source

$$N_i(z) = N_o \frac{z_o^2}{z^2}, \; z \gg z_o \tag{6.4}$$

As follows from Formula (6.4), the ion number density $N_i \approx 10\,\mathrm{cm}^{-3}$ will reach a distance $z = 1$ km from the ion source. Because a typical number density of ions in the troposphere is $N_i \approx 300\,\mathrm{cm}^{-3}$, one can conclude that along with intense ionization at altitudes 10−15 km, ionization due to hard radiation also proceeds at lower altitudes. This is confirmed by the total rate of ionization: that is, $4.5 \times 10^7\,\mathrm{cm}^{-2}\,\mathrm{s}^{-1}$ [2]. The ratio of this value to the maximum specific rate of ionization $30{-}40\,\mathrm{cm}^{-3}\,\mathrm{s}^{-1}$ [2–4] gives h $\sim$ 10 km; that is, ionization proceeds over the entire depth of the lower atmosphere, as we conclude from the analysis of Formula (6.4). One can also obtain that the absorption cross section of cosmic rays by air molecules at such altitudes is on the order of $10^{-25}\,\mathrm{cm}^2$. It should also be noted that the energy per formation of one ion pair in air is 30 eV. Based on the total rate of air ionization under the action of cosmic rays of $4.5 \times 10^7\,\mathrm{cm}^{-2}\,\mathrm{s}^{-1}$, one can obtain the energy flux $2.2 \times 10^{-10}\,\mathrm{W/cm}^2$ due to cosmic rays, which gives the power of cosmic rays for the total globe $P_{\cos} = 3 \times 10^8$ W. One can compare this with the total power of solar radiation $1.7 \times 10^{17}\,W$ that penetrates the Earth's atmosphere.

One can add to this that the used diffusion coefficient $D_L \sim 2 \times 10^4\,\mathrm{cm}^2/\mathrm{s}$ due to air convection also relates to particles located in air. In particular, for water molecules this value exceeds the diffusion coefficient of water molecules $D = 0.2\,\mathrm{cm}^2/\mathrm{s}$ [5] by several orders of magnitude, and this was used above. In reality, this small diffusion coefficient must be used for transport on small distances below l $\sim$ 30 cm. On larger distances the convective diffusion coefficients D_l or D_L must be used. The same relates to molecular ions, the mobility $2\,\mathrm{cm}^2/(\mathrm{V}\cdot\mathrm{s})$ of which corresponds to the ion diffusion coefficient at room temperature $D_i \approx 0.1\,\mathrm{cm}^2/\mathrm{s}$, and it must be used for ion transport on small distances, whereas the diffusion coefficients $D_l \sim 100\,\mathrm{cm}^2/\mathrm{s}$ and $D_L \sim 2 \times 10^4\,\mathrm{cm}^2/\mathrm{s}$ may be responsible for ion transport for large distances with corresponding sizes of vortices, l $\sim$ 30 cm and L $\sim$ 100 cm. Including the drift ion motion with the drift velocity $w = 2.6\,\mathrm{cm/s}$ under the action of the atmospheric electric field strength, one can introduce a specific size $r_o = D_i/w = 0.04$ cm, and ion propagation on distances below r_o is determined by ion diffusion with the diffusion coefficient D_i, whereas ion motion at larger distances $z \gg r_o$ proceeds under the action of the Earth's electric field. This takes place by neglecting the convection motion, that is, up to distances l $\sim$ 30 cm. Let us introduce a typical distance $R_o = D_L/w = 80$ m taking into account the convective motion of atmospheric air with ions. Propagation on distances $z \gg R_o$ results from the drift ion motion, and at lower distances of motion one can expect the diffusion character of motion. But convection is possible only at $z \gg 1 = 30$ cm, and because l $\sim R_o$, convective diffusion is not realized practically for ion motion. Thus, the drift character of ion motion takes place at $z \gg r_o$, possibly excluding a low range with ion transport due to convective diffusion. This analysis leads to the conclusion that molecular ions are formed in the lower Earth atmosphere under the action of solar cosmic rays and these ions move in the atmosphere as a result of diffusion or they drift in the atmosphere under the action of the Earth's electric field. Transport of molecular ions as a result of convection or under the action of the Earth's electric field is not of principle for formation of molecular ions in the troposphere.

The above analysis relates to a dry atmosphere, whereas electric atmospheric processes are determined by aerosols that form clouds. In this case an additional channel of decay of molecular ions arises due to ion attachment to aerosols. This process is identical to positive and negative ions. Subsequent collision of positive and negative ions leads to charge loss in the atmosphere, where ions are formed under the action of cosmic rays. Being guided by aerosol parameters in a typical cumulus cloud (2.3.2), we determine first the rate constant k_{at} of ion attachment to neutral aerosols of a cumulus cloud that is determined by the Smolukhowski formula [6]

$$k_{at} = 4\pi D_i r, \tag{6.5}$$

where D_i is the ion diffusion coefficient in air, r is an aerosol radius. On the basis of the above estimation $D_i = 0.1\,\mathrm{cm}^2/\mathrm{s}$ for the ion diffusion coefficient in motionless air, we obtain for the rate constant of ion attachment to the aerosol surface $k_{at} \sim 1 \times 10^{-3}\,\mathrm{cm}^3/\mathrm{s}$. Note that the rate of attachment of positive and negative ions to aerosols exceeds the rate of their recombination if the following criterion holds true

$$k_{at} N_p \gg \alpha N_i, \tag{6.6}$$

where N_p is the aerosol number density. This criterion gives $N_i/N_p \ll 10^3$ if the attachment process proceeds in a cumulus cloud, whereas in reality we have $N_p \sim N_i \sim 1\,\mathrm{cm}^{-3}$. Hence criterion (6.6) holds true in a cumulus cloud, and decay of molecular ions in a cloud results from attachment of ions to aerosols and subsequent recombination of positively and negatively charged aerosols in their collision. One can find from this a typical lifetime of molecular ions that penetrated inside a cumulus cloud.

A typical time of attachment of molecular ions to aerosols in a typical cumulus cloud is estimated as $\tau = (Nv\sigma)^{-1} \sim 0.02\,\mathrm{s}$, where $v \sim 3 \times 10^4\mathrm{cm/s}$ is a thermal velocity of molecular ions, and $\sigma \approx 2 \times 10^{-6}\,\mathrm{cm}^2$ is the cross section of an aerosol or the cross section of ion attachment to an aerosol. As can be seen, a typical time of ion attachment to aerosols in a cumulus cloud is less than a typical time of mutual neutralization of positive and negative molecular ions in collisions in dry air that is estimated as $\tau_{rec} \sim 300\,\mathrm{s}$. From this it follows for a thickness of penetration of molecular ions inside a cumulus cloud $1 \sim \sqrt{2D_i\tau_{at}} \sim 10\,\mathrm{cm}$ that is larger than the mean free path of ions in the atmosphere $\lambda \sim 0.06\,\mu\mathrm{m}$, but is significantly smaller than a typical cloud size. From this it also follows that decay of molecular ions results from ion-ion recombination if the number density of aerosols or dust particles in the atmosphere of a size of $10\,\mu\mathrm{m}$ is less than $1\,\mathrm{cm}^{-3}$ in the troposphere or less than $20\,\mathrm{cm}^{-3}$ at altitudes $11-15\,\mathrm{km}$ where the rate of formation of molecular ions is maximal.

6.2 Recombination of Positive and Negative Ions in Gases

Inasmuch as the negative charge of a low atmosphere is created by negative ions, decay of an atmospheric plasma at these altitudes may be connected with the process of mutual neutralization in collisions of negative and positive ions. The character of this process in air depends on the air pressure, thus one can divide the range of number densities of air molecules in three parts in accordance with the nature of this process for each range. Figure 6.1 represents this dependence.

At low pressures of a gas where ions are located, the ion recombination is described by the Thomson theory [7] and the rate of this process is proportional to the buffer gas density. Indeed, the three-body process proceeds according to the following scheme now

$$A^+ + B^- + C \rightarrow AB + C, \tag{6.7}$$

where C is an air molecule, and a compound system AB may be decayed in components. In this process particles A^+ and B^- combine to form a bound system, and the particle C carries away released energy. For formation of a bound state between A^+ and B^- in the course of their collision it is necessary to require that a distance between them must not exceed the critical radius b at which the interaction energy of ions is equal to their thermal energy T; that is, the critical radius equals $b = e^2/T$ for singly charged ions. One can define the rate constant K of the three-body process (6.7) from the balance equation for the rate of formation of compound system AB which has the form:

$$d[AB]/dt = K[A^+][B^-][C], \tag{6.8}$$

and $[X]$ is the number density of given sort particles.

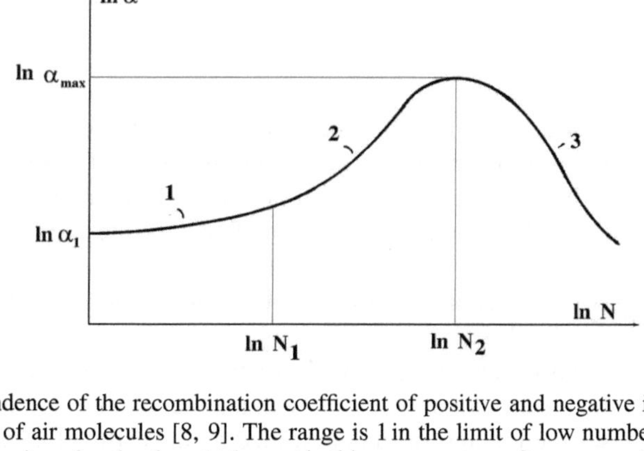

Fig. 6.1 Dependence of the recombination coefficient of positive and negative ions in air on the number density of air molecules [8, 9]. The range is 1 in the limit of low number densities of air molecules where air molecules do not take part in this process, range 2 corresponds to the Thomson theory, and range 3 relates to high number densities of air molecules where recombination is restricted by the mobility of positive and negative ions in air under the action of Coulomb interaction between them

According to the character of this process, the rate of conversion of particles A^+ and B^- into AB in three-body collisions (6.7) is the product of two factors, so that the first factor is the probability of location of a negative ion B^- in the critical region near A^+, that is equal to $[A^+]b^3$. The second factor is the rate $[C]v\sigma$ of collisions with the particle C which takes away an energy excess. Here v is a typical relative collision velocity, and σ is the cross section for the collision between C and either a positive A^+ or negative B^- ion, resulting in an energy exchange on the order of T. Assuming the masses of colliding particles to be of the same order of magnitude, we find this cross section to be comparable to the cross section for elastic collisions. This leads to the following rate of formation of particles AB

$$d[AB]/dt \sim [A^+[B^-]b^3[C]v\sigma$$

Comparing this expression and the definition of the rate constant of the three-body process (6.8), we obtain the following estimate for the rate constant [7]

$$K \sim vb^3\sigma \tag{6.9}$$

The criterion of this theory is that ions A^+ or B^- cannot simultaneously collide with particles C; that is, the mean free path of ions A^+ or B^- in a gas of particles C significantly exceeds a size b of the critical region. This criterion has the form

$$[C]\sigma b \ll 1, \tag{6.10}$$

where σ is a typical elastic cross section of ions A^+ or B^- with a particle C which is an air molecule now. In particular, taking the polarization interaction potential between an ion and molecule $U(R) = -\beta e^2/(2R^4)$, where R is a distance between them, and β is the molecule polarization, one can transform Formula (6.9) to the form

$$K \sim \frac{e^6}{T^3} \cdot \sqrt{\frac{\beta e^2}{\mu}}, \tag{6.11}$$

where μ is the reduced mass of an air molecule with a positive or negative ion.

Let us consider now another limiting case, recombination of positive and negative ions in a dense gas. In this case the approach of ions of different charge signs results from action of the field of a given ion on the ion of an opposite sign, and an approach of ions is braked due to gas resistance through ion collisions with gas atoms. As a result, the rate of ion-ion recombination is determined by ion mobilities under the action of another ion field.

Indeed, if an ion has the charge $\pm e$, and the distance between ions is R, the electric field of an ion at the location of another ion with another charge sign is $E = -e/R^2$, and this field causes the ion approach determined by the drift velocity

$$w = E(K_+ + K_-) = \frac{eE(K_+ + K_-)}{R^2}$$

This is valid for a dense gas if the criterion $R \gg \lambda$ holds true, and λ is the mean free path of ions in a gas.

To determine the decay rate for positive ions due to recombination, imagine a sphere of a radius R around the positive ion, and compute the number of negative ions entering this sphere per unit time. This is given by the product of the surface area of the sphere, $4\pi R^2$, and the negative ion flux $j = N_- w = N_- e(K_+ + K_-)/R^2$. From this we have the balance equation for the number density of positive ions:

$$\frac{dN_+}{dt} = -N_+ N_- 4\pi e(K_+ + K_-)$$

Comparing this with the definition of the recombination coefficient, we derive for this quantity the Langevin formula [10]

$$\alpha = 4\pi e(K_+ + K_-) \qquad (6.12)$$

This formula holds true for a dense gas, and we now combine all the limiting cases for recombination of positive and negative ions, and below we summarize various recombination cases of positive and negative ions in air.

Figure 6.1 gives the dependence of the recombination coefficient for positive and negative ions in air on the number density N of air molecules, and we divide the range of the molecule number density in three groups. At low number densities recombination of positive and negative ions in air has a pair character, and the cross section of their recombination at low collision energies is determined by Formula (5.3.2). Correspondingly, the pair recombination coefficient at low densities of air molecules is $\alpha_1 \geq \hbar^2/(m_e^2 \mu T)^{1/2}$, because $R_o \gg a_o$. Here m_e is the electron mass, μ is the reduced mass of the ions, T is the air (ion) temperature, and $a_o = \hbar^2/(m_e e^2)$ is the Bohr radius. The three-body process of ion-ion recombination (6.7) corresponds to the range 2, and the recombination coefficient for positive and negative ions in air is determined by Formula (6.11) $\alpha_2 \sim (Ne^6/T^3)(\beta e^2/\mu)^{1/2}$, where β is the polarizability of the air molecule, N is their number density, and μ is the reduced mass of the positive or negative ion and the air molecule. From this one can estimate the boundary number density N_1 of air molecules for transition from range 1 to range 2 of Fig. 6.1

$$N_1 \sim a_o (T/e^2)^{5/2} \beta^{-1/2}$$

Range 3 in Fig. 6.1 is described by the Langevin theory, and Formula (6.12) gives an estimate for the recombination coefficient in this range $\alpha_3 \sim e/(N\sqrt{\beta\mu})$, if we assume that the polarization interaction potential determines the ion-molecule scattering process acts. The transition between ranges 2 and 3 is associated with the number density

$$N_2 \sim T^{3/2}/(e^3\sqrt{\beta}),$$

and this number density corresponds to the largest recombination coefficient, $\alpha_{max} \sim e^4 \mu^{-1/2} T^{-3/2}$. The relevant physical process is pairwise ion-ion recombination if the cross section corresponds to elastic scattering of ions the Coulomb interaction between them. Thus, the maximum recombination coefficient has the same order of magnitude as the rate of elastic collisions of ions. Typical values for the quantities indicated in Fig. 6.1 for ions produced in air at room temperature can be cited as: $N_1 \sim 10^{17}$ cm^{-3}, $N_2 \sim 10^{20}$ cm^{-3}, $\alpha_1 \sim (10^{-9}-10^{-7})$ cm^3/s, and $\alpha_{max} \sim 10^{-6}$ cm^3/s.

6.3 Chemistry of Atmospheric Ions

The chemical activity of ions in air is higher than that for radicals and molecules because of a higher interaction of ions with molecules. Therefore chemistry of ions in atmospheric air is various and includes a series of effective processes that determine the atmospheric properties at various altitudes [11, 12]. Because the thermal velocity is $v_T \sim 10^4 - 10^5$ cm/s and the cross section of effective processes is $10^{-15}-10^{-13}$ cm^2, the rate constant of pair processes involving ions and air molecules is $(10^{-11}-10^{-8})$ cm^3/s in thermal collisions for a pair process. Table 6.1 contains the rate constants of some processes with ion participation in air. In this Table we start from processes of dissociative recombination that were analyzed in Sect. 5.3 and correspond both to electrons and ions. The rate constants of some processes of a given type at room temperature are taken from Table 6.1. Note a strong dependence of the recombination coefficient α on the electron energy as well as on the ion temperatures, and although at room temperatures of electrons and ions the data used are reliable, in an excited air plasma the accuracy of these data is lower.

The next group of processes relates to charge exchange with electron transfer from one colliding particle to another. The rate constant of this process is determined by the behavior of electron terms in the system of colliding particles at motionless nuclei. This process is effective if the ionization potentials for formation of ions in the incident and final states are nearby. The charge exchange process may be effective in thermal collisions if the ionization potential for formation of an incident ion is higher than that for the final ion.

Ion-molecular processes in an atmospheric plasma are various and may lead to formation of various ion types. Figure 6.2 represents the average number of basic ions in the lower part of the ionosphere where the negative charge belongs to electrons. As shown, there are four ion types at such altitudes. In contrast to ionospheric ions O^+, O_2^+ and N_2^+ which are formed as a result of photoionization of appropriate atoms and molecules, ions NO^+ in the ionosphere result from ion-molecular processes with participation of ions O^+, namely, processes 6 and 8 of Table 6.1 which are responsible for the formation of ions NO^+.

Processes of destruction of negative ions, in particular, processes 15–17 of Table 6.1, are of importance because they lead to formation of free electrons. Electron detachment from a negative molecular ion requires an energy which in thermal collisions may be taken due to chemical reactions; that is, chemical bonds are changed

Table 6.1 Processes involving collisions of positive and negative ions in air. The rate constants k of pair processes are given in cm^3/s and correspond to room temperature

Type	Number	Process	Rate constant, cm^3/s
Dissociative recombination	1	$e + N_2^+ \rightarrow N + N$	2×10^{-7} (Table 5.2)
	2	$e + O_2 \rightarrow O + O$	2×10^{-7} (Table 5.2)
	3	$e + N_4^+ \rightarrow N_2 + N_2^+$	4×10^{-7} (Table 5.2)
	4	$e + O_4^+ \rightarrow O_2 + O_2^+$	2.3×10^{-6}, T=205K [13]
Charge exchange	5	$N_2^+ + O_2 \rightarrow N_2 + O_2^+ + 3.51\,eV$	1×10^{-10} [14, 15]
	6	$O^+ + O_2 \rightarrow O + O_2^+ + 1.57\,eV$	2×10^{-11} [14]
	7	$N^+ + O_2 \rightarrow N + O_2^+ + 2.46\,eV$	1.4×10^{-10} [16]
	8	$NO + O_2^+ \rightarrow NO^+ + O_2 + 2.81\,eV$	4.4×10^{-10} [17, 18]
	9	$O^+ + NO \rightarrow O + NO^+ + 4.35\,eV$	$k < 1.3 \times 10^{-12}$ [19]
	10	$N_2^+ + H_2O \rightarrow N_2 + H_2O^+ + 2.97\,eV$	2.3×10^{-9}c [15]
	11	$O_2^- + O_3 \rightarrow O_2 + O_3^- + 1.67\,eV$	7×10^{-10} [20, 21]
Ion-molecular reactions	12	$N_2^+ + O \rightarrow NO^+ + N + 3.05\,eV$	1.4×10^{-10} [16]
	13	$O^+ + N_2 \rightarrow NO^+ + N + 1.1\,eV$	1.3×10^{-12} [14]
	14	$N^+ + O_2 \rightarrow O^+ + NO + 2.3\,eV$	4.5×10^{-10} [14]
Associative detachment of negative	15	$O^- + O \rightarrow e + O_2 + 3.6\,eV$	2×10^{-10} [22–24]
	16	$O^- + N \rightarrow e + NO + 5.1\,eV$	2×10^{-10} [22, 23]
	17	$O^- + N_2 \rightarrow e + N_2O + 0.2\,eV$	$k < 1 \times 10^{-14}$ [24, 25]

(continued)

Table 6.1 (continued)

Type	Number	Process	Rate constant, cm^3/s
Ions	18	$O^- + O_2(^1\Delta_g) \rightarrow e + O_3 + 0.5\,eV$	3×10^{-10} [24]
	19	$O^- + O_2 \rightarrow e + O_3 - 0.4\,eV$	$k < 1 \times 10^{-12}$ [25]
	20	$O_2^- + O \rightarrow e + O_3 + 0.6\,eV$	3×10^{-10} [22, 24]
	21	$O_2^- + N \rightarrow e + NO_2 + 4.1\,eV$	5×10^{-10} [22, 26]
	22	$O_2^- + O_2(^1\Delta_g) \rightarrow e + 2O_2 + 0.6\,eV$	2×10^{-10} [24]
	23	$O_3^- + O \rightarrow e + 2O_2$	1×10^{-10} [27]
	24	$OH^- + O \rightarrow e + H_2O + 0.9\,eV$	2×10^{-10} [22, 23]
	25	$OH^- + N \rightarrow e + HNO + 2.4\,eV$	$k < 1 \times 10^{-11}$ [23, 26]
Charge exchange dissociation	26	$N_3^+ + O_2 \rightarrow N_2 + N + O_2^+ 0.5\,eV$	7.4×10^{-11} [28]
	27	$N_4^+ + O_2 \rightarrow 2N_2 + O_2^+ 2.6\,eV$	3×10^{-10} [29–32]
	28	$N_4^+ + H_2O \rightarrow 2N_2 + H_2O^+ + 2.1\,eV$	2.4×10^{-9} [30, 33]
Exchange ion-molecular reactions	29	$O_2^+ \cdot N_2 + O_2 \rightarrow O_4^+ + N_2 + 0.2\,eV$	2.5×10^{-11}, $T = 80K$ [34]
	30	$O_2^+ \cdot N_2 + H_2O \rightarrow O_2^+ \cdot H_2O + N_2 + 0.5\,eV$	4×10^{-9} [35]
	31	$O_4^+ + H_2O \rightarrow O_2^+ \cdot H_2O + O_2 + 0.3\,eV$	1.8×10^{-9} [35, 36]
	32	$H_3O^+ \cdot OH + H_2O \rightarrow H_3O^+ \cdot H_2O + OH$	2.3×10^{-9} [35, 36]
	33	$NH_4^+ \cdot H_2O + NH_3 \rightarrow NH_4^+ \cdot NH_3 + H_2O$	1.2×10^{-9} [37]
	34	$O_3^- + NO_2 \rightarrow NO_3O^- + O_2$	2.8×10^{-10} [29]
	35	$O_4^- + O \rightarrow O_3^- + O_2$	4×10^{-10} [38]
	36	$O_2^- \cdot (H_2O)_2 + NO \rightarrow NO_3^- \cdot H_2O + H_2O$	2×10^{-9} [20, 21]

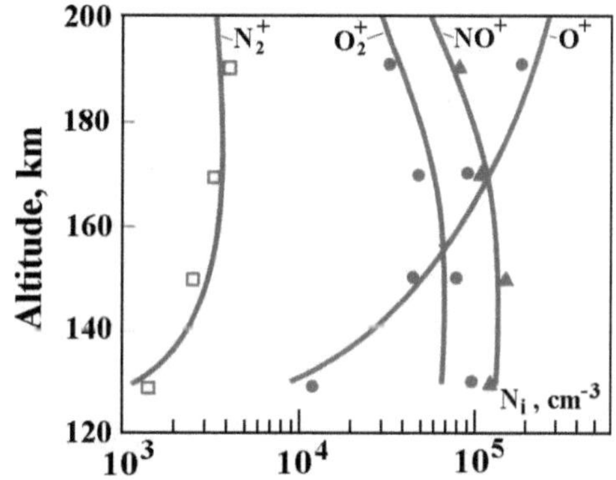

Fig. 6.2 Average number densities of simple ions in the lower day ionosphere [39]

between colliding particles. We restrict ourselves in Table 6.1 by processes that involve only nitrogen and oxygen compounds. In reality one can observe in the atmosphere a variety of processes with participation of impurities. In particular, ions in the lower atmosphere are connected mostly with impurities. In the lower atmosphere alkali metal positive ions and their compounds are observed, as well as chlorine negative ions and their compounds. The analysis in Sect. 3.4 exhibits that complex ions in the stratosphere include molecules of water, nitrogen, and sulfuric acids and other components. In addition, Table 6.2 contains the rate constants of three-body processes in air at room temperature and relates to nonexcited atomic particles. But an ionosphere as a nonequilibrium plasma system contains vibrationally excited molecules and electronically excited atoms and molecules that can change effective rate constants, especially in cases where these rate constants are small compared to gas-kinetic ones.

Three-body processes are important in the formation of complex ions. Complex ions are formed effectively in dense gases at low temperatures. As follows from Fig. 2.1, the favorable conditions are fulfilled in regions of the tropopause and mesopause. This possibility is confirmed by Fig. 3.1, and types of complex ions that are observed in the stratosphere are represented in Tables 3.3–3.5, and Fig. 3.6 gives the mass spectrum of complex ions under laboratory conditions. Complex ions are formed as a result of three-body processes. Some processes of this type involving nitrogen and oxygen components are given in Table 6.2 at room temperature.

6.4 Ions in the Upper Atmosphere

In contrast to ionization in the troposphere under the action of cosmic rays, ionization of molecules and atoms in the upper atmosphere at altitudes of 100–200 km (ionosphere) results from the photoionization process involving solar radiation in the vacuum ultraviolet spectrum. Figure 6.2 exhibits number densities of basic positive ions in the low ionosphere, where electrons carry a negative charge. Evidently, ions

Table 6.2 The rate constants of three-body processes involving positive and negative ions in air are expressed in cm^6/s and correspond to room temperature if the temperature is not indicated

Number	Process	Rate constant, cm^6/s
1	$O_2^+ + 2O_2 \to O_4^+ + O_2$	10^{-29}, T = 200 K [35, 40–44]
2	$CO^+ + 2CO \to CO^+ \cdot CO + CO$	1.3×10^{-28} [45–47]
3	$CO_2^+ + 2CO_2 \to CO_2^+ \cdot CO_2 + CO_2$	2.4×10^{-28} [42, 43, 48–50]
4	$NO^+ + 2NO \to NO^+ \cdot NO + NO$	5×10^{-30} [39, 51]
5	$CO_2^+ + 2CO \to CO_2^+ \cdot CO + CO$	3.5×10^{-28} [47]
6	$CO^+ \cdot CO + 2CO \to CO^+ \cdot (CO)_2 + CO$	3.9×10^{-30}, T = 280 K [46, 49]
7	$O_2^+ + H_2O + N_2 \to O_2^+ \cdot H_2O + N_2$	2.6×10^{-28} [34, 35]
8	$O_2^+ + H_2O + O_2 \to O_2^+ \cdot H_2O + O_2$	2×10^{-28} [35, 36, 41, 42, 52]
9	$NO_2^+ + H_2O + N_2 \to NO_2^+ \cdot H_2O + N_2$	5×10^{-28} [53]
10	$O^- + 2O_2 \to O_3^- + O_2$	9×10^{-31} [54–60]
11	$O_2^- + 2O_2 \to O_4^- + O_2$	4×10^{-31} [59, 61–63]
12	$O^- + CO_2 + O_2 \to CO_3^- + O_2$	2×10^{-28} [55, 64]
13	$O^- + 2CO_2 \to CO_3^- + CO_2$	2×10^{-29} [55]
14	$O^- + H_2O + O_2 \to O^- \cdot H_2O + O_2$	1.2×10^{-28} [63, 64]
15	$O_2^- + H_2O + O_2 \to O_2^- \cdot H_2O + O_2$	1.9×10^{-28} [59, 63, 64]
16	$O_3^- + 2N_2 \to O_3^- \cdot N_2 + N_2$	1.5×10^{-31} [65]
17	$O_3^- + H_2O + O_2 \to O_3^- \cdot H_2O + O_2$	2.4×10^{-28} [63, 64]
18	$CO_3^- + H_2O + O_2 \to CO_3^- \cdot H_2O + O_2$	1×10^{-28}[64]
19	$OH^- + CO_2 + O_2 \to HCO_3^- + O_2$	7.6×10^{-28}[64]
20	$OH^- + H_2O + O_2 \to OH^- \cdot H_2O + O_2$	2.5×10^{-28}[64]
21	$NO_2^- + H_2O + O_2 \to NO_2^- \cdot H_2O + O_2$	1.6×10^{-28}[64]
22	$NO_2^- + H_2O + NO \to NO_2^- \cdot H_2O + NO$	1.5×10^{-28}[51]
23	$O_2^- \cdot H_2O + H_2O + O_2 \to O_2^- \cdot (H_2O)_2 + O_2$	5×10^{-28} [59, 63, 64]
24	$OH^- \cdot H_2O + H_2O + O_2 \to OH^- \cdot (H_2O)_2 + O_2$	3.5×10^{-28} [64]

N_2^+, O_2^+, and O^+ are formed as a result of photoionization of nitrogen and oxygen molecules, as well as oxygen atoms, whereas ions NO^+ are formed from the atomic oxygen ion and molecular nitrogen ions in processes 9, 12, 13 of Table 6.1.

One can make a rough estimate for the number density of ionosphere ions. One can use the value $j_{ion} \sim 2 \times 10^{10}\,cm^{-2}\,s^{-1}$ for the flux of ultraviolet photons whose absorption by atmospheric molecules and atoms leads to their ionization. This radiation is absorbed effectively in an atmospheric layer of a thickness $\Lambda_N = 10{-}30\,km$, where the number density of absorbed molecules varies remarkably (see Fig. 2.5). The loss of electrons and ions proceeds from dissociative recombination, and the recombination coefficient is $\alpha = 4 \times 10^{-7}\,cm^3/s$ at room temperature for dominant molecular ion NO^+. The balance equation for the number density N_i of ions has the following form if they are formed in the photoionization process, and their recombination proceeds through dissociative recombination,

$$\frac{dN_i}{dt} = \frac{j_{ion}}{L} - \alpha N_i^2, \tag{6.13}$$

and we assume above that transformation of atomic oxygen ions in molecular ions according to processes 6, 9, and 13 of Table 6.1 proceeds quickly. From this we find a typical number density of ions in a quasineutral plasma ($N_e = N_i$)

$$N_i = \sqrt{\frac{j_{ion}}{\alpha \Lambda_N}} = (1-2) \times 10^5 \, \mathrm{cm}^{-3} \tag{6.14}$$

This corresponds to the maximum number density of ions in the atmosphere and is an understated value because the electron temperature remarkably exceeds room temperature at altitudes 120–200 km where maximal ionization of the atmosphere is observed under the action of solar radiation.

Accounting for low accuracy of an estimate (6.14) because of ignoring the temperature dependence for the coefficient of dissociative recombination, let us consider other aspects of ion formation. Let us denote the rate of photoionization

$$i = \frac{j_{ion}}{\Lambda_N}, \tag{6.15}$$

and the above rate of ionization in the region of maximal ionization this value is $i = 2 \times 10^4 \, \mathrm{cm}^{-3} \, \mathrm{s}^{-1}$. Note that this value due to troposphere ionization by cosmic rays at altitudes $11 - 15$ km is $i = (30–40) \, \mathrm{cm}^{-3} \, \mathrm{s}^{-1}$. The balance equation for the number density $[O^+]$ of oxygen atoms is

$$\frac{d[O^+]}{dt} = i - k_6[O^+][O_2], \tag{6.16}$$

where k_6 is the rate constant of process 6 of Table 6.1 which is equal at room temperature to $k_6 = 2 \times 10^{-10} \, \mathrm{cm}^3/\mathrm{s}$. Using this value and taking the number densities of atomic particles from Figs. 2.5 and 6.2, one can obtain $i = 3 \times 10^4 \, \mathrm{cm}^{-3} \, \mathrm{s}^{-1}$ using appropriate parameters at the altitude h = 170 km and $i = 1 \times 10^4 \, \mathrm{cm}^{-3} \, \mathrm{s}^{-1}$ at the altitude h = 200 km. One can see that these values are overestimated because they must not exceed the maximum rate $i_{max} \sim 10^4 \, \mathrm{cm}^{-3} \, \mathrm{s}^{-1}$. This means that the rate constant of the charge exchange process involving an oxygen atomic ion and oxygen molecule (process 6 of Table 6.1) is less in the ionosphere than that in air at room temperature.

Next, the balance equation for the number density of molecular ions $[NO^+]$ has the form

$$\frac{d[NO^+]}{dt} = k_{13}[O^+][N_2] - \alpha[NO^+]N_e, \tag{6.17}$$

where k_{13} is the rate constant of process 13 of Table 6.1 which is equal at room temperature to $k_{13} = 1.3 \times 10^{-12}\,\text{cm}^3/\text{s}$. Taking at altitudes above 120 km the number density of electrons as $N_e = [O^+] + [NO^+]$, one can obtain

$$\alpha[NO^+]([O^+] + [NO^+]) = k_{13}[O^+][N_2] \qquad (6.18)$$

Taking the value of the rate constant k_9 at room temperature and the number densities of nitrogen molecules, molecular ions NO^+ and oxygen atoms on the basis of data of Figs. 2.5 and 6.2, one can obtain $\alpha = 3 \times 10^{-8}\,\text{cm}^3/\text{s}$ at altitudes h = 170 and 200 km. Using the above parameters in Formula (6.13), one can estimate the number density of electrons that exceeds the estimate (6.14) by 2–3 times. This factor characterizes the accuracy of such estimations. We also give in Fig. 6.3 the number density of molecular ions and electrons from a source other than that used in Fig. 6.2, as well as for the other scales of altitudes and number densities. Comparison of these data also exhibits an indicated accuracy of these data. It is necessary to add that we consider photons to be directed under a zenith angle to the Earth's surface, and our consideration is also qualitative for this reason.

The information in Figs. 6.2 and 6.3 relates to a daytime atmosphere if solar radiation penetrates it. The ion number density of the nighttime atmosphere follows from the relation $\alpha N_i t \sim 1$, where t is the night duration. This equation takes into account that the number density of electrons and ions is determined by plasma destruction as a result of dissociative recombination and gives the estimate $N_i \sim 10^2 - 10^3\,\text{cm}^{-3}$ for the nighttime atmosphere. Note that in the above estimations we assume a typical time for establishment of the equilibrium for molecular ions to be small. According to (6.13) this time is estimated for the lower ionosphere as $\tau_{rec} \sim (\alpha N_e)^{-1} \sim (10-20)$ s. For atomic oxygen ions we assume above that a time of conversion of atomic ions in molecular ones is small compared to a time of dissociative recombination. This corresponds to the following criterion for the number density $[O_2]$ of oxygen molecules

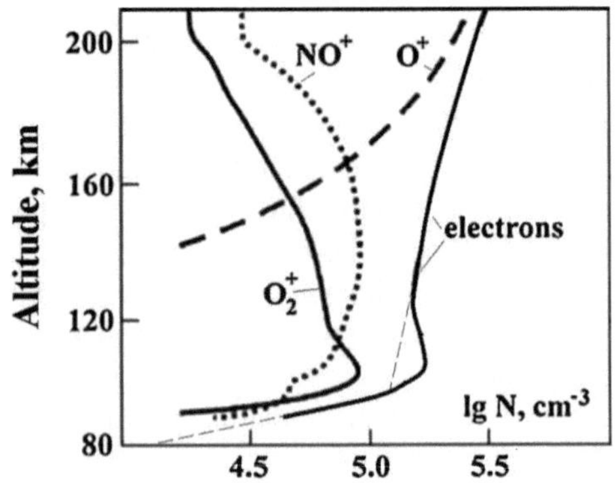

Fig. 6.3 Number density of electrons and ions in the lower part of the daytime ionosphere [66]

$$[O_2] \gg \frac{\alpha N_i}{k} \sim 5 \times 10^9\,\text{cm}^{-3}, \tag{6.19}$$

where $k = 2 \times 10^{-11}\,\text{cm}^3/\text{s}$ is the rate constant of conversion of oxygen atomic ions into molecular ones, that is, process 6 of Table 6.1. As follows from Fig. 2.5, the criterion (6.15) is valid for altitudes below approximately 150 km.

In this consideration we neglect transport of charged particles to other layers because a typical time τ_{dr} of this process is large compared to a recombination time τ_{rec}. We have for a typical time of gravitational drift or diffusion transport of ions accounting for that the drift velocity w_g of ions in the atmosphere coincides by an order of magnitude with that of atoms

$$\tau_{dr} = \frac{w_g}{\Lambda_N} \tag{6.20}$$

According to this formula the number densities of oxygen atoms and nitrogen molecules coincide at altitudes of 200 km where $N_a \approx 10^{10}\,\text{cm}^{-3}$, and then the drift velocity of ions is $w_g \sim 10^3\,\text{cm/s}$ on the basis of Formula (2.10). This gives for a typical time τ_{dr} of ion drift at these altitudes $\tau_{dr} \sim 3 \times 10^3$ s that significantly exceeds a typical recombination time. Because at lower altitudes the drift velocity is smaller, one can conclude that for considering altitudes below 200 km the process of dissociative recombination is responsible for the loss of ions.

The above number densities of ions refer to the daytime atmosphere. At night solar radiation does not penetrate the atmosphere, and the number density of ions is determined by electron-ion recombination. Hence, the ion number density of the nighttime atmosphere follows from the relation $\alpha N_i t \sim 1$, where t is the duration of the night. This gives the estimate $N_i \sim (10^2 - 10^3)\,\text{cm}^{-3}$ for the nighttime atmosphere. In the same manner one can determine the number density of ions at altitudes where solar radiation does not penetrate, but where electrons do not transform in negative ions. There the ion balance results from their drift (or diffusion) motion and dissociative recombination and has the form

$$w_g N_i / L \sim \alpha N_i^2, \tag{6.21}$$

where L is a distance of ion transport. From this we obtain on the basis of Formula (2.10)

$$\frac{N_i N_a}{L} \sim 5 \times 10^{19}\,\text{cm}^{-7} \tag{6.22}$$

Taking $L \sim \Lambda_N$, we have $N_i \sim 50\,\text{cm}^{-3}$ at an altitude of 110 km.

At altitudes where photoionization occurs, the negative charge of the atmospheric plasma comes from electrons. In the D-layer of the ionosphere solar radiation does not penetrate that is able to ionize air molecules. Therefore electrons attach to oxygen molecules in accordance with processes (5.3), and according to Formula (5.4) the rate of electron attachment $(K(O_2)[O_2] + K(N_2)[N_2])[O_2] = \kappa N_a^2$, where N_a is the total number density of air molecules and $\kappa = 1 \times 10^{-31}\,\text{cm}^6/\text{s}$. Assuming electrons

Table 6.3 The rate of decay of negative ions in D-layer of ionosphere [67]

Negative ion	decay rate, s^{-1}	character of ion destruction
O^-	1.4	Electron photodetachment
O_2^-	0.38	"—"
OH^-	1.1	"—"
NO_2^-	8×10^{-4}	"—"
O_3^-	0.47	Photodissociation
O_4^-	0.24	"—"
CO_3^-	0.15	"—"
CO_4^-	9.2×10^{-4}	"—"

to occur in the D-layer of the ionosphere as a result of ambipolar diffusion, we have the balance equation as $KN_a^2N_e = w_gN_e/L$. Taking, as earlier, $L = \Lambda_N$ and being guided by altitudes of approximately 100 km, we obtain from this

$$N_iN_a^2 \sim \frac{w_gN_i}{K\Lambda_N}, \tag{6.23}$$

that gives $N_a \sim 5 \times 10^{12}\,\mathrm{cm}^{-3}$. According to Fig. 2.4 this number density refers to an altitude h $\approx$ 100 km. Negative ions of the lower part of the ionosphere are destroyed as a result of their recombination with positive ions in pair collisions. This process proceeds according to the scheme $A^- + B^+ \to A + B$ and is characterized by a rate constant of about $\alpha \sim 10^{-9}\,\mathrm{cm}^3/\mathrm{s}$. The balance equation in this case is given again by Equation (6.21) from which it follows for the above values of parameters $N_iN_a \sim w_gN_a/(L\alpha) \sim 10^{16}\,\mathrm{cm}^{-6}$ which gives the number density of negative ions at the altitude of 100 km as $N_i \sim 2 \times 10^3\,\mathrm{cm}^{-3}$.

The D-layer of the ionosphere is roughly a boundary, so that ions of lower layers are impurity compounds, whereas above this boundary atmospheric ions include only nitrogen and oxygen. Various types of negative ions may be located in the $D-$layer of the ionosphere (see Table 6.3), and the lifetimes of these ions may be obtained on the basis of the rates of corresponding processes.

6.5 Ionization of Air by Cosmic Rays

Let us analyze the character of ionization of atmospheric air by cosmic rays under the assumption that cosmic rays consist of hard radiation and fast nucleons. The photon energies of several keV constitute the main contribution to ionization by X-rays, and the source of these photons is a solar corona plasma of temperature below 1 keV. These photon energies correspond to a tail of the plasma radiation spectrum where the radiative intensity decreases with increasing photon energies. Above we demonstrate that absorption of these photons by atmospheric air has a nonresonant character; that is, the absorption cross section drops monotonically with increasing photon energy. From this it follows that the rate of absorption and hence the ionization rate due to photons emitted by the solar corona decreases strongly with increasing altitude. Because the observed maximum of the ionization rate lies at altitudes of 11–15 km, this testifies to a not large contribution of solar photons to the rate of ionization of air molecules in the troposphere. One can add to this that the intensity of radiation of the solar corona in the range of keV photon energies has an irregular character and may be increased by several orders of magnitude during solar storms. In contrast to this, the rate of ionization of tropospheric air varies but does not vary as strongly in time (Fig. 3.12). Hence, ionization by X-rays is not the main channel of ionization in the Earth's troposphere, and penetration of fast atomic particles may be of importance for ionization of air molecules there.

Next, one can expect that the solar wind may be a source of cosmic rays that cause air ionization in the troposphere. Although protons of the solar wind cannot penetrate the atmosphere due to action of the Earth's magnetic field, they may be transformed in hydrogen atoms as a result of the charge exchange process at the atmosphere boundary. Indeed, according to Table 2.2 data, it is enough to transform approximately 0.4 % of the solar wind protons to provide the power of cosmic rays which causes ionization of air molecules in the troposphere. But the cross section of scattering of hydrogen atoms of energy of 1 keV on air molecules is on the order of $10^{-17}\,\mathrm{cm}^2$, and such collisions proceed at altitudes of 100–200 km. In particular, an aurora is realized at such altitudes and results from collisions of particles of the solar wind with air molecules in polar regions. As follows from this analysis, the solar wind cannot be responsible for tropospheric ionization by cosmic rays.

Evidently, the basic component of cosmic rays relates to fast particles. Assuming this particle to be singly charged, one can find its typical energy ε if it cannot penetrate at altitudes below $\sim$10 km. Assuming that a strong scattering results from the Coulomb interaction between an incident particle and nuclei of air molecules, one can find a particle energy ε for which the mean free path of an incident particle with respect to its Coulomb scattering on air molecules λ is compared with a typical altitude change $\Lambda_N \approx 6$ km at which the number density of air molecules varies remarkably. We have in accordance with Formula (3.7)

$$\lambda = \frac{1}{N\sigma}, \quad \sigma = \pi\rho^2, \quad \rho \sim \frac{Ze^2}{\varepsilon} \tag{6.24}$$

Here σ is the cross section of elastic Coulomb scattering for an incident singly charged particle and a nucleus of an air molecule, $N \sim 2 \times 10^{19}\,\mathrm{cm}^{-3}$ is a typical number density of nuclei of air molecules in the troposphere, and $Z = 7$ is the charge of nitrogen nuclei. From Formula (6.24) we obtain for a typical particle energy $\varepsilon \sim 10$ MeV. Particles of lower energies cannot go through the troposphere.

In addition, fast charged particles (protons) do not penetrate the Earth's atmosphere due to the action of the Earth's magnetic field. It is convenient to characterize the action of the magnetic field on a particle by the magnetic cutoff rigidity R_c that is proportional to the particle momentum and inversely proportional to its charge [68–70]. In this manner, the Earth's magnetic field deflects charged particles and, correspondingly, is working as a magnetic spectrometer in measurements of the flux of charged particles as a function of the altitude. In particular, the kinetic energy of protons that reach the Earth's surface at the equator must exceed 14 GeV [71] and the boundary energy for penetration of the particle in the atmosphere depends on the geographical latitude of a given locality. The particle rigidity is the boundary energy, so that if the particle energy is less than this value, the particle cannot penetrate the atmosphere because it is deviated by the Earth's magnetic field, and at polar latitudes $R_c = 0 - 0.6\,\mathrm{GeV}$. One can observe the cutoff of slow particles by variation of the detector thickness. Figure 6.4 gives the measured ionization rate for air molecules as a function of the altitude. One can see that the Earth with its magnetic field works like a mass spectrometer [72–75].

Thus, a typical energy of incident particles as a component of cosmic rays must be 10 GeV in order for these particles to penetrate the atmosphere. Note that particles of such energies are relativistic; that is, their velocity is close to the light speed c. In particular, the velocity of protons is equal to $0.43c$ at the energy of 100 MeV and to $0.996c$ at the energy of 10 GeV [78]. From experimental investigations of cosmic rays it follows that they have a galactic origin and a random magnetic field of a galactic space leads to isotropization of their motion, and the maximum of energy distribution function of cosmic particles is found in the energy range 100 MeV–10 GeV [79]. The energy density of cosmic rays in a galactic space $\sim 1\,\mathrm{eV/cm^3}$ is comparable to the energy density of stellar light $\sim 0.3\,\mathrm{eV/cm^3}$ [79]. In addition, cosmic rays are fully ionized.

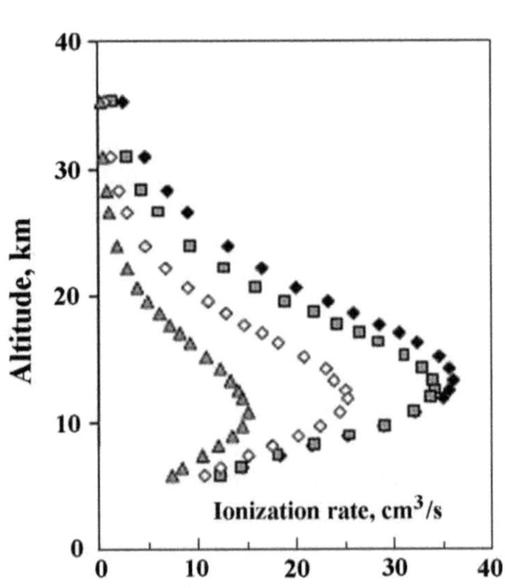

Fig. 6.4 Rate of air ionization rate by cosmic rays at different particle rigidities [76, 77]. ◆—$R_c = 0–0.6\,\mathrm{GeV}$, □—$R_c = 2.4–2.5\,\mathrm{GeV}$, ◇—$R_c = 6\,\mathrm{GeV}$, △—$R_c = 13.5\,\mathrm{GeV}$

Thus, from the above analysis it follows that the optimal energies of cosmic particles for their braking in the upper troposphere, where the maximum rate of air ionization due to cosmic rays are observed, is ~10 MeV and, on the other hand, a typical energy of cosmic particles penetrating the Earth's atmosphere exceeds or is on the order of 10 GeV. It is necessary to reconcile these conclusions. Let us compare the intensity of cosmic rays given in Fig. 6.4 with the observed rate of air ionization. Taking the total flux of formed electrons and ions in the troposphere as a result of ionization of air molecules by cosmic rays to be $4.5 \times 10^7\,\mathrm{cm^{-2}\,s^{-1}}$ [2] and the energetic cost of one electron-ion pair to be 30 eV, one can find the specific power of ionization by cosmic rays $P_{ion} \approx 2 \times 10^{-10}\,\mathrm{W/cm^2}$ (see Table 2.2). We now compare this value with that of the specific power of cosmic rays. From the spectrum given in Fig. 6.5 it follows that the particle flux with energies of 1 GeV and greater is about $1\,\mathrm{cm^{-2}\,s^{-1}}$ [79]. As can be seen, the energy flux of these cosmic rays exceeds $2 \times 10^{-10}\,\mathrm{W/cm^2}$. Hence, this energy flux can provide an observational rate of air ionization by cosmic rays if the particle energy is transferred to air molecules. Evidently, from comparing the data of Figs. 3.11 and 6.5 it follows that the energy flux in the atmosphere from fast protons and X-rays may be comparable. One can add to this that according to Fig. 6.6 although the energy flux of cosmic rays depends on the magnetic cutoff rigidity, relative values of this quantity depend on time in the identical manner. Indeed, Fig. 6.6 gives relative oscillations of the intensity as a result of neutron absorbtion for five stations with different magnetic cutoff rigidity, from $R_c = 1.1$ GeV for a Canadian station up to 12.9 GeV for a Peruvian station.

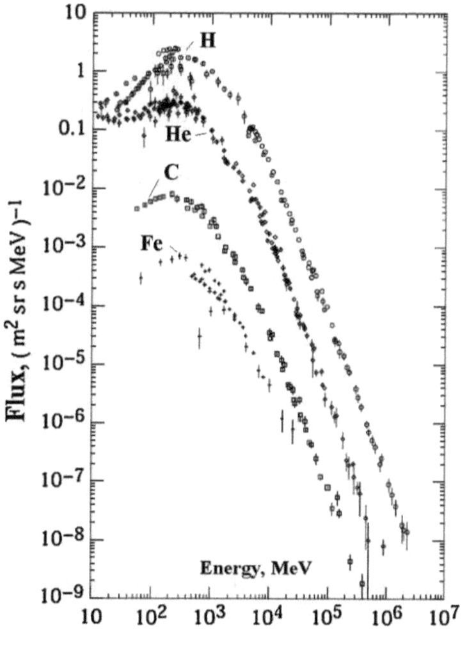

Fig. 6.5 Spectrum of different components of fast cosmic particles [79]

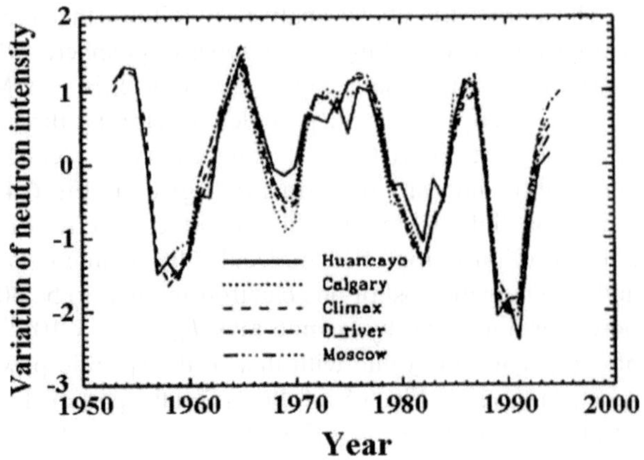

Fig. 6.6 Temporal relative variation of the intensity of cosmic rays measured at stations of different magnetic rigidity [80]

Note that solar radiation in the KeV range of photon energies is not the basic source of cosmic rays. Indeed, in this case solar flashes and other observational instabilities in the Sun's plasma must cause strong flux variations for radiation of various forms and may be accompanied by an increase of some orders of magnitude for the intensity of cosmic solar rays. Because of the absence of strong variations in ionization rates of atmospheric air, galactic cosmic rays make a significant contribution to the intensity of cosmic rays. The principal property of interaction of cosmic rays with atmospheric air is the maximum ionization at altitudes of 11–15 km (see Fig. 3.12). From this it follows that interaction between cosmic rays and air molecules results in nuclear reaction, and the cross section of this process is on the order of the cross section of nitrogen or oxygen nuclei. A part of the energy of fast protons is transferred to air molecules through secondary electrons and nucleons of MeV energies that resulted from nuclear reactions. Braking of forming particles proceeds through their inelastic collisions with air molecules [81], that is, excitation and ionization of air molecules due to Coulomb interaction of incident particles with bound electrons of air molecules. Thus, guided by braking of singly charged particles (protons or electrons) of MeV energies, we analyze their braking in air, taking into account that a typical energy transferred to bound electrons $\Delta\varepsilon$ is small compared to the proton energy, but it is large compared to the binding energies of electrons. Hence these electrons may be considered as free ones during a strong interaction with a fast particle. This allows summing the rate of the particle braking over transferred energies. As a result, we have the balance equation for the proton energy ε [82]

$$\frac{d\varepsilon}{dz} = -N_e \kappa, \qquad (6.25)$$

where z is the motion direction, N_e is the number density of bound electrons in air molecules, and the effective particle braking κ is given by

$$\kappa = \int \Delta\varepsilon d\sigma(\Delta\varepsilon) \tag{6.26}$$

Here $\sigma(\Delta\varepsilon)$ is the cross section of scattering of an incident particle with a transferred energy $\Delta\varepsilon$. Summation over molecule, atom, and ion excited states and usage the sum rule for the atom oscillator strengths leads to the following expression [82] for effective proton braking in the nonrelativistic case

$$\kappa = \frac{4\pi e^4 \ln \Lambda}{m_e v^2} = \frac{2\pi e^4 \ln \Lambda}{\varepsilon} \cdot \frac{M}{m_e}, \quad v \ll c \tag{6.27}$$

where e is the electron charge and the charge of an incident fast particle, M is the mass of an incident particle, v is the velocity of an incident particle, ε is its energy, m_e is the electron mass, and c is the light speed. The Coulomb logarithm is given by

$$\ln \Lambda = \ln \left(\frac{v}{v_o} \right)^2 \tag{6.28}$$

Here v_o is a typical velocity of bound electrons. In particular, in the case of the Thomas-Fermi atomic model we have $v_o \sim e^2 Z^{1/3}/\hbar$. Taking $v = c$ for air molecules ($Z = 7$), one can obtain $\ln \Lambda \approx 10$. If we consider the proton braking, at the energy $\varepsilon = 1\,\text{MeV}$, we have for the Coulomb logarithm $\ln \Lambda \approx 3$, and below we use this value for the Coulomb logarithm. The total number density of bound electrons N_e is given by

$$N_e = 2ZN_a, \tag{6.29}$$

where N_a is the number density of air molecules and Z is the average nuclear charge for air molecules.

Let us introduce the mean free path of a fast singly charged particle λ_{in} according to the formula:

$$\frac{d\varepsilon}{dz} = -\frac{\varepsilon}{\lambda_{in}}, \quad \lambda_{in} = \frac{\varepsilon}{N_e \kappa} = \frac{\varepsilon^2}{4\pi Z e^4 N_m \ln \Lambda} \frac{m_e}{M}, \tag{6.30}$$

where M is the particle mass. In particular, for protons we have:

$$\frac{\lambda_{in}}{\varepsilon^2} = 10 \frac{\text{cm}}{\text{MeV}^2} \tag{6.31}$$

From this we have the mean free path of protons $\lambda_{in} = 10\,\text{cm}$ in the upper troposphere at the energy of $\varepsilon = 1\,\text{MeV}$, if the braking is determined by excitation and ionization of air molecules. Note that within the framework of the Landau–Lifshitz model which leads to Formula (6.27) for effective braking it is based on the Born approximation for inelastic scattering of a charged particle on bound electrons. This approximation requires the fulfillment of the criterion $v \gg e^2/\hbar$ for valence electrons (the proton energy $\varepsilon \gg 50\,\text{keV}$) or of the criterion $v \gg Ze^2/\hbar$ for valence electrons (the proton energy $\varepsilon \gg 2\,\text{MeV}$) for ionization of internal electrons of air molecules.

One can compare this with the mean free path λ due to elastic scattering of a fast singly charged particle on nuclei of air molecules with the Coulomb interaction between them. The latter is defined by the formula (6.24):

$$\lambda = \frac{1}{2N_m\sigma}, \quad \sigma = \frac{\pi Z^2 e^4}{\varepsilon^2} \ln \Lambda \tag{6.32}$$

The ratio of these mean free paths equals

$$\frac{\lambda}{\lambda_{in}} = \frac{2M}{Zm_e} \tag{6.33}$$

As follows from this, one can ignore elastic scattering for protons, and the mean free paths for inelastic and elastic scattering are comparable for electron scattering.

It is important that fast protons or other charged nuclei with energy above 10 MeV may react with nuclei of air molecules as a result of a nuclear short-range interaction. One mechanism of nuclear reactions consists in formation of a compound nucleus in the course of the collision of two nuclei [83, 84], although the physics of nuclear reactions is richer (e.g., [85–88]) and various products may result from nuclear reactions. The first laboratory nuclear reaction was realized by Rutherford in 1919 [89] with only the participation of nitrogen nuclei

$$^4He +^{14} N \to p +^{17} O \tag{6.34}$$

Next, the spread geochronological method (e.g., [90, 91]) involves nitrogen and is based on discovery of radioactive carbon [92–94]. The nuclear process proceeds according to the scheme

$$n +^{14} N \to p +^{14} C \tag{6.35}$$

The lifetime of the isotope ^{14}C is 5730 years, and a certain equilibrium concentration of these isotopes in the atmosphere is determined by the flux of cosmic neutrons. If a carbon-contained compound is located underground, the equilibrium concentration of this isotope decreases in time, and its value allows the determination of a time during which this compound is found underground. One can expect the realization of the above mechanism of nuclear reactions involving protons of GeV energies through the formation of a compound nucleus that provides an effective energy transfer from fast protons to air molecules. In order to obtain the maximum ionization at altitudes of 10–15 km, the cross section of a nuclear reaction must be approximately 70 mb or $7 \times 10^{-26}\,\text{cm}^2$. The cross section of the nuclear process (6.34) is characterized by the cross section 130 mb at the collision energy of 13 MeV [95]; that is, the above requirement is practically fulfilled for an indicated collision energy. But at GeV proton energies along with nuclear reaction formation of showers of elementary particles (as π-mesons) are possible as a result of proton interaction with nitrogen or oxygen nuclei. Therefore, an observational character of ionization of air molecules in collisions with fast protons may result from nuclear reactions of cosmic rays with nuclei of air molecules, but we do not have certain information for this problem.

We now consider the first stage of lightning under conditions where the electric field arises between a cloud and the Earth's surface. We are guided by a typical electric field strength of $E = 200\,\mathrm{V/cm}$ in air in thunderstorm weather that is less than the breakdown electrical field strength $30\,\mathrm{kV/cm}$ in dry atmospheric air. Therefore, a random seed electron in a space between a cloud and Earth is not able to cause an electrical breakdown. Various versions were considered for processes that provide electrical breakdown [96], but their analysis leads to contradictions. The only mechanism of the initial breakdown stage is realized due to cosmic rays and is represented in Fig. 4.12. In this case fast protons or electrons propagate in the atmosphere and cause its ionization. These protons and electrons may accelerate under the action of an electric field and then they are called runaway electrons, as suggested by Wilson [97] in 1924.

Runaway electrons occur if they have high energy and move in the direction of an external electric field such that the power obtained by a test electron from the electric field exceeds the power scattered as a result of collisions with air molecules [98–101]. At the beginning, this mechanism was formulated for a thermonuclear hydrogen plasma and was based on a sharp decrease in the electron-proton cross section with increasing electron energy [102] for this case. Indeed, the Coulomb cross section of scattering of charged particles decreases sharply with increasing collision energy. In the case of other atoms or molecules the energy loss of a propagating electron follows from inelastic collisions, but the cross section of inelastic electron-molecule collisions decreases with an increasing electron energy. As a result, electrons take energy from an electric field and are accelerated.

If an electron is moving in air, one can include in consideration the energy loss due to inelastic electron scattering on air molecules. The balance equation according to Formula (6.25) has the form

$$\frac{d\varepsilon}{dz} = eE - N_e\kappa \tag{6.36}$$

From this it follows that an electron is accelerated under the action of an external electric field if its energy $\varepsilon > \varepsilon_r$, where the runaway energy ε_r is given by

$$\varepsilon_r = \frac{4\pi Z e^4 \ln \Lambda N_m}{eE} \tag{6.37}$$

In particular, under given parameters ($E = 200\,\mathrm{V/cm},\ N_m = 1 \times 10^{19}\,\mathrm{cm^{-3}}$) the runaway energy equals $\varepsilon_r \approx 500\,\mathrm{MeV}$ for a moving proton, and for a singly charged nitrogen nucleus $\varepsilon_r \approx 40\,\mathrm{MeV}$. From this it follows that the concept of runaway charged particles is not suitable for this case in this classical form.

Let us analyze the problem of streamer origin from another standpoint. We are guided by a number of electrons $n_s \sim 10^8$ and the number density of electrons $N_e \sim 10^{13}\,\mathrm{cm}^{-3}$ in accordance with [103]. Assuming that an energy $\varepsilon_i = 30\,\mathrm{eV}$ is consumed on formation of one electron-ion pair, as takes place in gas discharge, one can expect that the energy $\varepsilon_t \sim 3\,\mathrm{GeV}$ is required for streamer origin. Although this is comparable to the initial cosmic particle energy ∼10 GeV during a cosmic shower, this energy is distributed over a large distance, and neutral elementary particles (γ-rays, neutrons, neutrinos) carry an essential part of this energy. Therefore an energy consumed under the action of the electrical field is important for streamer creation. Therefore we consider a proton or a singly charged nitrogen ion with an energy of 1–10 MeV that moves at the beginning at a small angle to the electric field; its trajectory is shown in Fig. 6.7.

In the regime under consideration a fast charged particle ionized air in accordance with Formula (6.36), and in accordance with the Landau–Lifshitz character of particle braking (Formula (6.27)), free electrons are formed in the course of particle braking with an energy ∼100 eV, as follows from the Born approximation. Secondary electrons transfer their energy to excitation and ionization of molecules and are thermalized in the end as a result of 10^2–10^3 collisions with molecules. Because the mean free path of electrons in air at a given number density $N \sim 10^{19}\,\mathrm{cm}^{-3}$ is $\lambda \sim 10^{-3}\,\mathrm{cm}$, thermalized electrons are located in a tube of a radius $r \sim 0.01\,\mathrm{cm}$ with the fast particle trajectory as an axis. Thermalized electrons attach to oxygen molecules, and a typical time of this process is $\tau_{at} = 80\,\mu\mathrm{s}$ instead of (5.5) at room temperature. From this one can find a critical number of electrons n_{cr} which follows from the balance equation and accounts for formation of electrons under the action of the electric field

$$\varepsilon_t = eEw_e n_{cr}\tau \qquad (6.38)$$

Taking an electric field strength $E = 200\,\mathrm{V/cm}$ and the electron drift velocity $w_e \sim 10^6\,\mathrm{cm/s}$ at this electric field strength [104], one can obtain $n_{cr} \sim 10^8$ which can be formed during a time τ. Because of excited air in the region of the electron location, we assume effective ionization of air in a stepwise manner.

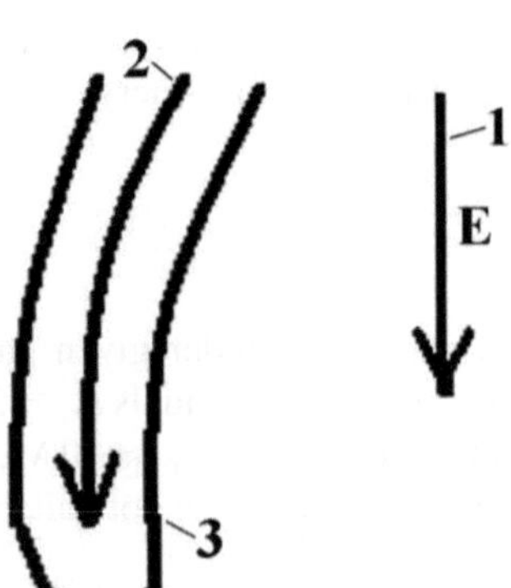

Fig. 6.7 Character of creation of a dense electron ensemble in propagation of a fast singly charged particle through atmospheric air: (*1*) direction of an electric field, (*2*) propagation of a fast particle, (*3*) region with free electrons

As can be seen, secondary electrons lead to the origin of a lightning channel that must contain a number n_s of electrons. Subsequently these electrons form a streamer. We assume that a fast particle creates excited air and the ionization process proceeds in this air similar to that in gas discharge. Then taking a typical energy $\varepsilon_i = 30\,\mathrm{eV}$ consumed on average on formation of an air electron-ion pair, we obtain for a typical length l for its formation as a result of motion of a secondary electron under the action of an electric field

$$l = \frac{\varepsilon_i}{eE}, \tag{6.39}$$

that is, $l \sim 0.2\,\mathrm{cm}$ for a proton. In a manner L one secondary electron leads to formation of n electrons that is given by

$$n = \exp\left(\frac{L}{l}\right) \tag{6.40}$$

Next, we obtain the total number of electrons n_L that are formed after a manner L with accounting for braking of a fast particle and multiplication of a secondary electron:

$$n_L = \frac{L}{\lambda_{in}}\frac{\varepsilon}{\varepsilon_i}\exp\left(\frac{L}{l}\right) \tag{6.41}$$

This formula gives $n_L = 4 \times 10^6$ at $L = 10\,\mathrm{cm}$ and $n_L = 1 \times 10^9$ at $L = 20\,\mathrm{cm}$, but a path time $\tau_L \sim L/v_p \sim 10^{-8} s$ is not enough for electron attachment to oxygen molecules ($\tau_L \ll \tau_{at}$).

We also estimate parameters of air heating up to temperatures where electron attachment is absent. This heating is determined by electron drift, and the heat balance equation has the form

$$eEw_eN_e\tau_h = c_p \Delta T N \tag{6.42}$$

Because air heating with its temperature change $\Delta T \sim 1000\,\mathrm{K}$, taking the electric field strength $E = 200\,\mathrm{V/cm}$, the number density of air molecules $N = 10^{19}\,\mathrm{cm}^{-3}$, the electron drift velocity at these parameters $w_e = 10^6\,\mathrm{cm/s}$, and the heat capacity per molecule at low temperatures $c_p = 7/2$, one can obtain $N_e\tau_p \sim 2 \times 10^7\,\mathrm{s/cm}^3$, which means air heating proceeds longer than the streamer formation.

We are guided above by fast protons as particles that create initial ionization at the origin of the lightning channel. In the case of harder particles the braking effect is stronger. As a demonstration, we now consider braking of bare nitrogen nuclei (the nuclear charge $Z = 7$, the nuclear mass $M = 14$) for which Formula (6.31) has the form

$$\frac{\lambda_{in}}{\varepsilon^2} = 0.02\frac{\mathrm{cm}}{\mathrm{MeV}^2}, \tag{6.43}$$

and the mean free path for the nitrogen nucleus at an energy $\varepsilon = 10\,\mathrm{MeV}$ is equal to $\lambda_{in} = 2\,\mathrm{cm}$. In this case the nitrogen nucleus creates 3×10^5 secondary electrons, and multiplication of each secondary electron leads to the formation of $\sim 10^4$ electrons under the action of the electric field, according to Formula (6.39), and Formula (6.40) gives for the total number of formed electrons $n \sim 3 \times 10^9$. This corresponds to the number density of electrons $N_e \sim 3 \times 10^{15}\,\mathrm{cm}^{-3}$ which is enough for formation of a streamer. But even in this case air heating in an ionization tube is a few Kelvin if all the released energy is consumed on air heating in the tube. Therefore a nonequilibrium character of ionization processes may provide destruction of negative ions in the ionization region.

Thus, from experiments [105] on Tien Shan station (Fig. 4.12) it follows that the origin of electrical breakdown in the atmosphere, lightning, is induced by cosmic rays. The above analysis shows that the basis of formation of a conductive lightning channel is air ionization by cosmic rays that gives secondary electrons as a result of motion of a charged nucleus with MeV energy, and this nucleus may be formed in a nuclear reaction involving cosmic rays. Multiplication of secondary electrons proceeds under the action of an external electric field; that is, the basic energy for air ionization is taken from an electric field. The importance should be noted of a nonequilibrium of the air state in the course of air excitation and ionization in collisions of secondary electrons with air molecules. This emphasizes the roughness of fulfilled estimations. Briefly, the first lightning origin stage is penetration of cosmic rays in the form of protons with GeV energies that collide with nuclei of air molecules as a result of nuclear reactions. Braking of products of such reactions, protons and nuclei of MeV energies, results in air ionization, so that secondary electrons are multiplied later under the action of the electric field. The energy obtained from the electric field for air exceeds the energy scattered by fast particles by some orders of magnitude.

References

1. H. Israel, *Atmospheric Electricity* (Keter Press Binding, Jerusalem, 1973)
2. M.A. Ruderman, J.W. Chamberlain, Planet Space Sci. **23**, 247 (1975)
3. M. Nicolet, Planet Space Sci. **23**, 637 (1975)
4. A. Kryvolutsky et al., Phys. Chem. Earth **27**, 471 (2002)
5. N.B. Vargaftik, *Tables of Thermophysical Properties of Liquids and Gases*, 2nd edn. (Halsted Press, New York, 1975)
6. M.V. Smolukhowski, Zs. Phys. **17**, 585 (1916)
7. J.J. Thomson. Philos. Mag. **47**, 334 (1924)
8. B.M. Smirnov, *Introduction to Plasma Physics* (Mir, Moscow, 1977)
9. B.M. Smirnov, *Physics of Ionized Gases* (Wiley, New York, 2001)
10. P. Langevin, Ann. Chem. Phys. **8**, 245 (1905)
11. H.S.W. Massey, J. de Physique. **40**, C7 (1979)
12. A.D. Danilov, *Chemistry of Ionosphere* (Plenum Press, New York, 1970)
13. M.T. Leu, M.A. Biondi, R. Johnson, Phys. Rev. **7A**, 292 (1973)
14. F.D. Findley, C.J. Fortin, D.R. Snelling, Chem. Phys. Lett. **3**, 204 (1969)
15. B.M. Smirnov, *Physics of Ionized Gas* (Nauka, Moscow, 1978), p. 118. in Russian
16. F.C. Fehsenfeld, E.E. Ferguson, J. Chem. Phys. **56**, 3066 (1972)
17. W.C. Lindinger, et al. *J. Geophys. Res.* **79**, 4753 (1974)
18. W.C. Lindinger, et al. *J. Geophys. Res.* **80**, 3725 (1975)

19. F.C. Fehsenfeld, E.E. Ferguson, Radio Sci. **7**, 113 (1972)
20. D.L. Albritton, Atom. Data Nucl. Data Tables **22**, 3 (1978)
21. C.G. Freeman, P.W. Harland, M.J. McEwan, Int. J. Mass Spectrom. Ion Phys. **28**, 377 (1978)
22. E.E. Ferguson, F.C. Fehsenfeld, A.L. Schmeltesopf, J. Chem. Phys. **45**, 1884 (1966)
23. E.E. Ferguson, F.C. Fehsenfeld, A.L. Schmeltesopf, Adv. Chem. **80**, 83 (1969)
24. F.C.Fehsenfeld, et al. *Can. J. Chem.* **47**, 1793 (1969)
25. J.L. Moruzzi, J.W. Ekin, A.V. Phelps, J. Chem. Phys. **48**, 3070 (1968)
26. F.C. Fehsenfeld, E.E. Ferguson, J. Chem. Phys. **51**, 3512 (1969)
27. L. Thomas, P.M. Gondhalekar, M.R. Bowman, J. Atmos. Terr. Phys. **35**, 397 (1973)
28. W. Lindinger, et al. *J. Chem. Phys.* **64**, 3720 (1976)
29. D.B. Dunkin et al. *J. Chem. Phys.* **54**, 439 (1971)
30. D. Smith, N.G. Adams, T.M. Miller, J. Chem. Phys. **69**, 308 (1978)
31. W. Lindinger, J. Chem. Phys. **68**, 2607 (1978)
32. S. Matsuoka, H. Nakamura, T. Tamura, J. Chem. Phys. **75**, 681 (1981)
33. V.M. Bierbaum, F. Kaufman, J. Chem. Phys. **61**, 3804 (1974)
34. N.G. Adams et al. *J. Chem. Phys.* **52**, 3133 (1970)
35. C.J. Howard, J. Chem. Phys. **57**, 3491 (1972)
36. F.C. Fehsenfeld, M. Mosesman, E.E. Ferguson, J. Chem. Phys. **55**, 2115 (1971)
37. F.C. Fehsenfeld, E.E. Ferguson, J. Chem. Phys. **59**, 6272 (1973)
38. F.C. Fehsenfeld, E.E. Ferguson, D.K. Bohme, Planet Space Sci. **17**, 1759 (1969)
39. L.J. Puckett, M.W. Teague, J. Chem. Phys. **54**, 2564 (1971)
40. D.A. Durden, P. Kebarle, A. Good, J. Chem. Phys. **50**, 805 (1969)
41. A. Good, D.A. Durden, P. Kebarle, J. Chem. Phys. **52**, 222 (1970)
42. C.E. Young, W.E. Falconer, J. Chem. Phys. **57**, 918 (1972)
43. L.W. Sieck, R. Gorden, J. Nat. Bur. Std. **74A**, 315 (1974)
44. B.R. Rowe et al. *J. Chem. Phys.* **80**, 4915 (1984)
45. M. Meot-Ner, F.H. Field, J. Chem. Phys. **61**, 3742 (1974)
46. R.L. Horton, J.L. Franklin, B. Mazzeo, J. Chem. Phys. **62**, 1739 (1975)
47. K.R. Jennings, J.W. Headley, R.S. Mason, Int. J. Mass. Spectrom. Ion. Phys. **45**, 315 (1982)
48. J.F. Paulson, F. Dale, R.L. Mosher, Nature **204**, 377 (1964)
49. L.W. Sieck, Int. J. Chem. Kinet. **10**, 335 (1978)
50. A.B. Rakshit, P. Warneck, *Zs. Naturforsch.* **34A**, 1410 (1979)
51. W.C. Lindinger, L.J. Puckett, Phys. Rev. **186**, 116 (1969)
52. D. Edelson, W.E. Falconer, J. Chem. Phys. **53**, 4295 (1970)
53. F.C. Fehsenfeld, C.J. Howard, A.L. Schmeltesopf, J. Chem. Phys. **63**, 2835 (1975)
54. D.S. Burch, R. Geballe, Phys. Rev. **106**, 188 (1957)
55. J.L. Moruzzi, A.V. Phelps, J. Chem. Phys. **45**, 4617 (1966)
56. J.H. Whealton, S.B. Woo, Phys. Rev. Lett. **20**, 2137 (1968)
57. L.G. McKnight, Phys. Rev. **2A**, 762 (1970)
58. R.M. Snuggs et al. *Phys. Rev.* **3A**, 487 (1971)
59. J.D. Payzant, P. Kebarle, J. Chem. Phys. **56**, 3482 (1972)
60. J.H. Schummers et al. *Phys.Rev.* **7A**, 689 (1973)
61. L.G. McKnight, J.M. Sawina, Bull. Am. Phys. Soc. **15**, 434 (1970)
62. L.G. McKnight, J.M. Sawina, Phys. Rev. **4A**, 1043 (1971)
63. J.L. Pack, A.V. Phelps, Bull. Am. Phys. Soc. **16**, 214 (1971)
64. F.C. Fehsenfeld, E.E. Ferguson, J. Chem. Phys. **61**, 3181 (1974)
65. E.E. Ferguson, Can. J. Chem. **47**, 1815 (1969)
66. D. Bilitza, Radio Sci. **36**, 261 (2002)
67. J. Wisemberg, G. Kokarts, J. Geophys. Res. **85**, 4642 (1980)
68. C. Störmer, *Polar Aurora* (Claredon Press, Oxford, 1955)
69. D.J. Cooke, J.E. Humble, M.A. Shea et al. *Nuovo Chim.* **C14**, 213 (1991)
70. D.F. Smart, M.A. Shea, Adv. Space Res. **36**, 2012 (1955)
71. G.A. Bazilevskaya et.al, in *Planetary Atmospheric Electricity*, ed. by F. Leblanc et.al. (Heidelberg, Springer, 2008), p. 149

72. J.A. Simpson, E. Hungerford, Phys. Rev. **77**, 847 (1950)
73. J.A. Simpson, Ann. Rev. Nucl. Part. Sci. **33**, 323 (1983)
74. G.A. Bazilevskaya, M.B. Krainev, V.S. Makhmutov, J. Atmos. Sol.-Ter. Phys. **62**, 1577 (2000)
75. G.A. Bazilevskaya, Space Sci. Rev. **94**, 25 (2000)
76. H.V. Neher, J. Geophys. Res. **76**, 1637 (1971)
77. H.V. Neher, J. Geophys. Res. **72**, 1527 (1967)
78. http://engineering.dartmouth.edu/~d76205x/research/Shielding/docs/Mewaldt-96.pdf
79. http://www.int.washington.edu/PHYS554/winter-2004/chapter8-04.pdf
80. H. Svensmark, E. Frils-Christensen, J. Atmos. Terr. Phys. **59**, 1225 (1997)
81. H. Bethe, Ann. Phys. **5**, 325 (1930)
82. L.D. Landau, E.M. Lifshitz, *Quantum Mechanics* (Pergamon Press, Oxford, 1965)
83. J.M. Blatt, V.E. Weisskopf, *Theoretical Nuclear Physics* (Wiley, New York, 1952)
84. V.F. Weiskopf, Rev. Mod. Phys. **29**, 174 (1950)
85. N.A. Jelley, *Fundamentals of Nuclear Physics* (Cambridge University Press, Cambridge, 1990)
86. G.R. Sattler, *Introduction to Nuclear Reactions* (Oxford University Press, New York, 1990)
87. J.S. Lilley, *Nuclear Physics* (Chichester, Wiley, 2001)
88. R.L. Murray, K.E. Holbert, *Nuclear Energy*, 7th edn. (Elsevier, Oxford, 2015)
89. E. Rutherford, Philosoph. Mag. **37**, 581 (1919)
90. G. Faure, *Principles of Isotope Geology* (Cambridge University Press, Cambridge, 1986)
91. A.P. Dickin, *Radiogenic Isotope Geology* (Cambridge University Press, Cambridge, 1995)
92. S. Ruben, M.D. Kamen, W.Z. Hassid, Science **90**, 510 (1939)
93. S. Ruben, M.D. Kamen, Phys. Rev. **57**, 549 (1940)
94. S. Ruben, W.Z. Hassid, M.D. Kamen, J. Am. Chem. Soc. **62**, 3443 (1940)
95. A.M. MacLeod, J.M. Reid, Proc. Phys. Soc. **87**, 437 (1966)
96. J.R. Dwyer, M. Uman, Phys. Rep. **534**, 147 (2014)
97. C.T.R. Wilson, Proc. Cambr. Philos. Soc. **22**, 34 (1924)
98. H. Dreicer, Phys. Rev. **115**, 238 (1959)
99. H. Dreicer, Phys. Rev. **117**, 329 (1960)
100. A.V. Gurevich, ZhETF **39**, 1296 (1960)
101. A.V. Gurevich, G.M. Milikh, R. Roussel-Dupre, Phys. Lett. A **165**, 463 (1992)
102. E.M. Lifshitz, L.P. Pitaevskii, *Physical Kinetics* (Pergamon Press, Oxford, 1981)
103. H. Raether, Arch. Electrotech. **34**, 49 (1940)
104. J. Dutton, J. Chem. Phys. Ref. Data **4**, 577 (1975)
105. A.V. Gurevich, A.N. Karashtin, V.A. Ryabov et al. *Usp. Phys. Nauk* **179**, 779 (2009)

Chapter 7
Processes in Dissociated Air

Abstract Atmospheric air is dissociated both at high altitudes and during propagation of lightning current through air. Various aspects of atmospheric dissociated air are analyzed. A conducted expanded lightning channel contains a hot almost equilibrium plasma. The analysis is represented for various channels in the lightning energetic balance. Processes involving atomic oxygen and ozone in excited air are considered.

7.1 Propagation of Electric Current Through Conductivity Lightning Channel

Charge transfer from the lower edge of the cumulus cloud to the Earth's surface includes several stages [1–5]. We consider now creation of the lightning conductivity channel as a result of propagation of the preionization wave: the arrow-like leader. In accordance with observational data and measurements, a leader consists of separated segments of length approximately 50 m, and the average velocity inside each segment is 2×10^8 cm/s. After passage of each segment the leader stops, and the average velocity of leader propagation is 1×10^7 cm/s. The statement [1–6] that the leader transfers a charge of 5 C is not correct. At the next stage, when an electric current passes through the conductivity lightning channel, we combine subsequent short stages of propagation of a strong current including recurrent stroke, in one stage with a duration time of 70 μs where a typical current strength is 30 kA. This corresponds to a transferred charge 2 C, the velocity of signal propagation is 1×10^{10} cm/s, and a typical channel radius is 1 cm [3]. The quiet steady current stage follows after this stage with a current strength 100 A and a time of duration 50 ms, that corresponds to the transferred charge 5 C at this stage. After this again a recurrent stroke arises with a current strength 15 kA and a duration time 50 ms, that corresponds to a transferred charge 1 C, and the last stages may repeat several times. We assume lightning to consist of 4 flashes on average, an average transferred charge to be approximately 25 C, and the total lightning duration 0.2 s. We use the above observed and measured lightning parameters [1, 3, 7] below. In addition, we take an average cloud electric potential with respect to the Earth to be $U = 60$ MV [8] and a distance

B.M. Smirnov, *Microphysics of Atmospheric Phenomena*,
Springer Atmospheric Sciences, DOI 10.1007/978-3-319-30813-5_7

cloud-Earth to be $L = 3\,\text{km}$, that corresponds to the average electric field strength $E = U/L = 200\,\text{V/cm}$. This is two orders of magnitude less than the breakdown electric field strength (about 30 kV/cm) for dry air.

In considering propagation of the lightning electric current through the atmosphere where the conductive lightning channel is created, we assume the channel to have a round cross section. Our task is to estimate parameters of an air plasma located in a conductive lightning channel on the basis of average measured lightning parameters. When the conductive lightning channel is formed and an electric current propagates along it, the air temperature of the conductive channel reaches the temperature $T = 20{,}000{-}30{,}000\,\text{K}$ according to spectroscopic measurements [1–3, 7]. Note the reliability of spectroscopy measurements fulfilled starting in 1868 [9] which give typical values of the air temperature as well as the number density of ions located in the conductive lightning channel because they are confirmed by various spectroscopic measurements, relate to various spectral lines, and were fulfilled by various scientists. In particular, the number density of electrons (and ions) is determined on the basis of shifts of spectral lines for excited atoms or molecules under the action of an electric field created by ions (the Holzmark effect [10]). This method allows one to determine reliably the number density of ions and, correspondingly, the number density of electrons.

Because of a high conductivity of the conductive lightning channel, it contains a plasma where the charge transfer has a collective character, so that the electric field propagates with light speed and is transferred over a large distance during a time

$$\tau_f \sim \frac{1}{\omega_p} = \frac{C}{\sqrt{N_e}}, \tag{7.1}$$

where ω_p is the plasma frequency, N_e is the number density of electrons, and $C = 2 \times 10^{-5}\,\text{cm}^{-3/2} \cdot \text{s}$. At the temperature $T = 20{,}000\,\text{K}$ and atmospheric pressure $p = 1\,\text{atm}$ the number density of electrons is estimated as $N_e \sim 10^{18}\,\text{cm}^{-3}$ in accordance with the data of Fig. 5.15, and according to formula (7.1) a typical transfer time of an electric signal is estimated as $\tau_f \sim 10^{-14}\,\text{s}$; that is, it is small compared to other plasma times. The charge transfer results from the shift of electrons as a whole with the drift velocity of electrons $w_e \sim 10^7\,\text{cm/s}$. In this case transfer of a charge $Q = (1-2)\,\text{C}$ at the recurrent stroke stage proceeds during a time $\tau_Q \sim 1\,\mu\text{s}/S$, where S is the cross section of the conductive channel expressed in cm^2. Because a radius of the conductive lightning channel is $r \sim 1\,\text{cm}$ [3], a typical charge transfer time τ_Q at the recurrent stroke stage is relatively small. Hence, charge transfer in the conductive lightning channel proceeds analogously to that in conductors. Namely, electrons are shifted as a whole along the entire conductor, and a typical charge transfer time is independent of a conductor length.

Note that most of a transferred charge proceeds at the weak continuous current stage [3]. In the course of current propagation, air of the conductive lightning channel is heated and expands at atmospheric pressure of surrounding air. Heat transport across the conductive lightning channel is of twofold importance. First, it establishes parameters of heat release during a pass of pulses of strong electric currents (recurrent strokes). Second, heat-removing determines a radius of the conductive lightning channel. Let us determine a radius of the conductive channel after the first recurrent stroke on the basis of the above parameters, where the electric current strength is equal to $I = 30\,\text{kA}$ at this stage, and its duration is $\tau = 70\,\mu\text{s}$. At a typical air temperature $T = 20{,}000\,\text{K}$ the specific conductivity of an air plasma is $\Sigma = 3 \times 10^4\,\text{S/m}$. From the relation

$$I = \pi r^2 \Sigma E \tag{7.2}$$

we obtain the electric field strength $E = 200\,\text{V/cm}$ for a radius of the conductive channel $r \approx 0.3\,\text{cm}$. The same estimation for a radius of the conductive lightning channel follows from the analysis of thermal expansion of this channel

$$r^2 = 4\chi\tau, \tag{7.3}$$

where the thermal diffusivity coefficient is equal to $\chi_a = 600\,\text{cm}^2/\text{s}$ for atmospheric air at the temperature $T = 20{,}000\,\text{K}$. Note that according to measurements [3] a radius of the conductive lightning channel is more; that is, expansion of a conductive channel during a current pass is not essential.

7.2 Conductivity of Dissociated Air

When a conductive channel of lightning is created, the current through the conductive channel depends both on its conductivity and radius. The air conductivity is determined by the electron concentration and hence grows with an increasing temperature. Therefore the temperature of air located in the conductive channel is large, and air is dissociated in this region. Hence the air conductivity Σ in the lightning conductive channel is determined by electron-atom and electron-ion scattering. In a general case the air conductivity is given by [11]

$$\Sigma = \frac{N_e e^2 \tau}{m_e} \tag{7.4}$$

Here N_e is the electron number density, m_e is the electron mass, and τ is a typical scattering time for electrons. Based on this formula, below we determine the conductivity of dissociated air being guided by the lightning conductive channel. There are two scattering mechanisms for electrons, namely, electron-atom scattering and electron-ion scattering, and the contribution of these scattering channels to the total conductivity is compared at the electron concentration $\sim 1\,\%$ [12]. In considering both scattering channels, we have for the scattering time $1/\tau = N_a v_T \sigma_{ea} + N_i v_T \sigma_{ei}$, where N_a is the number density of atoms, N_i is the number density of ions, v_T is a thermal electron velocity, σ_{ea} is the diffusion cross section of electron-atom scattering, and σ_{ei} is the cross section of electron-ion scattering. Because this analysis has a qualitative character, we assume the cross section of electron-atom scattering to be identical for nitrogen and oxygen atoms. Thus, accounting for both scattering mechanisms for electrons, we have the following expression for the conductivity of dissociated air

$$\Sigma = \left(\frac{1}{\Sigma_{ea}} + \frac{1}{\Sigma_{ei}} \right)^{-1}, \tag{7.5}$$

where Σ_{ea} is the conductivity due to electron-atom scattering, and Σ_{ei} is the conductivity due to electron-ion scattering. As can be seen, the air conductivity is determined by the scattering mechanism that gives the less partial conductivity. Therefore, electron-atom scattering determines the air conductivity at relatively small temperatures, whereas electron-ion scattering makes the main contribution to air conductivity at high temperatures.

At low temperatures the conductivity of dissociated air according to formula (7.4) is given by [11]

$$\Sigma_{ea} = \frac{N_e e^2 \tau}{m_e} \tag{7.6}$$

Guided by electron and gas temperatures $T_e \sim 10{,}000\,\text{K}$ in equilibrium dissociated air, or a typical electron energy of the order of 1 eV, we have for a typical cross section of electron-atom scattering [13] $\sigma_{ea} \approx 2 \times 10^{-16}\text{cm}^2$. This gives for the air conductivity at low temperatures:

$$\Sigma_{ea} = \frac{N_e \Sigma_o}{N_a \sqrt{T_e}} \tag{7.7}$$

Here N_e is the electron number density, N_a is the number density of atoms of dissociated air, and the parameter Σ_o in this formula is $\Sigma_o = 2 \times 10^6\,\text{S/m}$, and the electron temperature T_e is expressed in eV. (The Siemens unit is $S = 1/\Omega$ [14], where Ω is ohm). If the conductivity of dissociated air is determined by electron-ion scattering, this is determined by formula [15, 16]:

$$\Sigma_{ei} = \frac{\Sigma_1 T_e^{3/2}}{\ln \Lambda} \tag{7.8}$$

Here $\Sigma_1 = 1.9 \times 10^4\,\text{S/m}$, the electron temperature is expressed in eV, and the Coulomb logarithm $\ln \Lambda$ is taken as $\ln \Lambda = 7$. This transforms formula (7.8) to the form

$$\Sigma_{ei} = 2.7 \cdot 10^3 T_e^{3/2}\,\text{S/m} \tag{7.9}$$

Figure 7.1 gives the temperature dependence for the conductivity of dissociated air. The transition from the mechanism of electron-atom scattering to the mechanism of

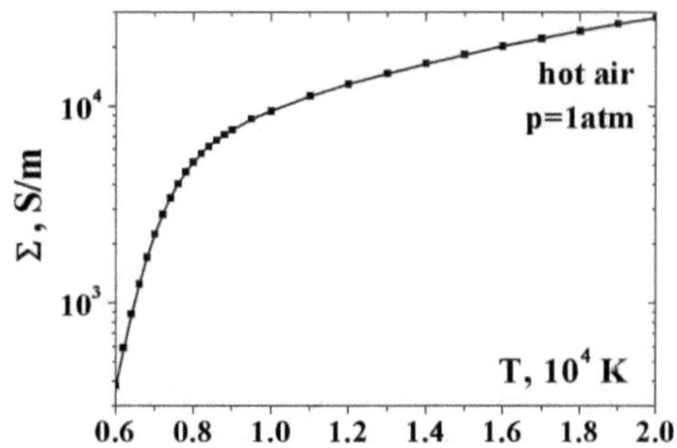

Fig. 7.1 Total conductivity coefficient of dissociated air due to electron-atom and electron-ion scattering

electron-ion scattering for atmospheric air proceeds at the temperature $T = 7300\,\mathrm{K}$, and the conductivity of atmospheric air is $\Sigma = 3 \times 10^3\,\mathrm{S/m}$ at this temperature. In addition, air is dissociated under these conditions.

When an electric current propagates through the lightning conductive channel, air is heated and expands. The velocity of propagation of a thermal wave is small compared with the sound speed; that is, in the course of thermal processes the pressure is identical in a space. Below we consider propagation of a thermal wave in the absence of convection, so that air expansion is determined by the thermal conductivity of heated air. Because of a high air temperature in the lightning channel, air is dissociated both in the channel and close to it. Therefore in the following we determine the thermal conductivity coefficient κ of dissociated air that characterizes heat transport across the lightning channel.

This heat transport is determined by two mechanisms: by collisions between air atoms and as a result of electron-atom collisions. Because the heat transfer is determined by transport of different atomic particles, the total thermal conductivity coefficient κ is a sum of the thermal conductivity coefficients due to transport of atoms κ_a and electrons κ_e; that is,

$$\kappa = \kappa_a + \kappa_e \tag{7.10}$$

In determining the thermal conductivity coefficient κ_a due to atom transport, we use the model of hard spheres (e.g., [17]), and according to this model collision of two atoms is analogous to collision of hard balls. In this case we have for the thermal conductivity coefficient (e.g., [18, 19])

$$\kappa_a = \frac{75\sqrt{\pi T}}{64\sigma_g\sqrt{m}}, \tag{7.11}$$

where T is the temperature, m is the atom mass, and σ_g is the gas-kinetic cross section for collision of two atoms. Note that dissociated air consists of nitrogen and oxygen atoms in the ground states, $N(^4S)$ and $O(^3P)$, and the statistical weights of atom states are $g_N = 4$ and $g_O = 9$. There are a number of electron terms for interaction of each atom pair that is a number of nondegenerate electron states of the quasimolecule consisting of colliding atoms, and this number is large. Therefore a gas-kinetic cross section σ_g in formula (7.11) corresponds to averaging over many channels. At high temperatures this cross section corresponds to the approach of colliding atoms where their electron shells are intersected. Using the same consideration for collisions of two neon atoms, we obtain a gas-kinetic cross section for collisions of nitrogen or oxygen atoms approximately $\sigma_g = 19\,\text{Å}^2$ at room temperature. Correspondingly, if we take the gas-kinetic cross section for collisions of nitrogen or oxygen atoms as one half of gas-kinetic cross sections of nitrogen or oxygen molecules, we obtain approximately the same result. The temperature dependence for the gas-kinetic cross section has the form $\sigma_g \sim T^{-0.2}$.

In the analysis of heat transport in air, we operate with the thermal diffusivity coefficient χ that is an analogue of the diffusion coefficient for particle transport and is equal to

$$\chi = \frac{\kappa}{c_p N_a}, \tag{7.12}$$

where $c_p = 5/2$ is the heat capacity per atom at constant pressure and N_a is the atom number density. Figure 7.2 contains the thermal diffusivity coefficient of dissociated air owing to atom transport.

The thermal conductivity coefficient due to electron transfer in an equilibrium plasma with identical gaseous and electron temperatures ($T_e = T$) is given by [20]

$$\kappa_e = \frac{2c_e\sqrt{T}}{3\sqrt{\pi m_e}\sigma_{ea}} = \frac{0.532c_e\sqrt{T}}{\sigma_{ea}\sqrt{m_e}}, \tag{7.13}$$

where c_e is the electron concentration and σ_{ea} is the diffusion cross section of electron-atom scattering that is assumed to be independent of the collision velocity. We take this cross section to be $\sigma_{ea} = 3 \times 10^{-16}\,\text{cm}^2$ on the basis of data for electron scattering in various gases [13]. Formula (7.13) corresponds to low electron and ion concentrations where one can neglect electron-ion scattering. In order to take into account the contribution of electron-ion scattering to the thermal conductivity coefficient due to electrons, we use the analogy in electron-ion scattering for the

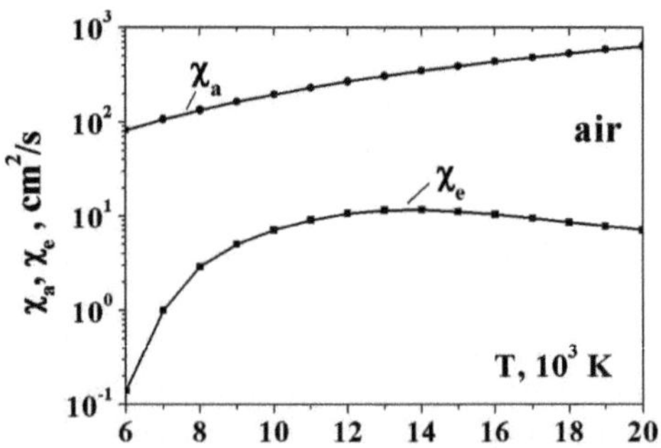

Fig. 7.2 Thermal diffusivity coefficient in dissociated air due to electron transfer and atom transfer

conductivity coefficient of this plasma and for the thermal conductivity coefficient. Then we multiply the thermal conductivity coefficient due to electrons by the factor $\Sigma_{ea}/(\Sigma_{ea}+\Sigma_{ei})$, where Σ_{ea}, Σ_{ei} are the air conductivities given by formulas (7.7) and (7.8), correspondingly. As a result, we have for the thermal diffusivity coefficient of ionized air owing to electron transfer

$$\chi_e = \frac{0.532\Sigma_{ea}c_e\sqrt{T}}{(\Sigma_{ea}+\Sigma_{ei})\sigma_{ea}\sqrt{m_e}}, \qquad (7.14)$$

Figure 7.2 contains the thermal diffusivity coefficient of dissociated air χ_e due to electron transfer. As shown, although a thermal velocity of electrons significantly exceeds that for atoms, the cross section of electron-ion scattering with Coulomb interaction is larger than that for electron-atom and atom-atom scattering. For this reason, the electron contribution to the thermal diffusivity coefficient is small in a wide range of temperatures.

7.3 Energy Balance for Conductive Lightning Channel

Let us estimate parameters of an air plasma of the conductive lightning channel if an electric current propagates through this channel. It is convenient to use formulas (7.2) and (7.3) and exclude the channel radius that gives the relation for a plasma temperature T in the form

$$\Sigma(T)\chi(T) = \frac{I}{4\pi E\tau} \qquad (7.15)$$

Substituting in formula (7.15) observational parameters of the lightning channel in a quiet lightning phase $I = 100\,\mathrm{A}$, $E = 200\,\mathrm{V/cm}$ and the total time of channel existence $\tau = 0.2\,\mathrm{s}$, we obtain $\Sigma(T)\chi(T) = 20\,\mathrm{S/(m\cdot cm)}$. From this follows the air temperature $T \approx 4600\,\mathrm{K}$ at the quiet stage, and the thermal diffusivity coefficient at this temperature is equal to $\chi = 50\,\mathrm{cm^2/s}$. From this one can find an average radius of the conductive lightning channel in the end as $r \approx 6\,\mathrm{cm}$, whereas according to photographs the channel radius is $r \approx 10\,\mathrm{cm}$ [21]. This estimation is rough because air in the conductive lightning channel is dissociated partially at the temperature $T = 4600\,\mathrm{K}$, and its equilibrium is not correct. One can add to this that air heating accumulates inside the channel in the course of propagation of new electric pulses. Nevertheless, the above estimation gives a correct temperature scale of the conductive channel, and it varies during propagation of the electric current through the channel.

Let us analyze the energetic balance in the lightning conductive channel when an electric current passes at the stage of the recurrent stroke. At this stage the conductive channel is created and the electric current propagates through this channel until the transferred positive charge to the channel base compensates the electric field of a cloud. The average measured parameters of lightning at this stage are [1–3, 7]: a transferred charge is $Q = 2\,\mathrm{C}$, the electric current is equal to $I = 30\,\mathrm{kA}$, a pulse duration is $\tau = 70\,\mu\mathrm{s}$, a channel radius is $r = 3\,\mathrm{cm}$, the electric field strength is $E = 200\,\mathrm{V/cm}$, and the temperature of a plasma of the conductive channel is $T = 20{,}000\,\mathrm{K}$. From this one can obtain the total specific energy due to an electric current pass $QE \approx 400\,\mathrm{J/cm}$, and this specific energy is spent in various heat release channels. Accounting for all the molecules in the conductive lightning channel to be dissociated, we obtain for the energy ε_T that is consumed for air heating, dissociation, and ionization of molecules,

$$\varepsilon_T = c_p T + \frac{D}{2} + p_i (J + 2c_p T) \tag{7.16}$$

Here $c_p = 5/2$ is the heat capacity for atoms, electrons, and ions; D is the molecule dissociation energy; p_i is the ionization probability; and J is the atom ionization potential. At the above air temperature $T = 2 \times 10^4\,\mathrm{K}$ the ionization probability is $p_i = 0.9$ in accordance with the data of Fig. 5.13, and we take a rough estimation for an energy $\varepsilon_T \approx 30\,\mathrm{eV}$ that is spent per formation of one electron-ion pair. Next, we take as plasma parameters in a conductive channel the temperature $T = 2 \times 10^4\,\mathrm{K}$ and atmospheric pressure $p = 1$ atm, that gives for the number densities of electrons and ions $N_e = N_i = 1.6 \times 10^{17}\,\mathrm{cm}^{-3}$, and for the atom number density, $N_a = 4.4 \times 10^{16}\,\mathrm{cm}^{-3}$. Correspondingly, the total number density of nuclei (atoms and ions) is equal to $N_a + N_i = 2 \times 10^{17}\,\mathrm{cm}^{-3}$, and taking a radius of the conductivity channel of a high temperature $r = 0.3\,\mathrm{cm}$, we obtain for the specific energy of the conductive lightning channel that is spent on air excitation

$$\pi r^2 \varepsilon_T (N_a + N_i) \approx 0.3\,\mathrm{J/cm}, \tag{7.17}$$

which is a small part of the input specific energy. Note that formula (7.16) takes into account the specific energy $p\Delta S$ (p is the air pressure and ΔS is a change of the cross section for the conductivity lightning channel of heated air). This specific energy is spent on expansion of the conductive lightning channel. But the contribution of the expansion energy to the total energy of air excitation (heating, dissociation, and ionization) does not exceed 10 %. Note that a shock wave arises when expansion of the conductive lightning channel proceeds with a velocity that exceeds the sound speed for surrounding cold air. But the specific energy consumed on formation of a shock wave is four orders of magnitude less than that released as a result of the pass of a lightning electric current.

We now determine the specific energy consumed on heating of the surrounding air near the conductive channel. Neglecting heat transfer due to electrons and internal degrees of freedom, we obtain for the specific heat energy

$$E_h = -2\pi r \kappa \frac{dT}{dr} \tau \sim 2\pi T \kappa \tau, \tag{7.18}$$

where τ is the time of the current pass. Formula (7.18) gives that the specific heat energy due to heat transfer in the surrounding air as a result of a pass of strong current pulses through the conductive channel is estimated as $E_h = 0.04\,\mathrm{J/cm}$, whereas the specific energy during a time between neighboring strong pulses is equal to $E_h = 4\,\mathrm{J/cm}$. As shown, these energy expenses are small compared to the total specific electric energy $IE\tau$ that is consumed during lightning flashes and in a quiet phase between neighboring flashes.

One more channel of energy consumption relates to emission of an air plasma. Based on the recombination character of emission of this plasma, one can find the specific energy E_{rad} consumed on radiation of the lightning conductive channel

$$E_{rad} = \pi r^2 P_{rad} \tau \tag{7.19}$$

Taking the temperature of the air plasma as $T = 2 \times 10^4\,\mathrm{K}$ and a typical radius of the conductive channel during the pulse to be $r = 0.5\,\mathrm{cm}$, we obtain for the specific radiation energy $E_{rad} = 20\,\mathrm{J/cm}$ on the basis of the radiation power due to recombination in a plasma [22]. One can see that this channel of energy consumption is also small compared to the total specific energy of lightning. From this it follows that the main part of the lightning energy is consumed during a charge entrance inside the Earth.

Thus, as follows from the analysis of the energetic balance of the conductive lightning channel, the specific total power of processes in the air plasma owing to air heating, dissociation, and ionization, and also due to heat transport in a surrounding atmosphere, radiation of lightning plasma is less than the specific electric power due to a lightning electric current. Hence the electric energy of lightning is released inside Earth. The above analysis allows us to describe general processes in the conductive lightning channel whose high temperature provides a high plasma temperature and, correspondingly, a high conductivity of the lightning channel during a pass of electric current through this channel, which expands in the course of current passing. The simple scheme under consideration does not allow an explanation of the complex character of a current pass. In reality, a current pass consists of several stages including the arrow-like leader, recurrent stroke, repeating phases of a strong electric current in the form of short pulses, and phases of a small, steady electric current. One can expect that Nature chooses an economic form of charge transfer, and, evidently, the convective character of air motion, as well as transport of electricity inside Earth. Indeed, typical Earth conductivity is $10^{-3}\,\mathrm{S/m}$ [3], so that some time is required for charge propagation inside Earth.

7.4 Atomic Oxygen and Ozone in the Upper Atmosphere

The atmosphere at not low altitudes is a nonequilibrium system because of the action of ultraviolet radiation from the Sun. Absorption of this radiation leads to formation of atomic oxygen as a result of photodissociation of oxygen molecules. As chemically active radicals, oxygen atoms may take part in various chemical processes, and therefore the Earth's atmosphere is a chemically active system. Chemical processes in atmospheric air depend on impurities that are present in air under certain conditions. These impurities are compounds of carbon, nitrogen, sulfur, and other atoms. Therefore chemical processes in atmospheric air are different depending on air admixtures, and atomic oxygen takes part in such chemical reactions that are accelerated under the action of solar radiation. In particular, the presence of organic molecules and oxygen radicals in the atmosphere causes photochemical smog [23] which results from interaction of organic radicals created under the action of solar radiation with ozone. One can see atmospheric chemistry reactions of oxygen atoms and ozone molecules, as well as involving atoms and molecules of atmospheric impurities of low concentrations that are excited by solar radiation. Atmospheric chemistry is not connected with atmospheric electricity and we consider this problem briefly. Based on elementary atmospheric processes involving oxygen atomic particles, below we analyze only processes related to atomic oxygen.

Atomic oxygen is one of the basic components of the ionosphere. Its formation in the atmosphere results from photodissociation of molecular oxygen due to absorption of solar radiation in the spectral range from 132 to 176 nm (or in the range of photon energies from 6 to 10.3 eV). This absorption proceeds at altitudes on the order of 100 km and is realized in two spectral ranges, namely, in the Schumann–Runge continuum and the Schumann–Runge band. Destruction of oxygen atoms at not high altitudes proceeds as a result of three-body processes

$$O + 2O_2 \to O_3 + O_2, \;\; O + O_2 + N_2 \to O_3 + N_2 \tag{7.20}$$

The rate constants of these processes are nearby and at room temperature their average value is $6 \times 10^{-34}\,\mathrm{cm^6/s}$ [24–32]. Accounting for the process of formation of oxygen atoms as a result of dissociation of oxygen molecules under the action of solar radiation, the process of loss of oxygen atoms in three-body attachment to an oxygen molecule, and the drift of oxygen atoms in lower layers of the atmosphere under the action of the Earth's gravitational field, we have the balance equation for the number density of oxygen atoms N_O in the form

$$w_g \frac{dN_O}{dh} = M(h) - KN_O N_a^2 \tag{7.21}$$

Here h is an altitude, w_g is the atom drift velocity according to formula (2.14), N_a is the number density of air molecules, and $M(h)$ is the number of events of photon absorption by oxygen molecules per unit volume and unit time that is given by

$$M(z) = \int j_\omega \sigma_\omega d\omega [O_2], \tag{7.22}$$

so that $[O_2]$ is the number density of oxygen molecules, j_ω is the flux of photons of a given frequency ω, and σ_ω is the cross section of photon absorption.

In analyzing the balance equation (7.21), we assume for simplicity that regions of formation and destruction of oxygen atoms are separated; that is, oxygen atoms are formed at high altitudes and are lost at low ones. Then starting from high altitudes, we obtain that the number density N_O of oxygen atoms increases with decreasing altitude, and the maximum of their number density N_{max} is attained at altitudes to which solar radiation in the range of the Schumann–Runge continuum and band cannot penetrate. Let us introduce the total flux of solar photons in the spectrum range

$$I = \int M(h)dh$$

Restricting by the Schumann–Runge continuum and band spectrum range, we obtain by integration of the partial flux of solar radiation over wavelengths below $\lambda \leq$ 176 nm, for the average photon flux penetrating into the atmosphere

$$I = \frac{1}{4}\int j_\omega d\omega = \frac{1}{4}\int j_\lambda d\lambda = 6 \times 10^{11}\,\mathrm{cm}^{-2}\,\mathrm{s}^{-1}$$

Solving (7.21) in this limiting case where one can ignore process (7.20), we determine the concentration of oxygen atoms $c_O = N_O/N_a$ at altitudes below which solar short-range radiation does not penetrate and photodissociation of oxygen molecules is absent:

$$w_g N_a c_O = I \qquad (7.23)$$

On the basis of formula (7.23) we obtain for the maximum of oxygen atoms $c_O \approx 0.04$. This concentration of oxygen atoms is established at altitudes below 90 km, until attachment of oxygen atoms to oxygen molecules is not significant. One can expect that this concentration of oxygen atoms corresponds to the maximum number density of oxygen atoms.

At lower altitudes the balance equation (7.21) for the number density of oxygen atoms takes the form

$$w_g \frac{dN_O}{dz} = K N_O N_a^2 c(O_2), \qquad (7.24)$$

where $c(O_2) = 0.2$ is the concentration of oxygen molecules, and due to convection the total number densities N of molecules is taken according to formula (2.3) in the form

$$N(h) = N_{max} \exp\left(\frac{h_{max} - h}{\Lambda_N}\right),$$

where $N_{max} = N(h_{max})$. Solution of (7.24) has the form

$$N_O(h) = N_{max} \exp\left(-\frac{\Lambda_N K N_a^3(h)}{3w_g N_a^3}\right) = N_{max} \exp\left(-\frac{N_a^3(h)}{N_*^3}\right), \qquad (7.25)$$

where

$$N_* = \left[\frac{3w_g N_a c(O_2)}{\Lambda_N K}\right]^{1/3} \qquad (7.26)$$

Taking the parameters of processes of formation of oxygen atoms and their attachment to oxygen molecules, we obtain on the basis of formula (7.26) $N_* = 3 \times 10^{13}\,\mathrm{cm}^{-3}$. This number density of air molecules corresponds to the altitude of 95 km, and the standard model of the atmosphere gives $c_O = 0.07$ at this altitude. This concentration corresponds to the above maximum number density of oxygen atoms and coincides by an order of magnitude with this estimation. But the standard atmospheric model leads to the maximum of the number density of oxygen atoms at the altitude of 105 km, and its concentration at this altitude is $c_O = 0.1$.

Let us analyze the results of the above model. This model relates to altitudes where the photodissociation process for oxygen molecules under the action of solar radiation is realized, and in reality this model may be considered for altitudes below 140 km. Next, this model assumes an altitude below which photodissociation ceases to be lower than an altitude where the process of the three-body process of attachment of an oxygen atom to an oxygen molecule becomes significant. But as follows from (7.21), both altitudes are identical, and the maximum of the number density of oxygen atoms according to this model corresponds to altitudes 90–95 km, whereas in reality this value is approximately 105 km (see Fig. 2.5). In addition, this model leads to a sharp decrease of the number density of oxygen atoms with decreasing altitude, but according to Fig. 2.6 such a sharp decrease of the number density is absent. An explanation of this discrepancy is that this model does not account for photodissociation of ozone molecules. One can add to this that the concentration of oxygen atoms at altitudes of the maximum N_O as an altitude function within the framework of standard atmosphere is approximately $c_O = 0.1$ according to Fig. 2.6 that is larger than the average value $c_O = 0.04$–0.07 which follows from the above estimations. This discrepancy testifies to the accuracy of the above model.

An average decay time for photodissociation of oxygen molecules (i.e., a time during which a test oxygen molecule is dissociated under the action of solar radiation), is given by

$$\tau_{dis} \sim \left[\frac{1}{4}\int \sigma_\omega dj_\omega\right]^{-1} \sim 2 \cdot 10^6\,\mathrm{s}, \qquad (7.27)$$

where the factor $1/4$ results from averaging the radiation flux over an angle between the Earth's surface and the direction of incident radiation. Thus, the lifetime of an oxygen molecule with respect to its dissociation under the action of solar radiation with photon energies in the range $\hbar\omega = 6$–$10\,\mathrm{eV}$ is on the order of 1 year.

The recombination rate of two oxygen atoms in three-body collisions decreases with increasing altitude because of a drop of the atmospheric density, but the concentration of atomic oxygen grows with increasing altitude. The number density of oxygen atoms becomes equal to the number density of oxygen molecules at altitudes between 100 and 120 km, and the number density of oxygen atoms equals that of nitrogen molecules at altitudes between 150 and 200 km (see Fig. 2.6). For comparison, we note that the average total number densities of nitrogen and oxygen molecules vary from $7 \times 10^{12}\,\text{cm}^{-3}$ to $3 \times 10^9\,\text{cm}^{-3}$, and the number density of atomic oxygen ranges from $6 \times 10^{11}\,\text{cm}^{-3}$ to $4 \times 10^9\,\text{cm}^{-3}$ as the altitude varies from 100 km up to 200 km. Thus, atomic oxygen becomes one of the basic components of the upper atmosphere at high altitudes.

At lower altitudes atomic oxygen is transformed in ozone as a result of processes (7.20). It is convenient to introduce the total number density of oxygen radicals N_{ox} as

$$N_{ox} = [O] + [O_3], \tag{7.28}$$

where $[O]$, $[O_3]$ are the number density of oxygen atoms and ozone molecules, respectively. The balance equation for the total number density of oxygen radicals by analogy with (7.21) has the form

$$w_g \frac{dN_{ox}}{dh} = M(h), \tag{7.29}$$

if we neglect the process of destruction of oxygen radicals in chemical processes. From this it follows that the number density of oxygen radicals N_{ox} increases monotonically with decreasing altitude such that its concentration is constant at low altitudes where solar radiation for dissociation of oxygen molecules does not penetrate. One can expect that transition from atomic oxygen to ozone proceeds at altitudes where the rates of drift of atomic oxygen or ozone is compared with the rate of the three-body process (7.20), so that the appropriate number density of air molecules N_* is given by formula (7.26); that is, $N_* = 3 \times 10^{13}\,\text{cm}^{-3}$. This number density of air molecules is realized at the altitude of 95 km. But photoprocesses lead to a decrease of the transient altitude. Approximately 97 % of the ozone amount is located between altitudes 15 and 55 km. Although the number density of ozone molecules varies significantly depending on the latitude, season, and time of day, we exclude these variations based on averaged parameters of ozone in atmospheric air in accordance with [33–37]. Note that the maximum number density of ozone molecules about $[O_3] = 7 \times 10^{12}\,\text{cm}^{-3}$ corresponds to the altitude range 20−25 km, and the maximum concentration of ozone molecules $c_{max} \sim 10^{-5}$ is observed at the approximate altitude of 35 km. We also have that the total number of ozone molecules in a column over the Earth's surface is $\int [O_3] dh \sim 2 \times 10^{19}\,\text{cm}^{-2}$. As follows from this, ozone is located in a layer of a thickness $\int [O_3] dh / [O_3] \approx 30\,\text{km}$.

Ozone is a metastable compound with respect to molecular oxygen, and decomposition of ozone molecules in pair collisions $2O_3 \to 3O_2$ is absent due to positions of electron terms of two interacted ozone molecules. Hence, ozone decomposition proceeds through slow channels [36] involving radicals with a low concentration in the atmosphere. We analyze below one such channel of ozone decay according to the Chapman cycle [38] or the oxygen cycle of ozone destruction. The main process of this cycle is

$$O + O_3 \rightarrow 2O_2 \tag{7.30}$$

The rate constant of this process averaged over data [24, 26, 30, 39–43] may be approximated through the Arrenius formula as

$$k = k_o \cdot \exp\left(-\frac{2230}{T}\right), \tag{7.31}$$

where the temperature T is expressed in K, and $k_o = 1.7 \times 10^{-11}\,\mathrm{cm}^3/\mathrm{s}$. It should be noted that atomic oxygen has a catalyzer role in ozone decay. Other catalyzers for ozone destruction exist and are realized in other cycles of ozone decay, namely, nitrogen, hydrogen, and chlorine cycles [44–50], may make the main contribution to the rate of ozone destruction depending on atmospheric impurities. Nevertheless, in the following we focus on the oxygen cycle only because it reflects general peculiarities of the decay process for atmospheric ozone. A small rate of the process of ozone destruction in the oxygen (Chapman) cycle is determined not only by a small rate constant of this process, but also because the maximum concentrations of atomic oxygen and ozone correspond to different altitudes.

Figure 7.3 contains the lifetime of ozone molecules as a result of decay in the process (7.30) according to [37] for standard atmosphere. Accounting for variations in the number densities of oxygen atoms and ozone molecules, we also determine the lifetime of ozone molecules at the altitude 50 km, where the temperature is $T = 271\,\mathrm{K}$ for the standard atmosphere. We have that the rate constant of the process (7.30) is equal according to formula (7.31) $k_d = 4.5 \times 10^{-15}\,\mathrm{cm}^3/\mathrm{s}$. Season, daytime, latitude, and longitude fluctuations of the number density of oxygen atoms are comparable with its value. We use the results of measurements according to which we have $[O] = 1 \times 10^9\,\mathrm{cm}^{-3}$ [51] and $[O] = 8 \times 10^9\,\mathrm{cm}^{-3}$ [52]. These number densities of oxygen atoms at altitude 50 km correspond to the lifetime of

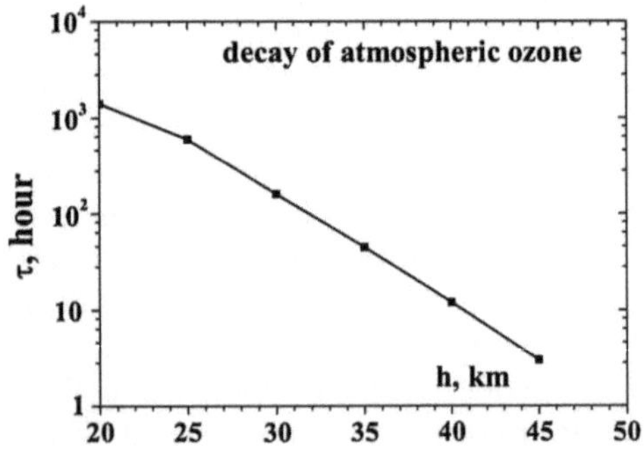

Fig. 7.3 Typical lifetime of an ozone molecule in atmospheric air depending on the altitude [37]

ozone molecules 60 and 8 h, respectively. As shown, these values exceed the results of Fig. 7.3 by an order of magnitude. This means that other circles of ozone destruction, rather than the oxygen circle (7.30), determine the rate of ozone decay under conditions of [37]. Another conclusion from this estimation is that this rate depends strongly on local atmospheric conditions which lead to significant variations of the number density of ozone molecules. In particular, at the altitude 50 km the number density of ozone molecules lies between $3 \times 10^{10}\,\mathrm{cm}^{-3}$ and $1 \times 10^{11}\,\mathrm{cm}^{-3}$ [52–55].

References

1. M.A. Uman, *Lightning* (McGrow Hill, New York, 1969)
2. M.A. Uman, *The Lightning Discharge* (Academic Press, New York, 1987)
3. V.A. Rakov, M.A. Uman, *Lightning, Physics and Effects* (Cambridge University Press, Cambridge, 2003)
4. YuP Raizer, *Physics of Gas Discharge* (Dolgoprudnyi, Intellect, 2009). (in Russian)
5. E.M. Bazelyan, YuP Raizer, *Physics of Lightning and Lightning Protection* (Moscow, Fizmatlit, 2001). (in Russian)
6. J.R. Dwyer, M. Uman, Phys. Rep. **534**, 147 (2014)
7. M.A. Uman, *About Lightning* (Dover, New York, 1986)
8. K. Berger, in *The Earth flash,* ed. by R.H. Golde. "Lightning." (Academic Press, London 1977) p. 119
9. R.A. Herschel, Proc. Roy. Soc. **15**, 61 (1868)
10. J. Holtsmark, Ann. der Physik. **58**, 577 (1919)
11. Ch. Kittel, *Introduction to Solid State Physics* (Wiley, New York, 1986)
12. B.M. Smirnov, *Plasma Processes and Plasma Kinetics* (Wiley, Berlin, 2007)
13. J. Dutton, J. Chem. Phys. Ref. Data **4**, 577 (1975)
14. http://en.wikipedia.org/wiki/Siemens
15. E.M. Bazelyan, YuP Raizer, *Spark Discharge* (CRC Press, Roca Baton, 1997)
16. L. Spitzer, *Physics of Fully Ionized Gases* (Wiley, New York, 1962)
17. B.M. Smirnov, *Principles of Statistical Physics* (Wiley, Berlin, 2006)
18. B.M. Smirnov, *Reference Data on Atomic Physics and Atomic Processes* (Springer, Heidelberg, 2008)
19. M. Capitelli, D. Bruno, A. Laricchiuta, *Fundamental Aspects of Plasma Chemical Physics* (Transport, Springer, New York, 2013)
20. B.M. Smirnov, *Physics of Ionized Gas.* (Moscow, Nauka, 1978) p. 118 (in Russian)
21. I.S. Stekol'nikov, *The Nature of Long Spark.* (Moscow, Izdat. AN SSSR,1960) (in Russian)
22. B.M. Smirnov, *Theory of Gas Discharge Plasma* (Springer, Heidelberg, 2014)
23. https://en.wikipedia.org/wiki/Smog
24. S.W. Benson, J. Chem. Phys. **42**, 2614 (1965)
25. T.G. Slanger, G. Black, J. Chem. Phys. **53**, 3717 (1970)
26. V.N. Kondrat'ev, *Rate Constants of Gas-phase Reactions* (Nauka, Moscow, 1971). (in Russian)
27. F. Stuhl, H. Niki, J. Chem. Phys. **55**, 4973 (1971)
28. R.E. Huie, J.T. Herron, D.D. Davies, J. Phys. Chem. **76**, 2653 (1972)
29. H. Niki, Can. J. Chem. **52**, 1397 (1974)
30. D.L. Baulch et al., J. Chem. Phys. Ref. Data **11**, 327 (1982)
31. C.L. Lin, M.T. Leu, Int. J. Chem. Kinet. **14**, 417 (1982)
32. D.L. Baulch et al., J. Chem. Phys. Ref. Data **13**, 259 (1984)
33. M.L. Salby, *Fundamentals of Atmospheric Physics* (Acad. Press, San Diego, 1996)
34. A.J. Krueger, R.A. Minznev, J. Geophys. Res. **84**, 6969 (1976)
35. H.U. Dütsch, Pure Appl. Geophys. **116**, 511 (1978)

36. S.P. Perov, AKh Khrgian, *The Modern Problems of Atmospheric Ozone* (Gidrometeoizdat, Leningrad, 1980). (in Russian)
37. J.H. Seinfeld, S.N. Pandis, *Atmospheric Chemistry and Physics* (Wiley, Hoboken, 2006)
38. S. Chapman, Mem. Royal. Meteorol. Soc. **3**, 103 (1930)
39. H.S. Schiff, Can. J. Chem. **47**, 1903 (1969)
40. D.C. Krezenski, R. Simonaitis, J. Heicklen. Int. J. Chem. Kinet. **3**, 467 (1971)
41. L.M. Crumb, F. Kaufman, J. Chem. Phys. **57**, 1270 (1972)
42. H.S. Johnston, Rev. Geophys. Space Phys. **13**, 635 (1975)
43. P.H. Wine et al., J. Phys. Chem. **87**, 3948 (1983)
44. J. Crutzen, Royal Meteorol. Soc. Quart. J. **96**, 320 (1970)
45. H.S. Johnston, Science **173**, 517 (1971)
46. P.J. Crutzen, J. Geophys. Res. **76**, 7311 (1971)
47. M.J. Molina, F.S. Rowland, Nature **249**, 810 (1974)
48. R.S. Stolarski, R.J. Cicerone, Can. J. Chem. **52**, 1610 (1974)
49. F.S. Rowland, M.J. Molina, *Rev. Geophys. Space Phys.***13**, l(1975)
50. B.M. Smirnov, Sov. Phys. Usp. **18**, 804 (1975)
51. A. Osepian, V. Tereschenko, P. Dalin, S. Kirkwood, Ann. Geophys. **26**, 131 (2008)
52. T. Shimazaki, A. Laird, Radio Sci. **1**, 23 (1972)
53. W. Swider, T.J. Keneshea, C.I. Foley, Planet. Space Sci. **26**, 883 (1978)
54. R. Rodrigo, J. Lopez-Moreno, M. Lopez-Puertas, F. Moreno, Planet. Space Sci. **34**, 723 (1986)
55. A. Osepian, S. Kirkwood, P. Dalin, Ann. Geophys. **27**, 577 (2009)

Chapter 8
Atmospheric Processes Involving Aerosols

Abstract It is shown that the mechanism of gravitation coagulation of aerosols due to different fall velocities is responsible for growth of charged aerosols and creation of electric currents in the atmosphere. Charging of aerosols results from collisions of water aerosols in different aggregate states, mostly graupels and ice particles. From the analysis of processes involving aerosols and observation data it follows that the average charge of aerosols is $Z = (25 - 30)e$. Electric fields occur in the troposphere because falling velocities of positively and negatively charged aerosols are different. Charged aerosols of a cloud are located in a plasma that is formed under the action of cosmic rays and prevents clouds from expansion. Fluxes of wet hot air from Earth to clouds which are a precursor of thunderstorms cannot increase the aerosol mass remarkably because of heat processes. Removing a charge from aerosols accelerates their growth and causes rain. In prethunderstorm weather the low edge of a cumulus cloud contains charged aerosols, and there is a negative charge of a lower density of molecular negative ions that are located between a cloud and Earth.

8.1 Aerosol Association in Earth's Atmosphere

Above we considered the key processes of a general chain in the development of atmospheric electric processes (see Fig. 4.4) which include processes of formation of charged aerosols as a result of collisions of neutral aerosols in different aggregate states, creation of electric fields in the atmosphere due to different velocities of gravitational fall for positive and negative aerosols, and electrical breakdown of the atmosphere. Below we consider these processes in detail. Aerosol formation proceeds from evaporated water at certain altitudes of the atmosphere with a not high temperature where the partial water pressure exceeds the saturated vapor pressure. Although water evaporates from the Earth's surface in the form of free molecules, at high altitudes with low atmospheric temperatures aerosols are formed, that is, liquid, snow, or ice micron-size particles. Because processes involving aerosols determine the character of charge formation and its separation in the atmosphere, which leads subsequently to atmospheric electrical breakdown in the form of lightning, in the following we study processes involving aerosols in the Earth's atmosphere.

B.M. Smirnov, *Microphysics of Atmospheric Phenomena*,
Springer Atmospheric Sciences, DOI 10.1007/978-3-319-30813-5_8

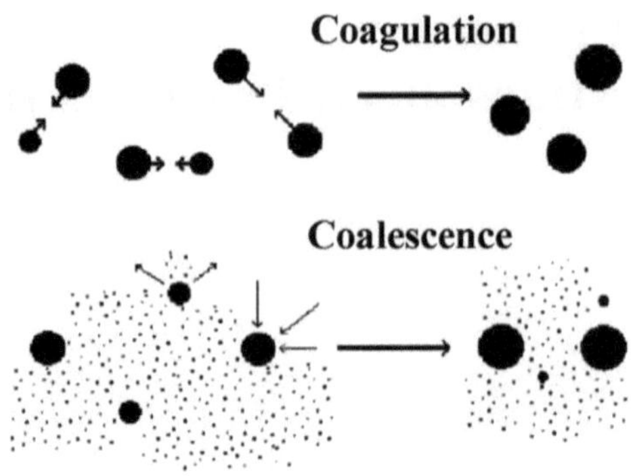

Fig. 8.1 Mechanisms of aerosol growth in atmospheric air. Coagulation results from association of two aerosols in one drop, and coalescence corresponds to attachment of individual molecules to large aerosols and evaporation of small aerosols [2, 3]

In considering aerosol growth, we use general principles of nucleation processes [1] as a basis. Two mechanisms of liquid aerosol growth in the atmosphere are given in Fig. 8.1 [2, 3], with the first being coagulation, where two liquid drops are joined after their contact, and the second nucleation mechanism being coalescence or Ostwald ripening [4, 5] resulting from the aerosol equilibrium with its vapor. Processes of molecule attachment to aerosols and evaporation of molecules from the aerosol surface lead to growth of large aerosols and vaporization of small aerosols. As a result, this leads to a continuous increase of the average aerosol size.

The diffusion mechanism of aerosol growth, which requires the validity of the criterion

$$r \gg \lambda \tag{8.1}$$

is realized at such aerosol sizes. Here λ is the mean free path of air molecules in air that is equal to $\lambda \approx 0.1\,\mu\text{m}$ [6] under atmospheric pressure. Hence the diffusion mechanism of growth holds true according to the criterion (8.1) and the rate constant of association of aerosols of radii r_1 and r_2 is equal to [7]

$$k_{as} = 4\pi(D_1 + D_2)(r_1 + r_2),$$

where D_1 and D_2 are the diffusion coefficients for colliding aerosols in air. The rate constant of association averaged over aerosol sizes is [2]

$$k_{as} = \frac{8T}{3\eta}, \tag{8.2}$$

where η is the viscosity coefficient for air. At room temperature this formula gives $k_{as} = 5.8 \times 10^{-10}\,\text{cm}^3/\text{s}$.

The equation for evolution of the average aerosol size $\overline{n}$ for this growth mechanism is given by [2]

$$\frac{d\overline{n}}{dt} = \frac{1}{2}k_{as}N_b = \frac{1}{2}k_{as}N_p\overline{n}, \tag{8.3}$$

where $N_b = N_p\overline{n}$ is the total number density of molecules in aerosols and N_p is the aerosol number density. For parameters of a typical cumulus cloud (2.24) this is equal to $N_b = 7 \times 10^{16}\,\mathrm{cm}^{-3}$. The solution of Equation (8.3)

$$\overline{n} = \frac{1}{2}k_{as}N_b t, \tag{8.4}$$

shows an increase of the average aerosol size in time. Let us introduce the aerosol doubling time τ_d during which the number of aerosol molecules is doubled, and the average aerosol radius becomes equal to $2^{1/3}r$ with respect to the initial average aerosol radius r. This leads to the aerosol number density $N_p/2$. In the case under consideration the doubling time is

$$\tau_d = \frac{2\overline{n}}{k_{as}N_b} = \frac{2}{k_{as}N_p} \tag{8.5}$$

Assuming that a cumulus cloud contains neutral aerosols, we find the doubling time for an aerosol size of a typical cumulus cloud $\tau_d = 2/(k_{as}N_p) = 40$ days. Inasmuch as this time exceeds the lifetime of a cumulus cloud, one can conclude that the diffusion mechanism of aerosol growth is not of interest for evolution of aerosols in a cumulus cloud.

The mechanism of gravitation association of aerosols is given in Fig. 8.2 and consists in the joining of two colliding aerosols at their contact if the collision of two aerosols is determined by a different velocity of aerosol fall in the gravitational field of the Earth. The rate constant of aerosol association is estimated as

$$k_{as} \sim \Delta v \cdot \sigma \sim \frac{\rho g r^4}{\eta}, \tag{8.6}$$

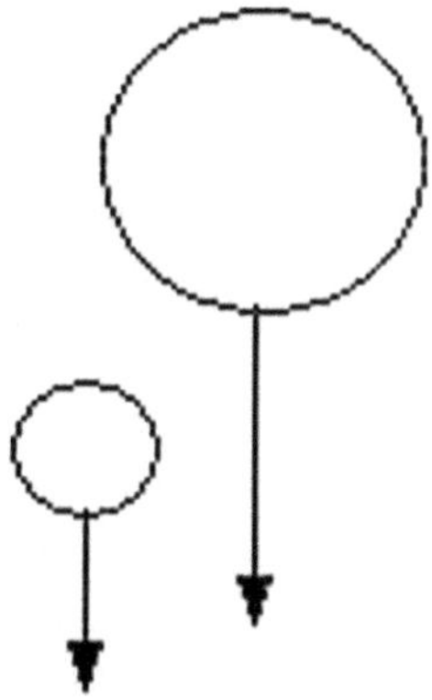

Fig. 8.2 Character of growth of liquid aerosols in atmospheric air under the action of the Earth's gravitational field. Heavy drops catch up light ones in the course of their falling and join with them in one drop as a result of their contact [3, 8]

where Δv is the difference of aerosol fall velocities in atmospheric air, $\sigma \sim r^2$ is a typical area of the projection of contacted aerosols on the Earth's surface, r is a typical aerosol radius, ρ is the density of an aerosol material (for water aerosols $\rho = 1\,\mathrm{g/cm^3}$), g is the free fall acceleration, and η is the air viscosity. The free fall velocity for a micron-size spherical particle in the gravitational field is equal to [7]

$$v_f = \frac{2\rho g r^2}{9\eta}, \tag{8.7}$$

Water droplets fall in atmospheric air under the action of their weight with the velocity

$$v_f = 0.012\frac{r^2}{a^2}, \tag{8.8}$$

where the fall velocity v_f is given in cm/s, and $a = 1\,\mu\mathrm{m}$. Averaging the rate constant of aerosol association taking into account the size distribution function of aerosols, one can obtain for the average rate constant [3]:

$$k_{as} = \frac{2r^4\rho g}{\eta}, \tag{8.9}$$

where r is the average aerosol radius. Comparing the rate constants of aerosol association in accordance with Formulas (8.2) and (8.9), one can conclude that the rates of coagulation and gravitation association of water aerosols are compared at $r \approx 1\,\mu\mathrm{m}$, and the gravitation mechanism of association dominates at larger aerosol sizes. In particular, for an average aerosol radius $r = 8\,\mu\mathrm{m}$ which is typical for a cumulus cloud, according to Formula (8.9) we have $k_{as} = 4.3 \times 10^{-6}\,\mathrm{cm^3/s}$. If aerosols are neutral, the lifetime of a cumulus cloud with respect to doubling of an aerosol size is $\tau \approx 8\,\mathrm{min}$, which is less than the lifetime of a cumulus cloud measured by hours. From this it follows that aerosols of a cumulus cloud are likely charged and this prevents aerosols from association. Assuming aerosol radii and charges to be identical, we find for the rate constant of aerosol association [3]:

$$k_{as} = \frac{2r^4\rho g}{\eta}\exp\left(-\frac{Z^2e^2}{2rT}\right) \tag{8.10}$$

instead of Formula (8.9) for association of neutral aerosols. In the case under consideration electrostatic repulsion of charged aerosols prevents them from the contact. One can represent Formula (8.10) for liquid water aerosols in the form

$$k_{as} = r^4\exp(-0.06Z^2/r), \tag{8.11}$$

where the rate constant of aerosol association k_{as} is measured in $10^{-9}\,\mathrm{cm^3/s}$, and an average aerosol radius is expressed in $\mu\mathrm{m}$. Note that an aerosol charge $Z \approx 27$ creates a weak field on its surface that under given conditions is equal to $E = Ze/r^2 \approx 6\,\mathrm{V/cm}$ and corresponds to the electric potential with respect to surrounding air $U = Ze/r \approx 5\,\mathrm{mV}$.

Let us determine the aerosol doubling time in a cumulus cloud for the gravitation association mechanism (Fig. 8.2), where the rate constant of aerosol association is given by Formula (8.10). Assuming that the number density of bound water molecules N_b does not vary in time, as well as the aerosol charge Z, we obtain for the doubling time τ_d on the basis of Formula (8.10) for the rate constant of aerosol association

$$\tau_d = \int_{n_1}^{n_2} \frac{2dn}{k_{as}N_b} = \frac{3\eta}{\rho g N_b r_W^3} \int_{r_2}^{r_1} d\frac{1}{r} \exp\left(\frac{Z^2}{2rT}\right) = \tau_o F[Z^2/(2rT)], \qquad (8.12)$$

where n_1, n_2 and r_1, r_2 are the initial and final average numbers of aerosol molecules and their average radii, and the parameters of the final state are given by

$$\tau_o = \frac{3\eta}{\rho g N_b r_W^3 r}, \quad F(x) = x\left[\exp(x) - \exp(x/2^{1/3})\right] \qquad (8.13)$$

On the basis of typical parameters of aerosols (2.24) in a cumulus cloud we have for a doubling time $\tau_o = 24\,\mathrm{min}$, and the function $F(x)$ is given in Fig. 8.3. If we take a typical lifetime of a cumulus cloud as between 3 and 12 h, one can obtain on the basis of Formula (8.12) for the corresponding aerosol charge

$$Z = 27 \pm 2 \qquad (8.14)$$

Thus the aerosol charge depends weakly on a typical time of aerosol association, and the repulsion electrostatic potential for aerosols of a charge $Z = 27$ and radius $r = 8\,\mu\mathrm{m}$ is equal to $Z^2e^2/(2r) = 760\,\mathrm{K}$, that exceeds a thermal aerosol energy. In particular, at room temperature $Z^2e^2/(2rT) = 2.6$ and $\exp[Z^2e^2/(2rT)] = 13$. Results of Fig. 8.4 confirm a weak dependence of the aerosol charge Z on a growth time, and an increase of the cloud lifetime by the order of magnitude leads to an increase of the aerosol charge by 15 %. This fact characterizes the reliability of

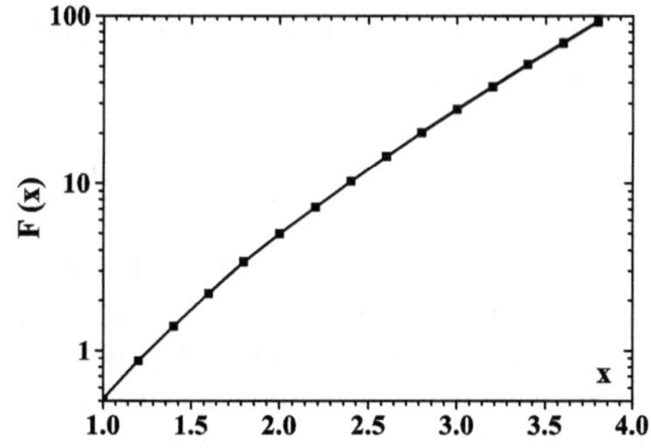

Fig. 8.3 Function $F(x)$ in accordance with Formula (8.13)

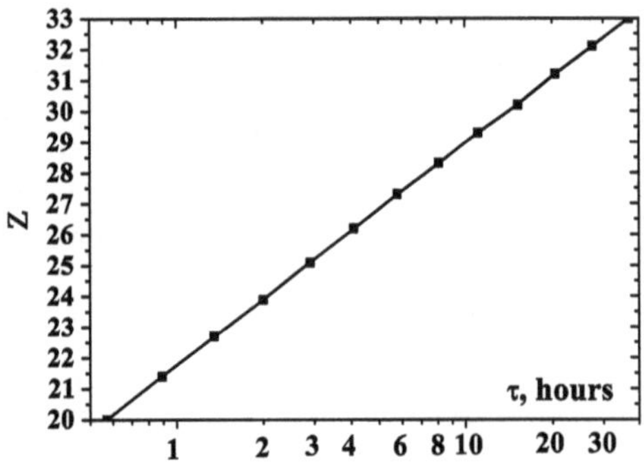

Fig. 8.4 Average aerosol charge Z for a typical cumulus cloud depending on a doubling time for a number of water molecules in an aerosol [9]

determination of the aerosol charge and confirms the presence of the aerosol stage of water circulation in the atmosphere.

Let us return to the concept where electricity transport in the atmosphere is a secondary phenomenon with respect to water transport, and the ratio of the transferred charge to the mass of transferred water is $1.4 \times 10^{-10}\,\mathrm{C/g}$. If a channel of water transport in the atmospheric water circulation corresponds to transfer of charged aerosols of a cumulus cloud, we obtain the ratio of a transported aerosol charge Ze to the water mass m in aerosols for this channel

$$\frac{Ze}{m} = 2 \times 10^{-9}\mathrm{C/g}, \tag{8.15}$$

and this is larger than that for the atmosphere on average. From this it follows that only a part of the water passed through the atmosphere becomes aerosols of cumulus clouds. This also confirms transfer of the atmospheric charge through aerosols.

Let us determine a typical thickness of a charged layer of a cumulus cloud. A typical electric field strength is determined by a charge per unit square of a cloud. On the other side, Formulas (2.24) and (8.14) give the charge number density for a cloud, and the combination of these quantities allows one to determine the thickness of the cloud's lower edge where a charge is located. Based on the observational value of the cloud voltage $U = 20$–$100\,\mathrm{MV}$ [10], we take for estimations the average voltage $U = 60\,\mathrm{MV}$, that at an altitude of the lower cloud edge $L = 3\,\mathrm{km}$ gives for the average electrical field strength $E = 200\,\mathrm{V/cm}$. From this we have for the charge density per unit cloud square:

$$\sigma = \frac{E}{4\pi} \approx 1 \times 10^{8}\frac{\mathrm{e}}{\mathrm{cm}^2} \approx 2 \times 10^{-11}\mathrm{C/cm}^2 \tag{8.16}$$

The charge density according to Formulas (2.24) and (8.14)is equal to $ZeN_p = 4 \times 10^{-15}\,\mathrm{C/cm}^3$. Note that according to airplane measurements [11, 12] the charge density in a cumulus cloud is $1\,\mathrm{C/km}^3$ or $1 \times 10^{-15}\,\mathrm{C/cm}^3$. The ratio of the surface charge density to its volume density gives for the thickness of a charged cloud layer $l \sim 100\,\mathrm{m}$ [3], which is less than a cloud size.

Taking the charge of the lower cloud edge to be equal to the charge of middle lightning $Q = 20\,\mathrm{C}$ [10, 13, 14], one can obtain that this charge is collected from an area

$$S = \frac{Q}{\sigma} \sim 100\,\mathrm{km}^2, \tag{8.17}$$

which corresponds to the above estimation. Thus, development of thunderstorm phenomena has a threshold character, and an observed minimum cloud size is $L \approx 3\,\mathrm{km}$ [10]. One can see that a lightning charge is collected from a large cloud area. But the conductivity of a cloud is low; that is, it is impossible to transfer a cloud charge to the conducting lightning channel during a time of lightning existence. This means that lightning evolution is accompanied by a change of electrical properties of a lower cloud edge as a result of passing to it a positive charge from the lightning channel.

8.2 Coalescence in Aerosol Plasma

According to the above analysis, the rate of association for identically charged aerosols drops sharply with an increasing charge. In the case of coalescence or Ostwald ripening [4, 5], where aerosols are located in a buffer gas, aerosol growth is determined by processes of molecule attachment to the aerosol surface and aerosol evaporation, and the rate of aerosol growth is independent of the aerosol charge. This system consisting of aerosols and their vapor is not an equilibrium, because the equilibrium between aerosols and vapor leads to an increase of the average aerosol size due to growth of large aerosols and vaporization of small aerosols.

The automodel form of the size distribution function for the coalescence process [15, 17, 18] is the principal peculiarity of this process and means that the size distribution function of aerosols depends on a nondimensional parameter, that is, the ratio of an aerosol size to its critical value, and does not depend on time directly, whereas the aerosol critical radius increases in time. The classical theory of coalescence [16, 19] relates to the diffusion mechanism of this process, whereas the above principal property also corresponds to the kinetic regime of coalescence [3, 9]. These two regimes differ by the molecule concentration in a buffer gas, and the approach of an attaching molecule and aerosol in the diffusion regime proceeds at diffusion molecule motion, whereas in the kinetic regime of coalescence the molecule approach results from a straightforward molecule trajectory. Below we consider the diffusion mechanism of coalescence being guided by growth processes for water drops in air.

According to the automodel character of aerosol growth, the size distribution function of aerosols has the form $f(u)$, where $u = n/n_{cr}$, n is a current number of aerosol molecules, and n_{cr} is the critical number of aerosol molecules; that is, the rates of molecule attachment and evaporation are equal at this number of aerosol molecules. Figure 8.5 represents this distribution function. Analyzing the equilibrium between free and bound water molecules, we use an expansion of the total binding

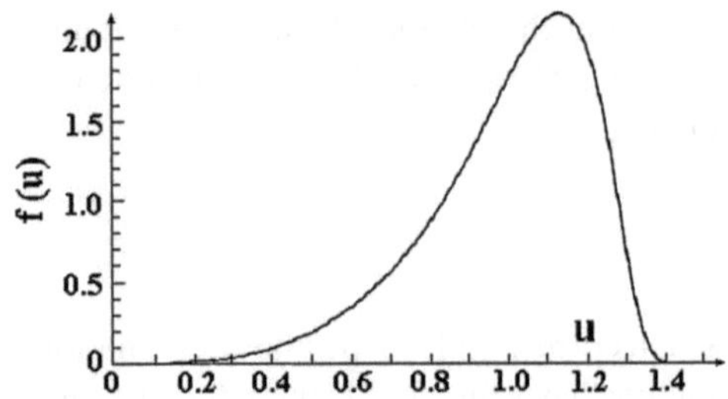

Fig. 8.5 Size distribution function of aerosols $f(u)$ for the diffusion mechanism of coagulation in accordance with Formula (8.1) [15, 16]

energy E_n of bound molecules in an aerosol consisting of n molecules over a small parameter $n^{-1/3}$ that has the form [20]

$$E_n = \varepsilon_o n - An^{2/3} \tag{8.18}$$

Here ε_o is the average binding energy per one water molecule in a macroscopic drop, A is the specific surface energy, and these parameters are equal for water drops [21] $\varepsilon_o = 0.43\,\text{eV}$, $A = 0.21\,\text{eV}$, where these parameters relate to the liquid aggregate state of aerosols at the temperature $T = 293\,\text{K}$. This gives the energy difference $\Delta\varepsilon(n)$ of water molecules in an aerosol consisting of n bound molecules compared to the limit of a large drop

$$\Delta\varepsilon(n) = \frac{dE_n}{dn} - \varepsilon_o = \frac{2A}{3n^{1/3}} \tag{8.19}$$

Let us consider the equilibrium between free and bound molecules for large aerosols if the criterion $A \ll T \cdot n^{1/3}$, holds true. Under equilibrium the relation between the number density of free molecules N_w and that in a saturated vapor $N_{sat}(T)$ has the form $N_w = N_{sat}(T)$. Taking for the saturated vapor pressure $N_{sat}(T) \sim \exp(-\varepsilon_o/T)$, where ε_o is the molecule binding energy for bulk water, we obtain for the critical aerosol size [7]

$$N_w = N_{sat}(T) \exp\left[\frac{\Delta\varepsilon(n_{cr})}{T}\right], \tag{8.20}$$

Taking for simplicity that each contact of a free molecule with the aerosol surface leads to molecule attachment, we obtain on the basis of the Smolukhowski formula [22] for the number of molecules attaching to an aerosol per unit time:

$$J_{at} = 4\pi Dr N_w, \tag{8.21}$$

where D is the diffusion coefficient for water molecules in air. The total change of the number of bound molecules for an aerosol of a given size is

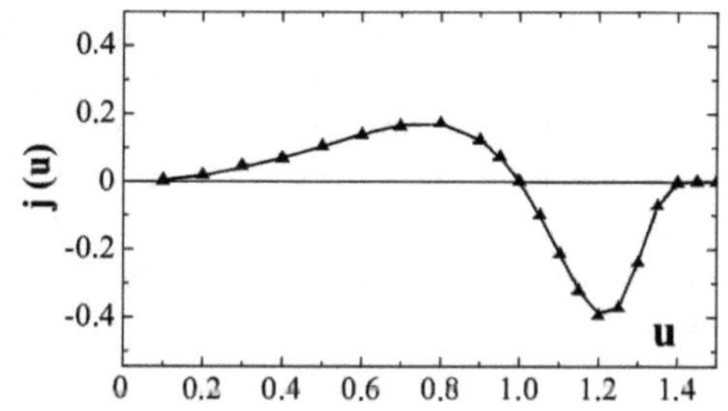

Fig. 8.6 Partial rate of molecule attachment to aerosols $j(u)$ [15, 16]

$$J = J_{ev} - J_{at} = J_o j(u), \; J_o = \frac{8\pi A r_W D N_m}{3T}, \; u = (n/n_{cr})^{1/3}, n < n_{cr} \quad (8.22)$$

where $J_{ev}(n,T)$ is the evaporation rate, and this formula relates to aerosols whose size is below the critical one. Figure 8.6 gives for the partial flux of molecules on the aerosol surface

$$j(u) = (1-u)f(u), \quad (8.23)$$

and from the equilibrium it follows that numbers of evaporated and attached molecules are equal; that is,

$$\int_0^1 (1-u)f(u)du = \int_1^{1.5} (u-1)f(u)du \quad (8.24)$$

From data of Fig. 8.6 we obtain for the total rate of molecule release for aerosols the size of which is below the critical one or for the rate of molecule attachment for aerosols of a larger size:

$$J = J_o \int_0^1 j(u)du = -J_o \int_1^{1.5} j(u)du = 0.085 J_o \quad (8.25)$$

From this we have the equation for the average aerosol size $\overline{n}$ that is connected with the critical aerosol size by the relation $\overline{n} = 1.13 n_{cr}$

$$\frac{d\overline{n}}{dt} = \frac{\overline{n}}{n_{cr}} \frac{J}{f(1)} = 0.056 J_o, \quad (8.26)$$

This gives the doubling time of growth of an aerosol size where the average number of aerosol molecules varies from $\overline{n}$ to $2\overline{n}$ [3, 9]:

$$\tau_d = \frac{36\overline{n}}{J_o} = \frac{4.3\overline{n}}{D r_W N_m} \frac{T}{A} \quad (8.27)$$

The processes of evaporation of molecules from the aerosol surface and attachment of molecules to aerosols are accompanied by heat processes. From this it follows that the aerosol temperature and that of surrounding air are different. This fact influences the total rate of aerosol evaporation or molecule attachment to aerosols. To characterize this process, we introduce the parameter Φ, which is the ratio of the partial rate of aerosol growth without accounting for the heat effect J to that with its account I:

$$I = \frac{J}{\Phi}, \tag{8.28}$$

where J is given by Formulas (8.25) and (8.22). The temperature dependence for the factor Φ for growth of water aerosols in air is given in Fig. 8.7 using parameters of water and its vapor [21], and the diffusion coefficient of water molecules in air at the temperature $T = 300\,\mathrm{K}$ and atmospheric pressure is given by $D = 0.2\,\mathrm{cm^2/s}$ [23]. Accounting for this effect leads to an increase of the doubling time of an aerosol size in Φ times compared to that according to Formula (8.27). Figure 8.8 gives the temperature dependence for the doubling time of an aerosol size in a cumulus cloud that according to Formula (8.27) is equal to

$$\tau_d = \frac{4.3\overline{n}}{\Phi D r_W N_m} \frac{T}{A} \tag{8.29}$$

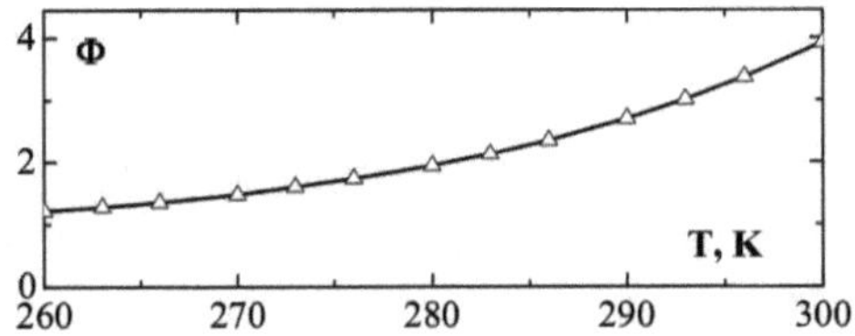

Fig. 8.7 Thermal factor $\Phi(T)$ that characterizes the rate of growth of water aerosols the size of which exceeds the critical one under the equilibrium of these aerosols with the water vapor in atmospheric air. Aerosol growth proceeds in the diffusion regime as a result of coalescence [3, 9]

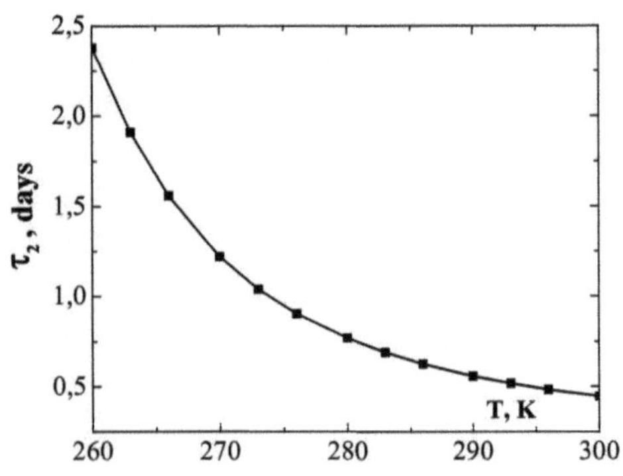

Fig. 8.8 Doubling time for an average number of aerosol molecules in aerosols of a cumulus cloud with parameters (2.24) in atmospheric air if aerosol growth is determined by the diffusion regime of coalescence [3, 9]

In analyzing the coalescence process, we assume a low time of equilibrium establishment between aerosols and vapor compared to a coalescence time. Let us determine a time of establishment of this equilibrium in the limit of large aerosols where the equilibrium number density of molecules near an aerosol N_o is equal to their number density N_{sat} in the saturated vapor. The number density of bound water molecules N_b in aerosols is given by

$$N_b - N_p \left(\frac{r}{r_W}\right)^3, \qquad (8.30)$$

where N_p is the aerosol number density and r is the average aerosol radius. Taking $N_b \gg N_{sat}$, we obtain from the equilibrium for the number density variations $\delta N_{sat} = -\delta N_b$, so that violation of the equilibrium does not influence the aerosol parameters. Then vapor relaxation is described by the equation

$$\frac{dN_w}{dt} = 4\pi Dr(N_{sat} - N_w), \qquad (8.31)$$

and the solution of this equation gives for the number density of free molecules:

$$N_w = N_{sat} - (N_{sat} - N_o)\exp(-t/\tau), \qquad (8.32)$$

As shown, a relaxation time or a time of establishment of the equilibrium between free molecules and aerosols is equal to

$$\tau = \frac{1}{4\pi Dr N_p} \qquad (8.33)$$

This time depends weakly on the air temperature where aerosols are located, and in the above consideration it is lower than a growth time for aerosols. In particular, for parameters of a typical cumulus cloud (2.24) in air at atmospheric pressure and the air temperature in the range 260–300 K this time ranges 0.5–0.4 s according to Formula (8.33).

Figure 8.9 contains typical times of aerosol growth and a time of equilibrium establishment between aerosols and saturated water vapor in atmospheric air for parameters of a typical cumulus cloud (2.24) and the air temperature 0 °C. Note that the number densities of free and bound water molecules are compared under these conditions. In particular, at the pressure 0.5 atm that corresponds to an altitude of standard atmosphere approximately 6 km, the specific water content in aerosols is 3.3 g/kg, whereas the content of free water molecules in the saturated vapor is equal to 7.5 g/kg.

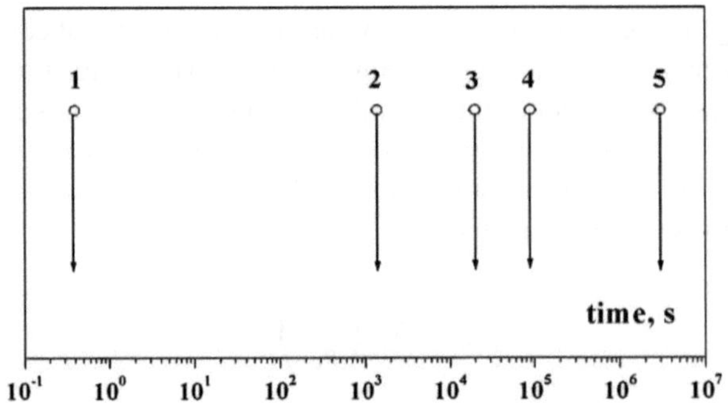

Fig. 8.9 Hierarchy of typical times in a system consisting of water aerosols and water vapor that are located in atmospheric air at the temperature $T = 0\,°\mathrm{C}$ and pressure 1 atm, if typical aerosol parameters correspond to a cumulus cloud (2.24): (1) equilibrium between free water molecules and aerosols, (2) gravitational growth of neutral aerosols, (3) gravitational growth of charged aerosols, (4) coalescence, (5) coagulation of neutral aerosols

8.3 Equilibrium Between Molecular Ions and Charged Aerosols

A specific situation occurs inside tropospheric clouds because air ionization by cosmic rays proceeds inside them, and forming molecular ions attach to aerosols, as we show in the analysis of Sect. 6.1. As a result, a specific aerosol plasma is formed, and for this reason a cloud conserves its volume more or less and does not scatter. Charged aerosols may change their charge owing to their attachment of molecular ions, and finally a charge distribution of aerosols is established. Our task now is to determine the charge distribution function of aerosols $f(Z)$, where Z is the aerosol charge. Assuming a typical aerosol charge to be large $Z \gg 1$, we represent the aerosol distribution function in the form

$$f(Z) = \frac{1}{Z_o\sqrt{\pi}} \exp\left(-\frac{Z^2}{Z_o^2}\right) \quad (8.34)$$

We use here the symmetry with respect to the charge sign, and this is a correct approximation of the charge distribution function if $Z_o \gg 1$. The average charge square for this distribution function is

$$\langle Z^2 \rangle = \int_{-\infty}^{\infty} Z^2 f(Z) dZ = \frac{Z_o^2}{2} \quad (8.35)$$

Molecular ions have charge $\pm e$ and if they attach to an aerosol, an aerosol charge changes by one. One can assume that the equilibrium takes place if the average charge square $\langle Z^2 \rangle$ does not vary at a given value Z_o.

The cross section for attachment of a single charged molecular ion to an aerosol of a charge Z is given by [24]

$$\sigma_{at} = \pi r^2 \left(1 - \frac{Ze^2}{r\varepsilon}\right), \tag{8.36}$$

where r is an aerosol radius and ε is the molecular ion energy. Because the aerosol charge is changed from Z to $Z + 1$ as a result of this process, the variation of the average charge square is $(\Delta Z)^2 = 2Z + 1$. Correspondingly, the equilibrium condition is

$$\left\langle \frac{dZ^2}{dt} \right\rangle = \langle (2Z + 1) N_i v \sigma_{at} \rangle, \tag{8.37}$$

where an average is made both over the aerosol charge and the Maxwell distribution for ion velocities. This average gives

$$\left\langle \frac{dZ^2}{dt} \right\rangle = \left\langle (2Z + 1) N_i v \pi r^2 \left(1 - \frac{Ze^2}{r\varepsilon}\right) \right\rangle = N_i \pi r^2 v \left(1 - \frac{Z_o^2 e^2}{r\varepsilon}\right) = N_i k_o \left(1 - \frac{2Z_o^2 e^2}{rT}\right) = 0, \tag{8.38}$$

where $k_o = \pi r^2 \sqrt{8T/\pi m}$ is the rate constant of ion attachment to a neutral aerosol, m is the ion mass, T is the ion temperature expressed in energetic units, and we assume that all the aerosols have an identical radius r. From the equilibrium condition we obtain for a typical aerosol charge

$$Z_o = \sqrt{\frac{rT}{2e^2}} \tag{8.39}$$

The criterion for the validity of the distribution function (8.34) is $Z_o \gg 1$ or

$$rT \gg e^2 \tag{8.40}$$

At the temperature $T = 300$ K this criterion has the form $r \gg 1\,\mu$m. For the average aerosol size $r = 8\,\mu$m according to (2.24) and $T = 300$ K Formula (8.39) gives $Z_o = 7$, and the criterion $Z_o \gg 1$ holds true. Note that this charging of cloud aerosols due to action of cosmic rays prevents clouds from fast scattering.

Let us consider another aspect of this problem and determine the number density of molecular ions N_i in a cumulus cloud. The balance equation for the ion number density has the form

$$\frac{dN_i}{dt} = \nu_{at} N_i, \tag{8.41}$$

if destruction of molecular ions results from attachment to aerosols, whereas their formation proceeds under the action of cosmic rays and is characterized by a rate

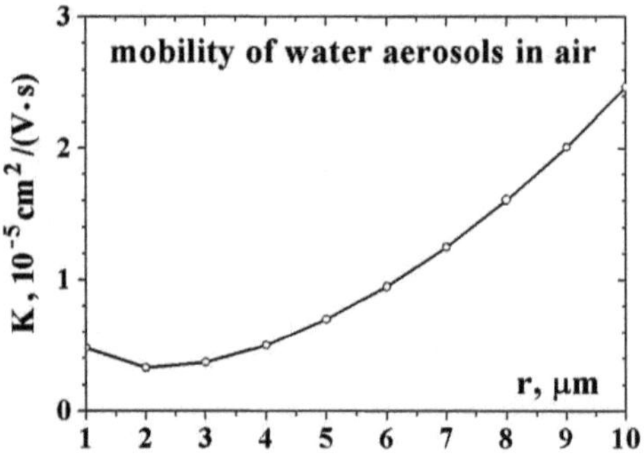

Fig. 8.10 Mobility of a growing water aerosol in a cumulus cloud under the assumption that this aerosol is singly charged at low sizes and joining of charged liquid aerosols leads to the aerosol charge (8.14) for the mean aerosol radius (2.24)

dN_i/dt. Here the rate of attachment of molecular ions to aerosols is given by the Smolukhowski formula (6.1.5) $\nu_{at} = 4\pi D_i r N_{aer}$. Taking the ion diffusion coefficient $D_i \approx 0.1\,\mathrm{cm}^2/\mathrm{s}$ at atmospheric pressure and room temperature and aerosol parameters in accordance with Formula (2.24), we obtain $\nu_{at} \sim 1\,\mathrm{s}^{-1}$. Then (8.41) gives $N_i \sim 10\,\mathrm{cm}^{-3}$ which is two orders of magnitude below this value in a dry atmosphere.

Thus, we have an aerosol plasma consisting of charged aerosols and also positive and negative molecular ions. The rate of collisions between aerosols in this plasma is determined by motion of charged aerosols under the action of external fields in air. Therefore below we consider the mobility of charged aerosols in weak electric fields. Figure 8.10 contains the dependence for the zero-field mobility of nano-size water aerosols on a number of water molecules constituting this aerosol. In this consideration an aerosol size is small compared to the mean free path of air molecules in air, and then particle motion proceeds in the kinetic regime [3] where each time a test aerosol may interact strongly with only one molecule. The mean free path of air molecules in air at atmospheric pressure and room temperature is approximately 0.1 μm. Therefore motion of a micron-size aerosol in atmospheric air is subject to another limiting case related to the diffusion regime of aerosol interaction with surrounding air where many molecules interact strongly with this aerosol simultaneously. Then the friction force F acted on a particle of a radius r that moves with velocity v is given by Stokes formula $F = 6\pi v r \eta$, where η is the air viscosity. Equalizing this force to the force ZeE from the electric field of a strength E (Z is the aerosol charge), we obtain for the aerosol mobility $K = v/E$,

$$K = \frac{Ze}{6\pi r \eta} \tag{8.42}$$

From this one can find the mobility of a singly charged water droplet of a radius r:

$$K = K_o Z \frac{a}{r}, \tag{8.43}$$

where $K_o = 4.6 \times 10^{-6}\,\mathrm{cm}^2/(\mathrm{V}\cdot\mathrm{s})$ at $a = 1\,\mu\mathrm{m}$. In the same manner one can obtain the mobility $K_o = 1.6 \times 10^{-5}\,\mathrm{cm}^2/(\mathrm{V}\cdot\mathrm{s})$ for an aerosol of a cumulus cloud of radius $r = 8\,\mu\text{m}$ according to Formula (2.24) and the charge $Z = 27$ according to Formula (8.14). This mobility corresponds to the drift velocity of such aerosols in air under the action of the Earth's electric field as $2 \times 10^{-5}\,\text{cm/s}$; that is, this velocity is respectively small. Figure 8.10 gives the dependence of the aerosol mobility on a size for charged aerosols.

If a negatively charged aerosol (water droplet) is located in atmospheric air, it falls down under the action of its weight, and the fall velocity is given by Formula (8.8). Along with this, a singly charged droplet rises under the action of the Earth's electric field, the strength of which is approximately $E = 1.3\,\text{V/cm}$, and the droplet mobility is given by Formula (8.54). The friction force and electric field force acted on the aerosol become equal at the droplet radius $r = 0.08\,\mu\text{m}$. Note that we are based on the diffusion regime of droplet motion that requires the criterion fulfillment

$$r \gg \lambda, \tag{8.44}$$

where λ is the mean free path of molecules in air. For atmospheric air this criterion is $r \gg 0.1\,\mu\text{m}$, so that the above result may be considered as an estimation.

Above we consider two types of an aerosol atmospheric plasma, so that in the first case water aerosols are charged in collisions with aerosols located in another aggregate state and then they fall down faster than aerosols of another charge sign. In the second case aerosols of a cloud are charged as a result of attachment to their molecular ions formed under the action of cosmic rays, and then the charge distribution function of aerosols is given by Formula (8.34), and the average aerosol charge is large if the criterion (8.40) holds true. At the temperature $T = 300\,\text{K}$ this criterion gives $r \gg 1\,\mu\text{m}$, and this and larger aerosol sizes in atmospheric air correspond to the diffusion regime of aerosol motion. In considering kinetics of aerosol recombination, we assume for definiteness all the cloud aerosols to be liquid. The recombination coefficient of a positive aerosol with a charge Z_+ and a negative aerosol with a charge Z_- in air under these conditions is given by the Langevin formula [25]:

$$\alpha = 4\pi e Z_+ Z_- (K_+ + K_-), \tag{8.45}$$

where $K_+,\ K_-$ are the mobilities of these singly charged aerosols. On the basis of Formula (8.42) one can transform this formula to

$$\alpha = \frac{2e^2}{3\eta} e^2 Z_+ Z_- \left(\frac{1}{r_+} + \frac{1}{r_-}\right), \tag{8.46}$$

Taking identical charges ($Z_+ + Z_- + Z$) and sizes ($r_+ = r_- = r$) of aerosols, we have from this

$$\alpha = \frac{4Z^2 e^2}{3r\eta} \tag{8.47}$$

One can compare this association rate constant with that for neutral aerosols according to Formula (8.2)

$$\frac{\alpha/2}{k_{as}} = \frac{Z^2e^2}{4rT} \gg 1, \tag{8.48}$$

where we use the factor $1/2$ for association of charged aerosols because under considering conditions each charged aerosol may recombine with an aerosol of another sign. In particular, for the charge distribution (8.34) that resulted from attachment of molecular ions to aerosols we have in accordance with (8.39) for the average size $\overline{Z}$

$$\overline{Z} = \frac{Z_o}{\pi} = \sqrt{\frac{rT}{2\pi e^2}}, \tag{8.49}$$

and the criterion (8.48) is not fulfilled because of a small aerosol charge. Hence the association rate constant for aerosols is given by Formula (8.2).

The charge distribution function of aerosols is given by Formulas (8.34) and (8.39) if aerosol charging results from attachment of molecular ions to an aerosol, where molecular ions are formed under the action of cosmic rays. In derivation of the charge distribution function (8.34), we ignore the process of aerosol association, and now we estimate the contribution of this process to aerosol charging. Let us denote the number density N_Z of the positive and negative charge, so that

$$N_Z = N_{aer} \int_0^\infty Zf(Z)dZ = \frac{N_{aer}Z_o}{2\sqrt{\pi}}, \tag{8.50}$$

where N_{aer} is the number density of aerosols, and, for definiteness, we extract aerosols that have a positive charge. Association of two aerosols that carry charges of different signs leads to a decrease of N_Z. Hence, the charge distribution of aerosols (8.34) with a typical charge (8.39) is valid if the rate of redistribution of aerosols over charges due to their association and loss a part of charges is small compared to the rate of formation of new charges dN_i/dt due to formation of molecular ions under the action of cosmic rays. One can represent this criterion in the form:

$$\frac{dN_i}{dt} \gg \frac{N_{aer}Z_o}{\tau_d}, \tag{8.51}$$

where we take $1/\tau_d$ as the rate of aerosol association, and τ_d is a time of doubling of an aerosol size; that is, a number of aerosol molecules will be doubled for this time. Taking $N_{aer} \sim 10^3\,\mathrm{cm}^{-3}$ according to (2.24) and $\tau_d = 8$ min for neutral aerosols of a radius $r = 8\,\mu\mathrm{m}$ on the basis of Formula (8.9), we obtain for the right-hand side of the relation (8.51)

$$\frac{N_{aer}Z_o}{\tau_d} \sim 10\,\mathrm{cm}^{-3}\mathrm{s}^{-1} \tag{8.52}$$

One can compare this value with the maximum rate of ion formation in the troposphere $dN_i/dt \approx 30 - 40\,\mathrm{cm}^{-3}\,\mathrm{s}^{-1}$. From this one can obtain the qualitative character of Formula (8.34) This creates a force that does not allow for clouds to expand quickly. Note that in contrast to the above aerosol plasma that contains water aerosols in different aggregate states and exists at temperatures below 0 °C, this description refers to clouds at lower altitudes where all aerosols are found in the same aggregate state, evidently in the liquid state.

We now analyze the influence of formation of molecular ions under the action of cosmic rays on the charge of aerosols located in the lower part of a cumulus cloud. In the absence of ion attachment to aerosols a typical aerosol charge is given by Formula (8.14) and, being expressed in electron charges, is equal approximately to $Z \approx 27$. Because this value follows from the lifetime of a cumulus cloud, we assume that this charge is also realized if air inside a cloud is ionized under the action of cosmic rays. When an equilibrium is established between charged aerosols and molecular ions, the number density of positive molecular ions N_+ becomes lower than the number density N_- of negative molecular ions. These number densities follow from the equilibrium

$$\frac{dN_i}{dt} = \nu_+ N_+ = \nu_o \left(1 + \frac{2Ze^2}{rT}\right) N_+, \quad \frac{dN_i}{dt} = \nu_o N_- \exp\left(-\frac{Ze^2}{rT}\right), \tag{8.53}$$

where the rate ν_o refers to neutral aerosols and is equal to

$$\nu_o = \pi r^2 \sqrt{\frac{8T}{\pi m}} N_{aer} \tag{8.54}$$

Here m is the mass of a positive or negative ion. From this we have for the ratio of the number densities of positive and negative ions

$$\xi = \frac{N_+}{N_-} = \sqrt{\frac{m_+}{m_-}} \frac{\exp(-x)}{1+2x}, \quad x = \frac{Ze^2}{rT}, \tag{8.55}$$

where m_+, m_- are masses of positive and negative ions, correspondingly. As can be seen, due to different number densities of positive and negative ions, the fluxes to an aerosol for molecular positive and negative ions of different charge signs are identical.

Let us apply these formulas for a typical cumulus cloud with parameters according to Formulas (2.24) and (8.14). For definiteness, we take room temperature of air and assume that basic ions are N_2^+ and O_2^-. We obtain $x = 0.19$, $N_+/N_- = 0.56$ and $\nu_o \approx 90\,\mathrm{s}^{-1}$. Next, in the case $dN_i/dt = 10\,\mathrm{cm}^{-3}\mathrm{s}^{-1}$, the number density of the positive charge is $N_+ = 0.08\,\mathrm{cm}^{-3}$, and the number density of the negative ions inside the cloud with charged aerosols is $N_- = 0.14\,\mathrm{cm}^{-3}$ under the above conditions.

Note that the total mass of water aerosols in a given volume is small compared to the mass of air molecules, and therefore the conductivity of wet air is determined by collisions of molecular ions with air molecules. In addition, the conductivity of this air is determined by the number density of molecular ions. As is seen, a typical number density of molecular ions in dry air is $\sim 10^3\,\mathrm{cm}^{-3}$, whereas according to the above estimations in a cumulus cloud it is $\sim 0.1\,\mathrm{cm}^{-3}$. From this it follows that the conductivity of dry air is larger than that of air in a cumulus cloud by four orders of magnitude. This disagrees with results of measurements [26] according to which the conductivity of cloudy air is 6 times lower than that for dry air. Evidently, this cloudy air differs from that of cumulus clouds.

8.4 Aerosols in Atmospheric Electricity

In considering the problem of atmospheric electricity, we use the bases of observational data, laboratory study, and elementary processes in atmospheric air which for the most part involve aerosols. Figure 8.11 gives a wide variety of aerosols that reside in the atmosphere, but only water aerosols are responsible for electrical processes in the atmosphere although the basis of these aerosols, the condensation nuclei, may be nonwater. The role of water aerosols is determined by the possibility of water to be condensed in the atmosphere, and its concentration significantly exceeds that for other air components with this property. For this reason the subsequent analysis is concerned with water aerosols only. Note that the property to be condensed leads to a nonuniform water distribution in the atmosphere and also to a nonstationary water distribution. On the basis of the average atmospheric parameters we obtain a qualitative description of atmospheric processes involving water aerosols.

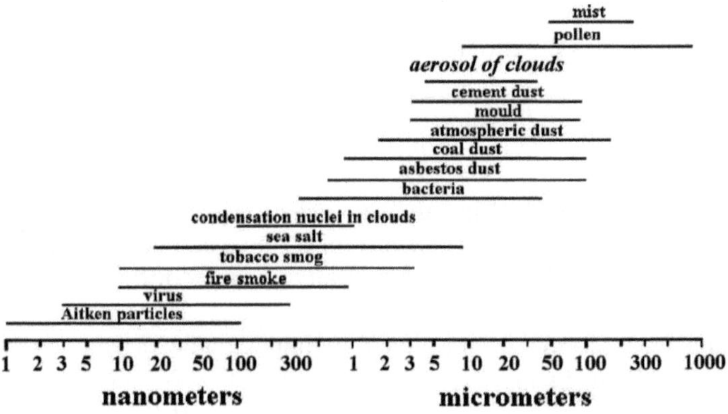

Fig. 8.11 Aerosols that are present in the atmosphere and their sizes. Aerosols that partake of processes of atmospheric electricity are represented in italics [27]

Aerosols are necessary for atmospheric electricity (see Chap. 4). The atmosphere contains various chemical impurities that can be the basis of aerosols. But in considering atmospheric electricity, we are guided by micron-size aerosols, and the total mass of atmospheric aerosols is a remarkable part of the total mass of atmospheric water vapor. Because other chemical components of this type are present in atmospheric aerosols with low concentrations, they will not take part in aerosol kinetics. In addition, ions and certain atoms and radicals may be nuclei of condensation; that is, they determine the character of aerosol growth at the first stage of growth, that is, at the stage of nano-size aerosols. But if an aerosol reaches a large size, these admixtures become nonessential. Therefore in considering micron-size water aerosols, one can ignore the presence of admixtures in them.

Atmospheric water consists of two components, free water molecules or a water vapor and bound water, or micron-size aerosols. An equilibrium between free and bound atmospheric molecules proceeds through processes of attachment of water molecules to aerosols and evaporation of water molecules from the aerosol surface. This equilibrium is established for a small time $\tau_{eq} \leq 1$ s. On the contrary, the system consisting of free atmospheric water and aerosols is unstable: if the atmospheric temperature is constant, in the end all the atmospheric aerosols will join in one condensed system. But this time is large compared to those under consideration; that is, the system of free atmospheric water and aerosols does not vary during such times.

Aerosols that take part in atmospheric electric processes constitute the clouds which are a basis of atmospheric electricity. But only a small part of clouds bears a relationship to atmospheric electricity. Hence, we only deal with cumulus clouds with the parameters of aerosols (2.24) in them. Because the temperature of the standard atmosphere decreases with increasing altitude at low altitudes (see Fig. 2.1), condensation processes proceed in an upper troposphere. This means that an evaporated water vapor ascends up and is partially converted in aerosols. During the first stage of this process, impurities determine the rate of condensation, for example [28]). But this stage gives a small contribution to a time of aerosol growth, and therefore we neglect this stage in the analysis of aerosol growth.

As follows from Fig. 4.2, the basic processes of atmospheric electricity include aerosol processes and consist in charging of aerosols as a result of their collisions. A subsequent process is a separation of charges that proceeds because of different falling velocities for positively and negatively charged aerosols. Inasmuch as the Earth is charged negatively, we assume the fast fall of negatively charged aerosols, whereas due to a low weight positively charged aerosols practically remain at altitudes of their charging. Note a variety of ways of aerosol charging (e.g., [29]). In particular, Frenkel [30] assumes a resonance character of formation of charged aerosols as a result of attachment of molecular ions to aerosols that contradicts the nature of this process. In reality, only the mechanism of aerosol charging [31] under consideration as a result of collisions of two aerosols in different aggregate states takes place. This charging mechanism has no alternative in atmospheric electricity, although some of these may be realized under other conditions.

Unfortunately, the charging process is not studied in detail in spite of the long space of time from the first understanding of its nature [31]. One reason is that the charging rate depends on whether this proceeds in the course of rime accretion or aerosol evaporation (e.g., [32–34]). In addition, gravitational fall of aerosols of a large mass may lead to a positive charging of the low cloud edge [35]. We use a simplified dipole model for charge distribution in a cloud with negatively charged aerosols that fall down faster than positive aerosols. This charge distribution corresponds to the so-called bipolar model [36–39] which differs from a real distribution (see Fig. 4.1) that is described better by the tripolar model [40–42] with three cloud regions where charges of a certain sign are located. Note that the total dipole moment of a cloud ranges from 20 to 100 C · km [43] at a typical distance between centers of the positive and negative charges of ∼1 km.

8.5 Charge Separation in Clouds

One of the stages of atmospheric electricity consists in separation of an atmospheric charge as a result of gravitational fall of ice and water aerosols. The fall velocity for aerosols follows from the equality of their weight and the friction force that is determined by the Stokes formula. From this we have for the fall velocity w of water micron-size drops in atmospheric air at the temperature $T = 300\,\mathrm{K}$ [2]

$$w = 0.012\left(\frac{r_o}{a}\right)^2, \tag{8.56}$$

where the fall velocity is expressed in cm/s, and $a = 1\,\mu\mathrm{m}$. Taking an average aerosol radius (2.24) $r_o = 8\,\mu\mathrm{m}$ and an average charge $Z = 27$ (8.14) we have the aerosol fall velocity $w = 0.7\,\mathrm{cm/s}$ on the basis of Formula (8.56) under the assumption of liquid aerosols, that leads to the following current density as a result of aerosol falling

$$i = ZewN_p w \tag{8.57}$$

This gives for the indicated parameters $i = 3 \times 10^{-15}\,\mathrm{A/cm^2}$, and the current $I = 1700\,\mathrm{A}$ in the Earth's atmosphere may be provided if approximately 10 % of the area of the Earth's surface is covered by cumulus clouds. It should be noted that above we neglect the drift velocity of aerosols due to convective motion which according to the criterion (2.22) is valid for $r_o < 1\,\mu\mathrm{m}$. In addition, an average time of water location in the atmosphere is 9 days [44], and during this time an aerosol passes in a manner that is compared to or exceeds the size of a cumulus cloud. As a result, we have a noncontradictory physical picture for processes of aerosol charging and charge separation in the atmosphere, which are the basis of physics of atmospheric electricity.

Fig. 8.12 Eruption of the Puyehue volcano in Chile in June 2011 with lightning inside eruption jets [45, 46]

As shown, atmospheric electric processes involving aerosols lead to negative charging of the Earth in collisions of aerosols located in different aggregate states. In other media, say, in atmospheres of other planets, particles of other types may be responsible for generation of electric charges. As for the Earth's atmosphere, electric processes are possible with participation of dust microparticles of any type, and lightning may be observed in sandstorms, in coal mines with coal dust, in jets of volcanic eruptions, and in an explosion of an atomic bomb. In particular, volcanic eruptions are accompanied often by lightning inside them, as demonstrated in Fig. 8.12, where eruption of the Puyehue volcano in Chile is represented. A minimum size for charge separation in volcano dust that leads to lightning creation is estimated as 500 m, whereas a minimum cloud size for a thunderstorm is 3 km [47]. In all these cases electricity in gas systems results from collisions of microparticles of various compounds. This mechanism is realized in atmospheres of other planets. Note that although electric processes in various cases are determined by the same nature of processes, parameters of these processes are different.

The Earth's charging as a result of falling of charged aerosols may proceed without lightning if charged aerosols reach the Earth's surface. Evidently, an upstream of air from the Earth's surface does not allow for clouds to reach the Earth's surface, and then the charge transfer proceeds in the form of lightning. Probably, this channel of charge transfer to the Earth directly by charged aerosols exists and is realized in mountains and oceans. Then high electric fields occur near the surface and cause corona discharges near conductive objects with the air glowing at their location. According to my mountaineering experience, it is dangerous to be on the ridge or high open places in prethunderstorm weather, where ice-axes, climbing irons, and other metallic objects are jingled testifying about propagation of electric currents through them. Glowing of oblong objects in prethunderstorm mist weather has been known for thousands of years under different names. This phenomenon was observed often by sailors in seas and oceans, and it is known as Saint Elmo's fire [48] (Saint Elmo is a protector of sailors). As follows from the above consideration, the character of electric phenomena in the Earth's atmosphere is more complex and richer than that following from simple schemes.

8.6 Mixing of Cloud Aerosols with Streams of Wet Air and Rain

Basic electric atmospheric processes proceed in clouds consisting of aerosols. Some examples of clouds are given in Figs. 8.13 and 8.14, and they demonstrate a nonuniform cloud structure with displacement of their elements due to convection that is ignored in this qualitative consideration. Cumulus clouds are of interest for the analysis of electric atmospheric phenomena. Aerosol parameters for these clouds are described by Formula (2.24), and these clouds are formed in regions of water condensation in the troposphere.

Let us analyze in detail atmospheric processes involving aerosols. The dry air mass is 5.1×10^{18} kg in the Earth's atmosphere, its specific weight near the air surface is $1.2\,\mathrm{kg/m^3}$ on average, and an average water mass in the atmosphere is equal to 1.3×10^{16} kg [50] which corresponds to the average specific weight of atmospheric water $3\,\mathrm{g/m^3}$. The latter coincides with the specific water weight in aerosols if one takes into account an altitude decrease of this quantity and corresponds to the partial water pressure of about $2Torr$, whereas even at the temperature $0\,°\mathrm{C}$ the saturated

Fig. 8.13 Cloud view from a flight [49]

Fig. 8.14 Clouds below mountains [53]

vapor pressure is $4.7 Torr$ [51]. One can conclude from this that water is found in the atmosphere mostly in the form of free molecules, and aerosols are formed in regions with a heightened water pressure and low temperature. The atmosphere temperature decreases with an increasing altitude up to the tropopause, that is, the atmospheric layer with a minimal temperature. The tropopause of a middle atmosphere is found at the altitude 9 km at the poles and 17 km at the equator [50]. The temperature of an average atmosphere at the Earth's surface is 15 °C, the temperature of the tropopause equals −55 °C, and the temperature gradient is −6.5 K/km [52]. On the basis of this we introduce a typical size by analogy with (2.1) (see Fig. 2.5):

$$\Lambda_T = \left(\left| \frac{dT}{T dh} \right| \right)^{-1}, \tag{8.58}$$

and $\Lambda_T \sim 10$ km near the Earth's surface. We use this value for the subsequent analysis.

The important stage of cloud evolution consists in mixing of a cumulus cloud with wet air streams from surface layers. This process is important for two reasons. These wet streams contain molecules and particles that obtain the aerosol charge in the processes of charge exchange. This process is accompanied by the charge exchange process with charge transfer to molecules, radicals, and small particles located in fluxes of warm air at the beginning. On the other hand, as a result of the charge exchange process, aerosols become neutral and grow in the process of gravitation coagulation (see Fig. 8.2). Therefore this process is a trigger for a fast growth of aerosols-droplets which creates rain.

In order to analyze the process of growth of liquid droplets in the course of creation of rain, we consider a simple example taking into account that a thunderstorm is realized at the water content in the atmosphere of $A_{cr} = 7\,\mathrm{g/kg}$ (grams of water per kilogram of air) as the threshold water content for a thunderstorm [47]. This water content corresponds to the saturated vapor pressure at the temperature 8 °C; that is, water aerosols are stable at lower atmospheric temperatures. The total number density of free and bound water molecules is $N = 3 \times 10^{17}\,\mathrm{cm}^{-3}$ at this water content and the total concentration of water molecules of $c = 1.1\,\%$ at the above content and temperature. Note that these parameters for bound water molecules in aerosols with average parameters (2.24) ($r = 8\,\mu\mathrm{m},\ N_p = 10^3\,\mathrm{cm}^{-3}$) are $N = 7 \times 10^{16}\,\mathrm{cm}^{-3}$ and $c = 0.27\,\%$; this corresponds to the content of bound water molecules in aerosols of $A_* = 1.7\,\mathrm{g/kg}$. This amount of free water molecules corresponds to the temperature of $-10\,°\mathrm{C}$. Because processes of growth and evaporation of aerosols are accompanied by energy release or absorption, injection of wet air in a cumulus cloud leads to a change of bound molecules in aerosols depending on the temperature of injected air. Subsequently this determines the amount of water that descends on Earth in the form of rain.

To analyze the character of mixing an air stream with air of a cumulus cloud, we consider certain conditions of this mixing. Let us take for definiteness the air temperature inside a cumulus cloud to be $T_1 = 258\,\mathrm{K}$ ($T_1 = -15\,°\mathrm{C}$) that corresponds to the content of the saturated vapor pressure $A_s = 1\,\mathrm{g/kg}$ (see Table 1), and to the total water content $A_1 = 2.7\,\mathrm{g/kg}$. Let T_2 be the air temperature in a stream. Taking the water content $A_{cr} = 7\,\mathrm{g/kg}$, we assume that a forming mixture contains equal amounts of air from a cloud and stream. Then at the first stage of mixing the temperature of the forming mixture is $(T_1 + T_2)/2$ and the water content is $A_{tot} = (A_{cr} + A_1)/2 = 4.9\,\mathrm{g/kg}$. At the second stage an excess of free water molecules attaches to aerosols leading to a temperature change. As a result, the pressure of free water molecules is equal to the saturated water pressure at the final mixture temperature T_f. This leads to the equations:

$$T_f = \frac{T_1 + T_2}{2} + \Delta T,\ \ \Delta T = \frac{\varepsilon}{C_p}\Delta A,\ \ A_s(T_f) = A_{tot} - A_* + \Delta A \qquad (8.59)$$

Here ΔT is a decrease of the mixture' temperature as a result of an increase ΔA of the content of bound molecules, ε is the specific energy for evaporation of water molecules from the aerosol surface, C_p is the air heat capacity, and $A_s(T_f)$ is the water content in air at an indicated temperature that corresponds to the saturated vapor pressure. In this case $\varepsilon/C_p = 0.96\,\mathrm{K/g/kg}$, and (8.59) allow us to determine the parameters of mixing under given conditions. Taking the stream temperature as $T_2 = 278 + \delta T$, we reduce the set of (8.59) to the form

$$A_s(268 + 0.5\delta T + \Delta T) = 3.2 + 1.04\Delta T, \qquad (8.60)$$

where the temperature is expressed in Kelvin. Using the data of Table 1, we find the dependence $\Delta T(T_2)$ or the dependence $A_b(T_2)$ that is given in Fig. 8.15.

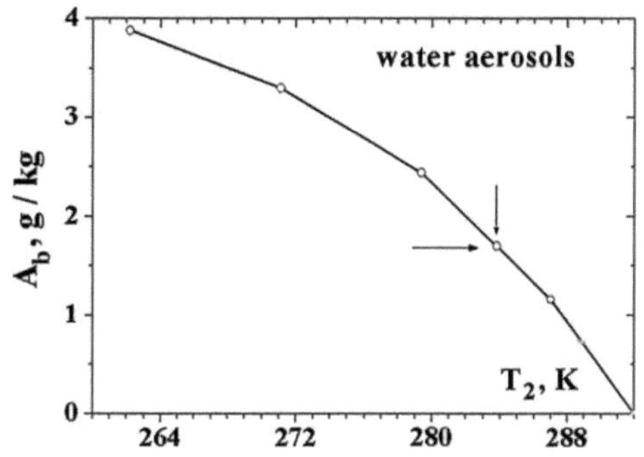

Fig. 8.15 Dependence of the content A_s of bonded water in aerosols located in atmospheric air at the initial temperature T_2 of the wet air jet under conditions given in the text. *Arrows* show the jet temperature at which mixing does not lead to change of the water content A_s in aerosols

Above we assume that aerosols are absent in the incident air jet. We also use a fast establishment of equilibrium between bonded and free water molecules in the formed mixture. Indeed, a typical time τ_{eq} for this equilibrium is given by the Smoluchowski formula

$$\frac{1}{\tau_{eq}} = 4\pi D_w r N_p \sim 1\,\mathrm{s}^{-1}, \tag{8.61}$$

where D_w is the diffusion coefficient of water molecules in air that at atmospheric air pressure is $D_w \approx 0.2\,\mathrm{cm}^2/\mathrm{s}$ [23], $r = 8\,\mu\mathrm{m}$ is the aerosol radius, and $N_p = 10^3\,\mathrm{cm}^{-3}$ is the initial number density of aerosols. Hence, the excess of water in air compared to the saturated vapor pressure is transformed in a condensed phase in the form of aerosols in the course of mixing of the jet of wet air and cold air with charged aerosols of a cumulus cloud.

The example that results is represented in Fig. 8.15 which demonstrates the character of mixing ascending air jets with the air of a cumulus cloud. One can see that an increase of the aerosol mass proceeds in a narrow range of jet temperatures. A small temperature of ascending air is required for this; that is, air of the jet is cooled on the way to a cumulus cloud. One can conclude from this demonstration that transport of water to a cumulus cloud is not necessary for the origin of rain. On the contrary, charge exchange of charged aerosols of a cumulus cloud in collisions with molecules and particles located in the jet is a trigger for subsequent growth of aerosols and the origin of rain. As a result of the charge exchange process, aerosols become neutral and associate in the process of gravitation association (Fig. 8.2). In addition, a charge of a cumulus cloud transfers to molecules and small dust particles. This increases the mobility of charged particles and facilitates the conditions for electrical breakdown in air because of transport of weakly ionized components in this air.

Let us consider the kinetics of aerosol growth if neutral aerosols are located in a cumulus cloud and they can associate in the process of gravitation association (Fig. 8.2). The balance equation (8.3) for aerosol growth has the form

$$\frac{d\overline{n}}{dt} = \frac{1}{2} k_{as} N_b,$$

where $\overline{n}$ is the average number of aerosol molecules, N_b is the number density of bound water molecules, and k_{as} is the rate constant of aerosol association for the gravitation mechanism of their growth which is given by Formula (8.9)

$$k_{as} = \frac{2r^4 \rho g}{\eta},$$

Let us represent this formula in the form

$$k_{as} = k_o \left(\frac{r}{a}\right)^4,$$

where $k_o = 1.1 \times 10^{-5}\,\text{cm}^3/\text{s}$ and $a = 10\,\mu\text{m}$. Along with the equation of aerosol growth, we use the equation of aerosol motion under the action of the weight force

$$\frac{dh}{dt} = w, \tag{8.62}$$

where h is the manner in which aerosol passes during its falling and w is the falling velocity given by Formula (8.7). Excluding time from these equations and assuming the total number density of bound molecules N_b to be independent of time, we obtain

$$\frac{dh}{dr} = const = 1.4 \times 10^6, \tag{8.63}$$

where $const$ is determined by initial conditions of this equation in accordance with (2.24). In particular, if a fall height is $h = 2\,\text{km}$, the final drop radius is $r = 1.4\,\text{mm}$, and a typical time of aerosol fall is $\tau_f \approx 25\,\text{min}$.

According to observations [54], a radius of rain drops near the Earth's surface is found in a range $0.1 - 9\,\text{mm}$. Next, Formula (8.7) is violated at large drop sizes, and the fall velocities for drops of radii 0.5 and 5 mm are equal, respectively, to 2 and 9 m/s, whereas according to Formula (8.7) these velocities are 2 m/s and 200 m/s, respectively. Indeed, at large Reynolds numbers the Stokes formula (8.7) is violated and the spherical drop shape is lost [54]; that is, the Stokes formula (8.7) may be used as an estimate. In addition, if aerosols of a cumulus cloud with parameters (2.24) are transformed in water at the Earth's surface, the thickness of a water layer at the Earth's surface Δ is proportional to the thickness L of a cumulus cloud, and

$$\frac{\Delta}{L} = 2\,\text{mm/km} \tag{8.64}$$

The rate of water precipitation during rain ranges from 2.5 mm/h up to 7.6 mm/h [54]. From this it follows that air moisture contributes to the process where wet air is mixed with a cumulus cloud [55]. Hence, convection of wet air and its interaction of jets of wet air with aerosols of a cumulus cloud is of importance for rain generation [56–58].

Thunderstorm weather is accompanied by a high air moisture that exceeds 7 g/kg [47] which corresponds to the saturated vapor pressure at the temperature 8 °C (see Table 1). At lower temperatures water aerosols are formed in this wet air, and according to Formula (2.24) the water content in aerosols of a typical cumulus cloud is 3 g/kg and their formation leads to an increase of the air temperature by 16 K. Below we estimate the rate of moisture conversion in raindrops during thunderstorm weather, if we start from aerosol parameters in a cumulus cloud (2.24) and assume air to penetrate quickly inside a cumulus cloud. The rate constant of attachment of water molecules to an aerosol of a radius r is determined by the Smolukhowski formula (6.5) [22]

$$k_{at} = 4\pi D_w r, \tag{8.65}$$

where the diffusion coefficient of water molecules in air $D_w \approx 0.2\,\mathrm{cm^2/s}$ [23], and the equation of aerosol growth has the form

$$\frac{dn}{dt} = k_{at} N_w \tag{8.66}$$

Here n is a typical number of water molecules in a drop-aerosol, and $N_w = 3 \times 10^{17}\,\mathrm{cm^{-3}}$ is the number density of water molecules that corresponds to the threshold density of water molecules 7 g/kg for thunderstorm weather [47]. This gives a typical time of aerosol growth

$$\tau = \frac{3r^2}{8\pi D N_w r_W^3}, \tag{8.67}$$

where $r = 8\,\mu\mathrm{m}$ is an aerosol radius and $r_W = 0.192\,\mathrm{nm}$ is the Wigner-Seitz radius for water, and this leads to a typical time of aerosol growth $\tau \approx 0.2\,\mathrm{s}$ under given conditions.

This also exhibits that aerosol growth as a result of attachment of free water molecules to an aerosol proceeds faster than that at the gravitation mechanism of aerosol growth. Indeed, comparison of the rate constants for gravitation growth (8.9) and for attachment of free water molecules to an aerosol (8.65) gives that they become equal at an aerosol radius $r = 0.1\,\mathrm{mm}$, whereas the mechanism (8.65) dominates at lower sizes. In addition, attachment of free water molecules to aerosols proceeds quickly if this process is permitted by thermal conditions.

Note that rain rises together with a lightning discharge in thunderstorm weather. Then after lightning flashes a positive charge it is transferred to charged aerosols, and at the beginning it is found in the form of positive ions. Collisions of positive ions with charged aerosols neutralize the aerosols and cause fast growth of water drops that leads to rain generation. Evidently, convective jets of warm wet air, as a thunderstorm precursor, delay attachment of positive molecular ions to aerosols due to thermal effects, so that they neutralize charged aerosols, and subsequent growth of water drops proceeds quickly. One can expect that a typical time of aerosol neutralization does not exceed 10 min, a typical time of rain generation. In the course of rain development, a part of the charge remains to be bound with water droplets. As a result, rainfall is accompanied by charge transport to the Earth. In particular, according to [59] rainfall of 25 mm/h is accompanied by a current density of $0.2\,\mathrm{fA/cm^2}$.

It is important to note that thunderstorm processes simultaneously include rain development and creation of air electric breakdown, lightning. Let us determine the electrical parameters of a cumulus cloud before these processes. According to observational data, the average electric cloud-Earth potential is $U = 50\,\mathrm{MV}$ and a cloud-Earth distance is $L = 3\,\mathrm{km}$ which corresponds to the electric field strength between the cloud and Earth due to a cloud charge as $E = 200\,\mathrm{V/cm}$. Assume a depth of a charged layer to be small compared to a cloud-Earth distance, as well as a distance between the centers of the positive and negative cloud charges. We use the above values for the number density of aerosols (2.24) and the average charge per aerosol $Z \approx 27$; we have for the number density of cloud charges $N_{ch} = 4\times 10^4\,\mathrm{cm}^{-3}$ which is close to an observational value [47] $N_{ch} = 5 \times 10^4\,\mathrm{cm}^{-3}$. Considering the charge of the lower cloud edge to be distributed over a plane surface, we obtain for the surface charge density

$$\sigma = \frac{E}{4\pi} = 0.5\,\mathrm{C/km}^2 \tag{8.68}$$

Because a charge of 20 C is transferred over the lightning channel on average, the size of a thunderstorm cloud is of several km.

Note that in this estimation we assume the lower cloud edge to be flat. In reality, development of cloud electric processes is accompanied by air convection; that is, this assumption is not fulfilled, and these data have an estimated character. For example, flight measurements in New Mexico [29, 60] gave the center of the negative cloud charge to be at an altitude of 7 Km with temperature of $-15\,°\mathrm{C}$. Correspondingly, typical velocities of charge transport are of several km/s. One can add to this the following description of thunderstorm development according to observational data [61]. A typical time of electric field generation is on the order of 30 min, and from radar measurements the first lightning flash occurs 10–20 min after the aerosol formation that follows. Next, charge generation and separation proceed in a region of a radius of 2 km. The negative charge density formed is $(5-30)\,\mathrm{C/km^3}$ and the rate of charge generation is on the order of $1\,\mathrm{C/(km^3\cdot min)}$. These values [61] correspond to those from lightning observations [14].

In considering an aerosol plasma of a cumulus cloud, we assume it to be uniform. But such a cloud structure is violated near the cloud boundary with dry air, especially in air vortices which result from convection and are accelerated if fluxes of hot air from the Earth's surface reach a cloud [62–64]. Because an aerosol plasma contains micron-size charged water aerosols and also positive and negative molecular ions inside it, negative molecular ions are ejected in dry air near the boundary of a cloud and dry air. Negative ions that are ejected downward move towards the Earth with the drift velocity estimated as $v_d = EK \sim 4\,\mathrm{m/s}$. The total time of cloud discharging under optimal conditions as a result of interaction with surrounding dry air is estimated as $\sim N_-/(dN_i/dt) \sim 2\,\mathrm{min}$.

Thus, a cumulus cloud before the thunderstorm is a specific physical object, where the lower cloud layer contains charged water aerosols, and molecular negative ions are present between the cloud and Earth underlayer. The number density of negative ions cannot exceed that inside an aerosol plasma, and therefore the ions do not influence the electric field in a space between the cloud and Earth underlayer. Indeed, according to the Poisson equation, the electric field strength gradient in a region of location of molecular negative ions is estimated as $dE/dh \sim 10^{-7}\,\mathrm{V/cm^2}$. These ions do not influence the formation of the conducting channel, however, they take part in charge transport through this channel at other stages.

In conclusion, we divide the processes in aerosol plasma between free and bound water molecules and the charge transfer process involving aerosols. Processes between free water molecules and aerosols establish an equilibrium between two phases, a gas phase, that is, free water molecules of a water vapor and a condensed phase (liquid and solid) in aerosols which proceeds quickly ($\tau \approx 0.2$ s under typical conditions). The process of aerosol growth leads to reconstruction of the condensed phase, and the process of charge transfer changes the conditions of this process.

References

1. I. Gutzow, J. Schmelzer, *The Vitreous State* (Springer, Berlin, 1995)
2. B.M. Smirnov, *Cluster Processes in Gases and Plasmas* (Wiley, Berlin, 2010)
3. B.M. Smirnov, *Nanoclusters and Microparticles in Gases and Vapors* (De Gruyter, Berlin, 2012)
4. W. Ostwald, Z. Phys, Chem. **22**, 289 (1897)
5. W. Ostwald, Z. Phys, Chem. **34**, 495 (1900)
6. B.M. Smirnov, Phys. Usp. **54**, 691 (2011)
7. B.M. Smirnov, *Clusters and Small Particles in Gases and Plasmas* (Springer, New York, 1999)
8. A.A. Lushnikov, V.N. Piskunov, DAN SSSR **228**, 642 (1976)
9. B.M. Smirnov, EPL **99**, 13001 (2012)
10. K. Berger, *The Earth Flash, in "Lightning"*, p. 119 ed. by R.H. Golde (Academic Press, London, 1977)
11. W. Gaskell, A.J. Illingworth, J. Latham, C.B. Moore, Q.J.R. Meteorol, Soc. **104**, 460 (1978)
12. H. Christian, Q. J. R. Meteorol. Soc. (e.a.) **106**, 159 (1980)
13. M.A. Uman, *Lightning* (McGrow Hill, New York, 1969)
14. V.A. Rakov, M.A. Uman, *Lightning, Physics and Effects* (Cambridge University Press, Cambridge, 2003)
15. I.M. Lifshitz, V.V. Slezov, Sov. Phys. JETP **35**, 331 (1958)
16. E.M. Lifshitz, L.P. Pitaevskii, *Physical Kinetics* (Pergamon Press, Oxford, 1981)
17. I.M. Lifshitz, V.V. Slezov, Fiz. Tver. Tela **1**, 1401 (1959)
18. I.M. Lifshitz, V.V. Slezov, J. Phys. Chem. Sol. **19**, 35 (1961)
19. V.V. Slezov, V.V. Sagalovich, Sov. Phys. Usp. **30**, 23 (1987)
20. S. Ino, J. Phys. Soc. Jpn. **27**, 941 (1969)
21. D.R. Lide (ed.), *Handbook of Chemistry and Physics*, 86th edn. (CRC Press, London, 2003–2004)
22. M.V. Smolukhowski, Z. Phys. **17**, 585 (1916)
23. N.B. Vargaftik, *Tables of Thermophysical Properties of Liquids and Gases*, 2nd edn. (Halsted Press, New York, 1975)
24. L.D. Landau, E.M. Lifshitz, *Mechanics* (Pergamon Press, London, 1980)
25. P. Langevin, Ann. Chem. Phys. **8**, 245 (1905)
26. W.D. Rust, C.B. Moore, Q.J.R. Meteorol, Soc. **100**, 450 (1974)

27. B.M. Smirnov, Phys. Usp. **184**, 1153 (2014)
28. C.F. Bohren, *Clouds in a Glass of Beer* (Wiley, New York, 1987)
29. C.P.R. Saunders, *Planetary Atmospheric Electricity*, p. 335, ed. by F. Leblanc et al. (Springer, Heidelberg, 2008)
30. J. Frenkel, *Theory of Phenomena of Atmospheric Electricity* (Gostekhizdat, Leningrad-Moscow, 1949). (in Russian)
31. S.E. Reynolds, M. Brook, M.F. Gourley, J. Met. **14**, 426 (1957)
32. J. Latham, B.J. Mason, Proc. R. Soc. **260A**, 523 (1961)
33. J. Latham, B.J. Mason, Proc. Roy. Soc. **260A**, 537 (1961)
34. E.R. Javaratne, C.P.R. Saunders, J. Hallett, Q. J. R. Meteorol. Soc. **109**, 609 (1983)
35. D.R. MacGorman, Bull. Am. Meteorol. Soc. (e.a.) **89**, 997 (2008)
36. G.C. Simpson, Phil. Trans. Roy. Soc. **A209**, 379 (1909)
37. C.T.R. Wilson, Proc. R. Soc. **92A**, 555 (1916)
38. C.T.R. Wilson, Proc. R. Soc. **37A**, 32D (1925)
39. G.C. Simpson, Proc. R. Soc. **A114**, 376 (1927)
40. M. Stolzenburg, W.D. Rust, B.F. Smull, T.C. Marshall, J. Gephys. Res. **103**, 14059 (1998)
41. M. Stolzenburg, W.D. Rust, T.C. Marshall, J. Gephys. Res. **103**, 14078 (1998)
42. M. Stolzenburg, W.D. Rust, T.C. Marshall, J. Gephys. Res. **103**, 14097 (1998)
43. M. Brook, T. Ogawa, *The Cloud Discharge, in "Lightning"*, p. 191, ed. by R.H. Golde (Academic Press, London, 1977)
44. http://en.wikipedia.org/wiki/water-vapor
45. http://news.nationalgeographic.com/news/2008/05/photogalleries/volcano-photos
46. http://www.theatlantic.com/infocus/2011/06/chiles-puyehue-volcano-erupts/10081/
47. C.B. Moore, B. Vonnegut, *Point-discharge Currents, in "Lightning"*, p. 51, ed. by R.H. Golde (Academic Press, London, 1977)
48. http://en.wikipedia.org/wiki/St.-Elmo' s-fire
49. http://www.bugaga.ru/pictures/1146729812-oblaka-v-fotografiyah.html
50. http://en.wikipedia.org/wiki/Atmosphere-of-Earth
51. http://en.wikipedia.org/wiki/Properties-of-water
52. http://en.wikipedia.org/wiki/Troposphere
53. http://ru.fishki.net/picsw/042011/27/post/mir/mir-029.jpg
54. http://en.wikipedia.org/wiki/Rain
55. http://en.wikipedia.org/wiki/Convection
56. W.R. Cotton, R.A. Anthes, *Storm and Cloud Dynamics* (Academic Press, San Diego, 1989)
57. K.A. Emanuel, *Atmospheric Convection* (Oxford University Press, New York, 1991)
58. R.A. Houze, *Cloud Dynamics* (Academic Press, San Diego, 1993)
59. B.J. Mason, *The Physics of Clouds* (Claredon Press, Oxford, 1971)
60. P.R. Krehbiel, *Studies in Geophysics*, pp. 90–113 (National Academy Press, Washington, 1986)
61. B.J. Mason, Q. J. R. Soc. **79**, 501 (1953)
62. J.M. Straka, *Clouds and Precipitation* (Butterworth, Woburn, 1989)
63. R.R. Rogers, M.K. Yau, *A Short Course in Cloud Physics* (Pergamon Press, Oxford, 1989)
64. W. Cotton, G. Bryan, S. van den Heever (eds.), Storm and cloud dynamics. Int. Geophys. **99**, 1–809 (2011)

Part III
Radiative Processes in the Earth Atmosphere

Chapter 9
Photoionization of Atmospheric Processes

Abstract The relation is obtained between the cross sections of photoprocesses and the altitude where corresponding photons penetrate. Parameters of photoprocesses involving vacuum ultraviolet photons and oxygen atmospheric components (O, O_2, O_3) are represented. The penetration depth for a photon of a given frequency is determined on the basis of the absorption cross section of corresponding processes. General principles of air ionization in the Earth's atmosphere under the action of solar radiation are analyzed. Penetration of X-rays from solar corona in the Earth's atmosphere is considered as a possible channel of the ionization process under cosmic rays.

9.1 Photoionization in Upper Atmosphere

An upper atmosphere at altitudes above 80 km contains electrons as a negative charge (see Fig. 3.1) that results from absorption of the vacuum ultraviolet radiation from the Sun. Figure 9.1 gives the altitudes below which solar radiation of a given wavelength does not penetrate. Note that the wavelength 102.6 nm corresponds to the threshold of the process [1]

$$O_2 + \hbar\omega \rightarrow O_2^+ + e \tag{9.1}$$

and the wavelength 98.0 nm is the threshold for ionization of nitrogen molecules [2]

$$N_2 + \hbar\omega \rightarrow N_2^+ + e, \tag{9.2}$$

and these values are close to the ionization potentials of these molecules. Indeed, the ionization potential of the nitrogen molecule N_2 is 15.58 eV that corresponds to the threshold wavelength $\lambda = 79.6$ nm and the ionization potential of the oxygen molecule O_2 is equal to 12.88 eV and corresponds to the threshold wavelength $\lambda = 102.6$ nm. Evidently, in the nitrogen case excitation of the nitrogen molecule leads to subsequent ionization in collisions with other molecules. Next, the ionization potential of the oxygen atom O is 13.61 eV and relates to the threshold wavelength $\lambda = 91.1$ nm, whereas the ionization potential of the nitrogen atom N 14.54 eV corresponds to the threshold wavelength $\lambda = 85.3$ nm. The above components may be

B.M. Smirnov, *Microphysics of Atmospheric Phenomena*,
Springer Atmospheric Sciences, DOI 10.1007/978-3-319-30813-5_9

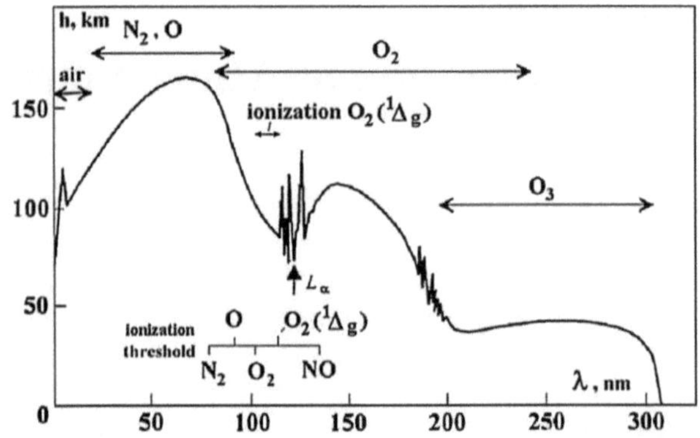

Fig. 9.1 Boundary altitudes for absorption of solar ultraviolet radiation at various altitudes and air components responsible for its absorption [3]

responsible for air ionization at altitudes above 100 km. In the following we consider the character of photoionization at penetration of short-range radiation in the atmosphere.

Absorption of solar radiation may change the altitude distribution of molecules. Indeed, this process is not essential for this distribution, however, the number density of atoms or molecules of a given sort $N(h)$ depending on an altitude h is given by the Boltzmann formula which may be approximated by Formula (2.3)

$$N(h) = N_o \exp\left(-\frac{h}{\Lambda_N}\right), \tag{9.3}$$

where the length Λ_N is introduced according to Formula (2.1). Let us consider the absorption process from the standpoint of incident photons. Let $I_\omega(h)$ be the intensity of incident photons with a frequency ω. Then the following equation describes photon penetration in the atmosphere

$$\frac{dI_\omega(h)}{dh} = -I_\omega \sigma_\omega N(h), \tag{9.4}$$

where h is the altitude, $N(h)$ is the number density of given molecules at an altitude h, and σ_ω is the photoionization cross section at a given frequency. Solution of this equation is

$$\ln\left[\frac{I_\omega(h)}{I_o}\right] = -\Lambda_N \sigma_\omega N(h), \tag{9.5}$$

where $I_o = I_\omega(\infty)$. Let us introduce the parameter h_o as

$$\Lambda_N \sigma_\omega N(h_o) = \ln 2, \tag{9.6}$$

and (9.5) takes the form

$$P(h, \omega) = \exp\left[-\ln 2 \cdot \exp(-x)\right], \tag{9.7}$$

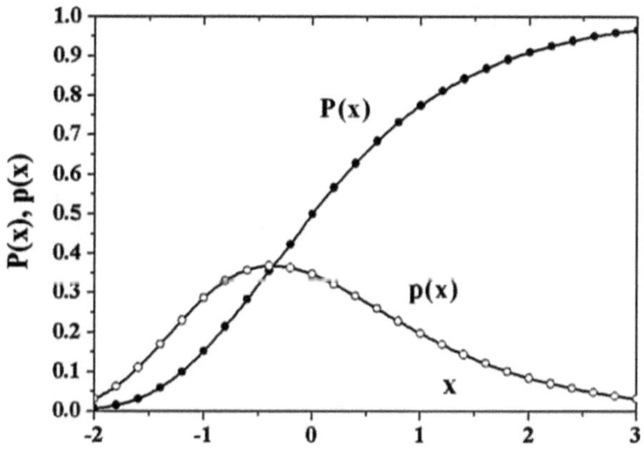

Fig. 9.2 Dependence of the surviving probability $P(x)$ for solar photons in the Earth's atmosphere and the derivative of this function $p(x)$ on the reduced altitude $x = (h - h_o)/\Lambda_N$

where $P(h, \omega) = P(x, \omega) = I_\omega(h)/I_o$ is the probability of photon surviving, and $x = [h - h_o(\omega)]/\Lambda_N$ is the reduced altitude. Figure 9.2 gives the dependence of the probability of photon surviving $P(x)$ as a function of the reduced altitude. As can be seen, the surviving probability is equal to 0.1 at the reduced altitude $x = -1.2$ and $P(x) = 0.9$ at the reduced altitude $x = 2$. From this it follows that photons of this frequency are absorbed mostly in the range of altitudes $\Delta h = 3.2\Lambda_N$. Along with the probability of surviving, Fig. 9.2 contains the derivative of this function $p(x) = dP(x)/dx$. Note that (9.6) for the altitude, where the strongest absorption takes place, may be represented in the form

$$\sigma_\omega \int_{h_o}^{\infty} N(h)dh = \ln 2 \qquad (9.8)$$

Solar radiation with wavelengths below 102.5 nm may lead to photoionization in the upper atmosphere. This radiation is emitted in the solar corona and the average flux of photons in this spectrum range is about $2.4 \times 10^{10}\,\mathrm{cm^{-2}s^{-1}}$. The significant part of the absorption spectrum is caused by the Lyman-series transitions of atomic hydrogen, including the Rydberg spectrum and the continuum. The contribution to the photon flux in the spectral range from 84 to 103 nm is about $1.3 \times 10^{10}\,\mathrm{cm^{-2}s^{-1}}$ on average. A contribution to this spectrum range also follows from the transition of *C III* at 99.1 nm with the photon flux $9 \times 10^{8}\,\mathrm{cm^{-2}s^{-1}}$, the transition of *C III* at 97.7 nm with the photon flux $4.4 \times 10^{9}\,\mathrm{cm^{-2}s^{-1}}$, the transition of an oxygen ion *O V* at 63 nm with the photon flux $1.3 \times 10^{9}\,\mathrm{cm^{-2}s^{-1}}$, the transition of *He I* at 58.4 nm with $1.3 \times 10^{9}\,\mathrm{cm^{-2}s^{-1}}$, and Lyman transition of *He II* at 30.4 nm with the flux $7.7 \times 10^{9}\,\mathrm{cm^{-2}s^{-1}}$. In addition, the maximum rate constant for the generation of electrons as a result of photoionization processes occurs at an altitude of about 160 km and is $4 \times 10^{15}\,\mathrm{cm^{-3}s^{-1}}$.

Note that the maximum ionization takes place at the altitude 160 km, and the number density of oxygen molecules varies from $2 \times 10^{12}\,\mathrm{cm}^{-3}$ at an altitude 100 km up to $2 \times 10^{8}\,\mathrm{cm}^{-3}$ at an altitude of 200 km (see Fig. 2.6). This corresponds to a typical length $\Lambda_N = 6 - 40$ km according to Fig. 2.5. Taking the total number density of molecules at the altitude 160 km to be $N = 1 \times 10^{11}\,\mathrm{cm}^{-3}$, we obtain from formula (9.6) for a typical cross section of photoionization $\sigma_\omega \approx 5 \times 10^{-18}\,\mathrm{cm}^2$. The photoionization cross section in the wavelength range is $\sigma_\omega \sim 1 \times 10^{-18}$–$1 \times 10^{-17}\,\mathrm{cm}^2$ at wavelengths of 130–170 nm where ionization proceeds under the action of radiation.

9.2 Absorption of Solar Radiation in Atmosphere

The above analysis gives general principles of absorption of solar radiation propagated through the Earth's atmosphere. With an altitude decrease, the absorption of solar radiation for a given wavelength increases because of increasing air density, and total absorption takes place in layers according to (9.4) where the maximum degree of absorbed photons at a given frequency is observed. We now determine the total rate of absorbed intensity at a given point integrated over photon frequencies. We have for the total power of an absorbed photon energy $\Phi(h)$ at a given altitude h as

$$\Phi(h) = \int_0^\infty d\omega I_\omega p(h,\omega), \;\; p(h,\omega) = \frac{dP(h,\omega)}{dh}, \tag{9.9}$$

and we use (9.4) according to which

$$p(h,\omega) = \frac{dP(h,\omega)}{dx} = -P(h,\omega)\Lambda_N \sigma_\omega N(h) \tag{9.10}$$

Figure 9.2 contains the derivation of the probability of photon surviving $p(x,\omega)$ as a function of the reduced altitude. Let us take for definiteness the dependence of the absorption cross section σ_ω on the photon frequency ω as

$$\sigma_\omega = \sigma_{\omega_o}\left(\frac{\omega_o}{\omega}\right)^\gamma \tag{9.11}$$

Let us assume that absorption at a given space point h is centered around the frequency ω_o in accordance with Formula (9.6)

$$\Lambda_N \sigma_\omega N[h_o(\omega_o)] = \ln 2,$$

and $h = h(\omega_o)$. On the basis of Formulas (9.3) and (9.6) we have

$$\frac{h_o(\omega) - h_o(\omega_o)}{\Lambda_N} = \ln\left(\frac{\omega_o}{\omega}\right)^\gamma = \gamma \ln\left(\frac{\omega_o}{\omega}\right) \tag{9.12}$$

Let us take the frequency dependence for intensity of incident radiation I_ω in the form

$$I_\omega = I_{\omega_o} \cdot \left(\frac{\omega_o}{\omega}\right)^\alpha \tag{9.13}$$

This leads to the rate of absorbed radiation at an altitude h

$$\Phi(h) = \Phi_o \int_0^\infty \frac{d\omega}{\omega_o} \left(\frac{\omega_o}{\omega}\right)^\alpha P(h,\omega)\sigma_\omega N[h_o(\omega)] \exp\left[-\frac{h - h_o(\omega)}{\Lambda_N}\right] \tag{9.14}$$

where

$$\Phi_o = \frac{I_{\omega_o}\omega_o}{\Lambda_N} \tag{9.15}$$

From the relation

$$\sigma_\omega \cdot N[h_o(\omega)] = \sigma_{\omega_o} \cdot N[h_o(\omega_o)] \tag{9.16}$$

we have

$$\frac{h_o(\omega) - h_o(\omega_o)}{\Lambda_N} = \ln\left[\frac{\sigma_\omega}{\sigma_{\omega_o}}\right] \tag{9.17}$$

Taking the frequency dependence for the absorption cross section according to Formula (9.11), we obtain from Formula (9.17)

$$P(h,\omega) = \exp\left[-\ln 2\left(\frac{\omega_o}{\omega}\right)^\alpha\right] \tag{9.18}$$

Substituting this relation in Formula (9.14), we obtain

$$\Phi(h) = \Phi_o(\ln 2)^{1-(1-\alpha)/\gamma}\Gamma\left(\frac{1-\alpha}{\gamma}\right) \tag{9.19}$$

Let us represent this dependence in the form

$$\Phi(h) = \Phi_o F(\alpha,\gamma) \tag{9.20}$$

Function $F(\alpha,\gamma)$ is represented in Fig. 9.3.

The above analysis allows us to determine an altitude where solar radiation of a given frequency is absorbed according to Formula (9.4) and is given in Fig. 9.4. In the first case it is assumed that only oxygen molecules partake of the absorption process; in the second case we assume for simplicity that the absorption cross sections for oxygen and nitrogen molecules are identical.

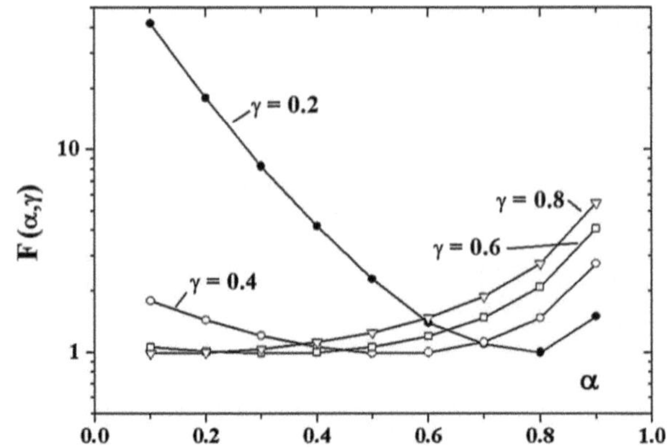

Fig. 9.3 Function $F(\alpha, \gamma)$ in accordance with Formulas (9.19) and (9.20)

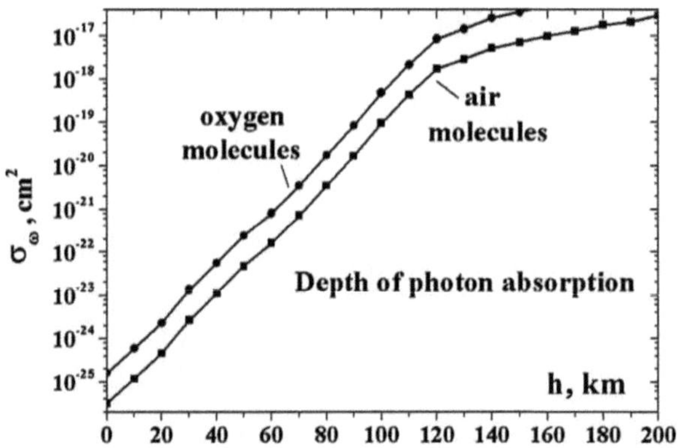

Fig. 9.4 Absorption cross section at a given frequency for solar radiation if this radiation is directed perpendicular to the Earth's surface and is absorbed mostly at a given altitude according to (9.4). Solar radiation may be absorbed by oxygen molecules only (the *upper curve*) or air molecules if we assume the absorption cross section to be identical for nitrogen and oxygen molecules (the *lower curve*)

9.3 Excited Atoms and Molecules in Atmospheric Air

As follows from the above analysis, absorption of solar radiation at a given altitude is determined by an appropriate cross section for a given altitude in accordance with Fig. 9.1. As follows from Fig. 9.1, absorption of solar radiation at atmosphere altitudes below 100 km is determined by oxygen and ozone molecules with the wavelengths below the ionization threshold of these molecules; that is, this absorption leads to dissociation of these molecules. The absorption cross sections for molecules and atoms as basic air components in the ionosphere are measured and represented in [4–7] and other papers and books. In particular, Fig. 9.5a contains the cross section

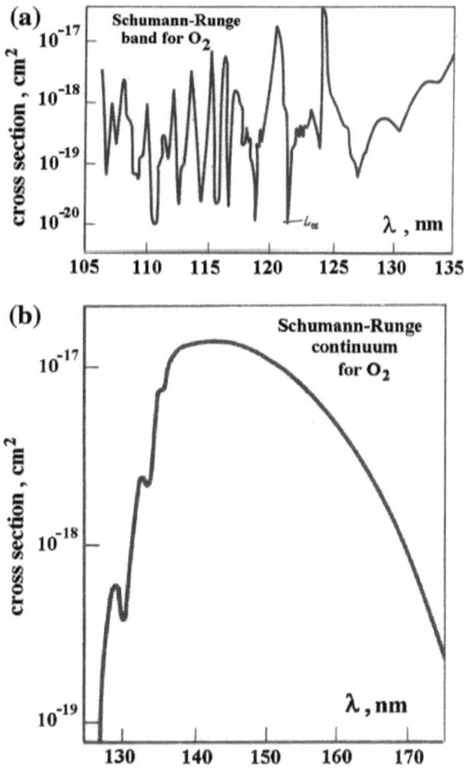

Fig. 9.5 Absorption cross section of the oxygen molecule for the range of the Schumann–Runge band (**a**) and Schumann–Runge continuum (**b**) according to measurements [8]

in the Schumann–Runge band that corresponds to radiative transitions $X^3\Sigma_g^- \to C^3\Delta_u$, $X^3\Sigma_g^- \to A^3\Sigma_u^+$, $X^3\Sigma_g^- \to B^3\Sigma_u^-$, and Fig. 9.5b gives the cross section in the Schumann–Runge continuum. These transitions include both transitions in bound states and in states of continuous spectra.

The process of absorption of molecular oxygen by solar radiation leads to formation of atomic oxygen which becomes an important component of the atmosphere. In particular, the Schumann–Runge continuum in the spectral range of 132–176 nm (the photon energy lies in the range of 6–10.3 eV) is described by the process

$$\hbar\omega + O_2 \to O(^1D) + O(^3P) \qquad (9.21)$$

As shown, along with formation of oxygen atoms in the ground state $O(^3P)$, this process leads to formation of excited oxygen atoms $O(^1D)$ that are of importance in atmospheric chemistry. The Schumann–Runge continuum is responsible for generation of atomic oxygen at altitudes higher than 120 km. At lower altitudes, up to $h \approx 8$ km, the generation of atomic oxygen is accomplished mostly by dissociation of oxygen molecules due to the weak Herzberg continuum in the wavelength range of 200–250 nm, where the absorption cross section is 10^{-24}–10^{-23} cm^{-3} [9, 10].

Oxygen atoms are formed as a result of photodissociation of oxygen molecules that attach subsequently to oxygen molecules in lower atmospheric layers, so that ozone molecules are formed. Ozone molecules located in the atmosphere absorb a remarkable part of ultraviolet solar radiation and this process may lead to the formation of excited oxygen atoms and molecules. In particular, the threshold wavelength for the process

$$\hbar\omega + O_3 \to O(^1D) + O_2(^1\Delta_g) \tag{9.22}$$

is 310 nm. Hence, absorption of solar radiation in the Earth's atmosphere may be accompanied by formation of electronically excited atoms and molecules, and these atomic particles determine the atmospheric chemistry to a great degree.

Above we consider dissociated air formed under action of an electric field in the lightning channel where air is transformed in an equilibrium plasma of a high temperature. We now consider another situation in the upper Earth atmosphere where air is excited under the action of solar radiation. Absorption of solar radiation leads to formation of radicals and excited components in the atmosphere and also results in partial dissociation of atmospheric air. In the first place, absorption of solar radiation is determined by oxygen components. Moreover, as a result of these processes atomic oxygen is formed, and in the ionosphere, that is, at altitudes above 100 km, the number density of oxygen atoms exceeds that of oxygen molecules (see Fig. 2.6). Along with atomic and molecular oxygen, atmospheric air contains ozone, and these three oxygen components absorb a basic part of solar radiation. Therefore below we only consider processes with participation of oxygen components.

Ozone O_3 is an important atmospheric component that absorbs ultraviolet radiation and prevents the Earth from penetration of this radiation on its surface. Absorption of atmospheric ozone proceeds in the Chappins band with wavelengths 400–600 nm and in the Hartley band with wavelengths below 320 nm. The cross section of absorption in the Hartley band is represented in Fig. 9.6 and according to (9.22) absorption by an ozone molecule with photon wavelengths below 300 nm may lead to formation of an excited oxygen atom $O(^1D)$. Indeed, Fig. 9.7 represents

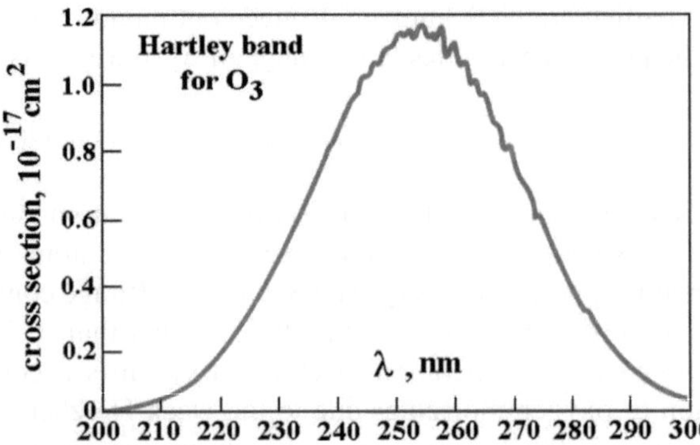

Fig. 9.6 Absorption cross section of the ozone molecule in the range of the Hartley band and continuum according to measurements [11]

Fig. 9.7 Quantum yield of excited oxygen atoms $O(^1D)$ as a result of photodissociation of an ozone molecule: ▲—[12], *dark circles*—[13], ▽, □, △—[14] at temperatures $T = 221, 258$ and 293K correspondingly, *solid curve*—[15], *dotted curve*—[16], *open circles*—[17], +—[18]

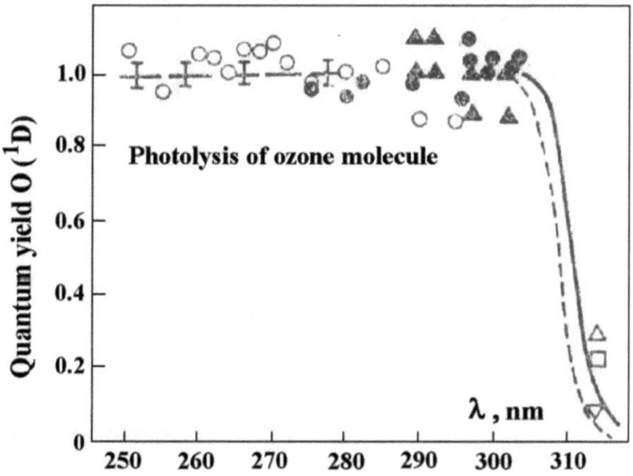

the quantum yield of photolysis of the ozone molecule that confirms formation of excited oxygen atoms as a result of absorption of an ozone molecule in the Hartley band region. Note that excited atoms are characterized by a heightened chemical activity and therefore the air chemistry is complicated in the region of their formation.

The ozone density and the number density of its molecules vary significantly depending on the geographic coordinates, season, and time of day. Averaging over these parameters, one can obtain that approximately 97 % of ozone is located between altitudes 15 and 55 km [19–22]. The maximum number density of ozone molecules about $5 \times 10^{12}\,\mathrm{cm}^{-3}$ corresponds to the altitude range 20−25 km, and the maximum concentration of ozone molecules $c_{max} \sim 10^{-5}$ corresponds to the altitude of approximately 35 km. Next, the total number of ozone molecules in a column over the Earth is $\int N(O_3)dh \approx (2-3) \times 10^{19}\,\mathrm{cm}^{-3}$ ($N(O_3)$ is the number density of ozone molecules, and h is the altitude). On the basis of data of Fig. 9.6 for the absorption cross section of an ozone molecule one can conclude that the optical depth of an atmosphere at wavelengths about 250 nm is equal to a few hundreds. The Earth's atmosphere is optically thick also at wings of the Hartley band, and therefore ozone protects the Earth's surface from the longwave part of ultraviolet radiation.

The radiation flux from the Sun in the range from 115 up to 310 nm is given in Fig. 9.8 at a distance from the Sun that corresponds to the Sun-Earth distance [23, 24]. When this radiation penetrates inside the Earth's atmosphere, it is absorbed by atmospheric ozone. Integration over wavelengths gives the radiation power per unit area in the ultraviolet spectrum range that is about 1 % from the total specific power of solar radiation penetrating inside the Earth's atmosphere. In spite of a small part of this ultraviolet radiation, it is important because of a strong action on living organisms.

As a result of absorption of ultraviolet radiation by ozone molecules, excited oxygen atoms and molecules are formed. According to measurements [12, 15–18, 25], the process (9.22) with formation of excited oxygen atoms $O(^1D)$ and excited molecules $O_2(a^1\Delta_g)$ proceeds. If the wavelength of radiation lies in the

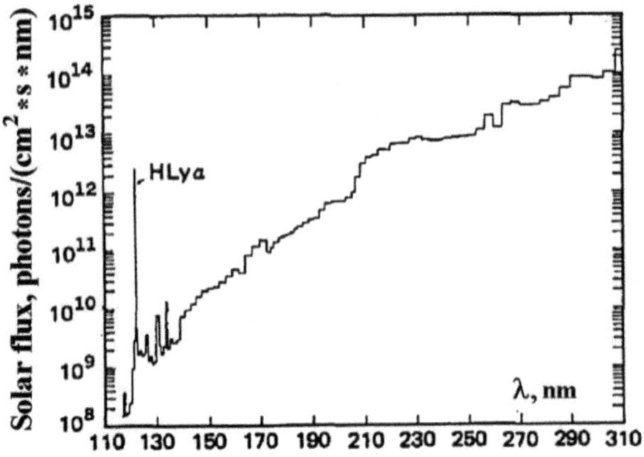

Fig. 9.8 Spectrum of the energy flux for solar photons at Earth level [23, 24]

range 254–310 nm, and if the wavelength is less, excited molecules $O_2(b^1\Sigma_g^+)$ and atoms $O(^1D)$ may be formed. Note that the threshold wavelength for the latter process is equal to 260 nm. Absorption of solar radiation by ozone molecules and subsequent collision processes involving air molecules lead to formation of metastable atoms and molecules with a large radiative lifetime. These atoms and molecules have the same structure of the electron shell as those in the ground electron state. Therefore the radiative transition in the ground state is forbidden and the rate of radiative transitions is small compared to dipole radiative transitions. Radiation parameters for metastable states of oxygen and nitrogen atoms and molecules are given in Table 9.1.

Because of the large lifetime of metastable atoms and molecules, their quenching in a dense atmosphere results from collisions with molecules of atmospheric air. In Tables 9.2 and 9.3 we give the rate constants of quenching of nitrogen and oxygen metastable atoms and molecules in collisions with air molecules. As a result, the maximum of the number densities for metastable atoms or molecules corresponds to high altitudes with low number density of quenched molecules. Next, according to the data of Tables 9.2 and 9.3, the rate constant of quenching for metastable molecules

Table 9.1 Radiative lifetime of metastable nitrogen and oxygen atoms and oxygen molecules [26–31]

Atom, molecule	Excitation energy, eV	Lower state of transition	λ, nm	τ, s
$N(^2D_{5/2})$	2.38	$N(^4S)$	520.2	1.4×10^5
$N(^2D_{3/2})$	2.38	$N(^4S)$	519.9	6.1×10^4
$N(^2P)$	3.58	$N(^2D_{5/2,3/2})$	1040, 1041	12
$O(^1D)$	1.97	$O(^3P)$	630	140
$O(^1S)$	4.19	$O(^1D)$	577.7	0.8
$O_2(a^1\Delta_g)$	0.98	$O_2(X^3\Sigma_g^-)$	1270	3×10^3
$O_2(b^1\Sigma_g^+)$	1.64	$O_2(X^3\Sigma_g^-)$	761.9	12

Table 9.2 Rate constants of quenching of metastable atoms and molecules in collisions with nitrogen molecules at room temperature

Atom,molecule	k_q, cm^3/s
$N(^2D)$	$(1.5 \pm 0.5) \times 10^{-14}$ [33–36]
$N(^2P)$	6×10^{-14} [34]
$O(^1D)$	$(6 \pm 2) \times 10^{-11}$ [37–42]
$O(^1S)$	$<1 \times 10^{-17}$ [43]
$O_2(a^1\Delta_g)$	$<3 \times 10^{-21}$ [44]
$O_2(b^1\Sigma_g^+)$	$(1.9 \pm 0.3) \times 10^{-18}$ [45–53]

Table 9.3 Rate constants of quenching of metastable atoms and molecules in collisions with oxygen molecules at room temperature

Atom, molecule	k_q, cm^3/s
$N(^2D)$	$(8 \pm 3) \times 10^{-12}$ [33–36, 54]
$N(^2P)$	$(4 \pm 1) \times 10^{-12}$ [35, 36]
$O(^1D)$	$(5 \pm 2) \times 10^{-11}$ [37, 38, 40, 42, 55–57]
$O(^1S)$	$(3 \pm 1) \times 10^{-13}$ [30, 33, 43, 45, 48, 58, 59]
$O_2(a^1\Delta_g)$	$(1.9 \pm 0.3) \times 10^{-18}$ [44, 48, 52, 60–62]
$O_2(b^1\Sigma_g^+)$	$\sim 1 \times 10^{-16}$ [37, 45, 47, 48, 51, 52, 63]

$O_2(a^1\Delta_g)$ by oxygen and nitrogen molecules is lower than that for other metastable atoms. This means that the number density of these metastable molecules in a dense atmosphere is higher than that for other metastable atomic particles in atmospheric air and, on the other hand, the maximum at middle altitudes. As follows from Fig. 9.9 [32], where a typical distribution of the number density of oxygen molecules over the altitude is given, the maximum in the number density corresponds to altitudes 50−60 km. At these altitudes one can ignore quenching of metastable molecules in collisions with electrons. In addition, the number density of metastable oxygen molecules $O_2(a^1\Delta_g)$ in the atmosphere depends on the latitude and zenith angle. Hence, one can consider the data of Fig. 9.9 as estimates. In particular, the maximum number density of these molecules N_* is $N_* \sim 10^{10}\,cm^{-3}$, and a typical number of these molecules per atmospheric column is $\int N_*(h)dh \sim 10^{16}\,cm^{-2}$.

The data in Fig. 9.9 for a typical altitude profile of the number density of metastable oxygen molecules $O_2(a^1\Delta_g)$ was measured in Reference [32] in the Canadian atmosphere. It should be noted that variations in the atmospheric number densities of metastable oxygen molecules $O_2(a^1\Delta_g)$ are larger than those for other excited atomic particles in atmospheric air, but depend on conditions (season, time of day, geographical coordinates). Hence, Fig. 9.9 indicates only the order of magnitude for this value. Nevertheless, Fig. 9.9 exhibits that the number density is compared with that for ozone molecules at high altitudes. Therefore these molecules may take part in the absorption of solar radiation.

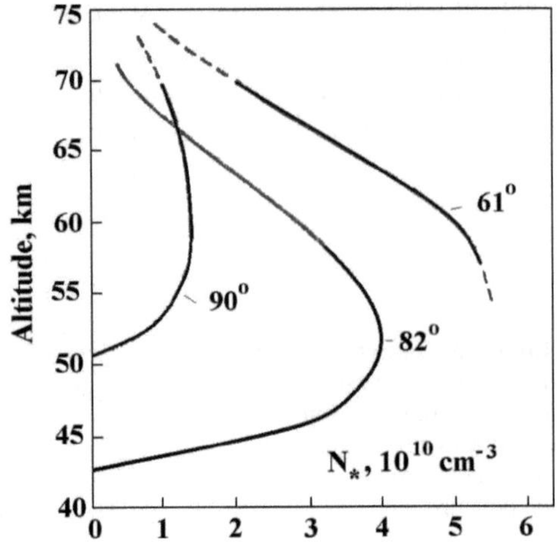

Fig. 9.9 Altitude distribution for the number density N_* of metastable molecules $O_2(a^1\Delta_g)$ in a daytime atmosphere at different zenith angles of the Sun [32]

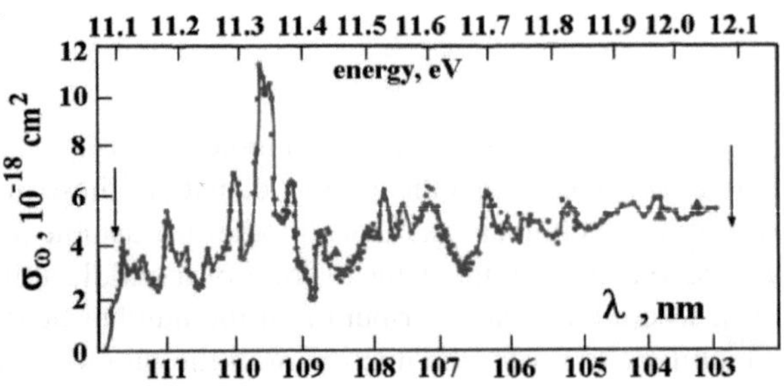

Fig. 9.10 Cross section of absorption of vacuum ultraviolet radiation by metastable oxygen molecules $O_2(a^1\Delta_g)$ according to measurements [64]

Figure 9.10 [64] gives the cross section of photoionization for metastable oxygen molecules $O_2(a^1\Delta_g)$, and this process gives an introduction to absorption of solar radiation in the thermosphere and stratosphere at altitudes below 60 km. Indeed, taking the absorption cross section by these molecules to be $\sigma_\omega \sim 5 \times 10^{-18}\,\text{cm}^2$, as follows from Fig. 9.10, one can obtain for the optical density of the atmosphere $u_\omega = \sigma_\omega \int N_*(h)dh/\cos\alpha$, where α is the incident angle for solar radiation. As shown, the optical density due to ionization of oxygen molecules $O_2(a^1\Delta_g)$ is $u_\omega \sim 1$. Next, metastable atoms and molecules of nitrogen and oxygen located in the atmosphere influence their optical characteristics. In particular, radiation emitted by metastable oxygen molecules $O_2(a^1\Delta_g)$ as a result of the process

$$O_2(a^1\Delta_g) \rightarrow O_2(X^3\Sigma_g^+) + \hbar\omega \qquad (9.23)$$

gives an essential contribution to radiation of the daytime atmosphere [65–67]. This band of infrared radiation is concentrated near the wavelength 1.27 μm, and its intensity is equal to about 2×10^7 Ra [65, 68] ($1\text{Ra} = 10^6$ photons/(cm$^2 \cdot$ s)). Radiation of metastable atoms gives a contribution to nighttime radiation.

In addition, these radiative transitions determine the glowing of an aurora. Namely, a red aurora color follows from the transition $O(^1D) \to O(^3P) + \hbar\omega$, whereas its green color is caused by transitions $N(^2D) \to N(^4S) + \hbar\omega$ and $O(^1S) \to O(^3P) + \hbar\omega$. An aurora occurs when a flow of protons of the solar wind penetrates into the atmosphere. The charge exchange of these protons and their deceleration takes place in the E- and F-layers of the ionosphere. The subsequent chain of elementary processes leads to formation of excited atoms whose radiation is observed from the Earth as an aurora. Because of a large radiative lifetime of metastable atoms and molecules, their radiation occurs at high altitudes only because of quenching in collisions with air molecules at low altitudes. The probabilities for an excited atom to radiate and to be quenched are determined by the parameter $(k\tau)^{-1}$, where τ is the radiative lifetime of an excited atom, and k is the quenching rate constant. An excited oxygen atom $O(^1S)$ is quenched in the upper atmosphere mainly by atomic oxygen with the rate constant of $k = 7 \times 10^{-12}\,\text{cm}^3/\text{s}$, and the threshold number density of oxygen atoms is $2 \times 10^{11}\,\text{cm}^{-3}$ which corresponds to an altitude approximately 115 km. In the case of $O(^1D)$ the main quenching mechanism proceeds with participation of nitrogen molecules, and its rate constant of the quenching process is approximately $k = 6 \times 10^{-11}\,\text{cm}^3/\text{s}$, and the threshold number density of nitrogen molecules is approximately $10^8\,\text{cm}^{-3}$. If the number density of quenching particles is lower than the above values, radiation of the corresponding excited atoms may be significant, and this relates to the atmosphere altitude of about 170 km.

References

1. P.M. Dehmer, W.A. Chupka, J. Chem. Phys. **62**, 4525 (1975)
2. V.L. Carter, J. Chem. Phys. **56**, 4195 (1972)
3. H. Friedman, in *Physics of the Upper Atmosphere*, ed. by J.A. Rateliffe (Academic Press, New York, 1960)
4. P.W. Langhoff, A. Gerwer, C. Asaro, B.W. McKoy, Int. J. Quantum Chem. **16, S13**, 645 (1979)
5. R.M. Goody, Y.L. Yung, *Atmospheric Radiation: Theoretical Basis* (Oxford University Press, New York, 1989)
6. J.A. Fennelly, D.G. Torr, At. Data Nucl. Data Tables **51**, 321 (1992)
7. J. Berkowitz, J. Phys. B **30**, 583 (1997)
8. K. Watanabe, E.C.Y. Inn, M. Zelikoff, J. Chem. Phys. **21**, 1026 (1953)
9. R.W. Ditchburn, P.A. Young, J. Atmos. Terr. Phys. **24**, 127 (1962)
10. V. Hasson, R.W. Nicholls, J. Phys. B **4**, 1789 (1971)
11. M. Griggs, J. Chem. Phys. **49**, 857 (1968)
12. I.T.N. Jones, R.P. Wayne, Proc. R. Soc. **391A**, 870 (1970)
13. C.L. Lin, W.B. DeMore, J. Photochem. **2**, 161 (1973)
14. S. Kuis, R. Simonaitis, J. Heicklen, J. Geophys. Res. **80**, 1328 (1975)
15. I. Arnold, F.J. Comes, G.K. Moortgat, Chem. Phys. **2**, 211 (1977)
16. D.L. Philen, R.T. Watson, D.D. Davis, J. Chem. Phys. **67**, 3316 (1977)
17. P.W. Fairchild, E.K.C. Lee, Chem. Phys. Lett. **60**, 36 (1978)
18. O. Kajimoto, R.J. Cvetanovic, Int. J. Chem. Kinet. **11**, 605 (1979)
19. M.L. Salby, *Fundamentals of Atmospheric Physics* (Academic Press, San Diego, 1996)
20. A.J. Krueger, R.A. Minznev, J. Geophys. Res. **84**, 6969 (1976)
21. H.U. Dütsch, Pure Appl. Geophys. **116**, 511 (1978)

22. J.H. Seinfeld, S.N. Pandis, *Atmospheric Chemistry and Physics* (Wiley, Hoboken, 2006)
23. U.S. Standard Atmosphere (U.S. Government Printing Office, Washington, 1976)
24. M. Ackerman, in *Mesospheric Models and Related Experiments*, ed. by G. Fiocco (Springer, New York, 1971)
25. P.M. Scott, R.J. Cvetanovic, J. Chem. Phys. **54**, 1440 (1971)
26. R.H. Garstang, J. Res. Natl. Bur. Stand. **68A**, 61 (1964)
27. E.C. Zipf, Can. J. Chem. **47**, 1863 (1969)
28. C. Nicolaides, O. Sinanoglu, P. Westhaus, Phys. Rev. **4A**, 1400 (1971)
29. A. Corney, O.M. Williams, J. Phys. **5B**, 686 (1972)
30. K. Schofield, J. Photochem. **9**, 55 (1978)
31. K. Schofield, J. Phys. Chem. Ref. Data **8**, 723 (1979)
32. J.C. Haslett, L.R. Megill, H.I. Schiff, Can. J. Phys. **47**, 235 (1969)
33. G. Black et al., J. Chem. Phys. **51**, 116 (1969)
34. C.L. Lin, F. Kaufman, J. Chem. Phys. **55**, 3760 (1971)
35. D. Husain, L.J. Kirsch, J.R. Wiesenfeld, Faraday Discuss. Chem. Soc. **53**, 201 (1972)
36. D. Husain, S.K. Mitra, A.N. Young, J. Chem. Soc. Faraday Discuss. Part II **70**, 1721 (1974)
37. J.F. Noxon, J. Chem. Phys. **52**, 1852 (1970)
38. R. Gilpin, H.I. Schiff, K.H. Welge, J. Chem. Phys. **55**, 1087 (1971)
39. R.F. Hampson et al., J. Phys. Chem. Ref. Data. **2**, 267 (1973)
40. R.F. Heidner, D. Husain, J. Wiesenfeld, J. Chem. Soc. Faraday Trans. Part 2 **69**, 927 (1973)
41. J.C. Tully, J. Chem. Phys. **62**, 1893 (1975)
42. J. Davidson et al., J. Chem. Phys. **64**, 57 (1976)
43. R.A. Young, G. Black, J. Chem. Phys. **47**, 2311 (1967)
44. F.D. Findley, C.J. Fortin, D.R. Snelling, Chem. Phys. Lett. **3**, 204 (1969)
45. F. Stuhl, K.H. Welge, Can. J. Chem. **47**, 2311 (1967)
46. L. Wailace, D.M. Hunten, J. Geophys. Res. **73**, 4813 (1968)
47. T.P. Izod, R.P. Wayne, Proc. R. Soc. **308A**, 81 (1968)
48. F. Stuhl, K.H. Welge, Can. J. Chem. **49**, 1870 (1969)
49. S.J. Arnold, M. Kubo, E.A. Ogryzlo, Adv. Chem. Ser. **77**, 133 (1969)
50. R.J. O'Brien, G.H. Myers, J. Chem. Phys. **53**, 3832 (1970)
51. S.V. Filseth, A. Zia, K.H. Welge, J. Chem. Phys. **52**, 5502 (1970)
52. K.H. Becker, W. Growth, U. Schurath, Chem. Phys. Lett. **8**, 259 (1971)
53. L.R. Martin, R.B. Cohen, J.F. Schatz, Chem. Phys. Lett. **41**, 394 (1976)
54. T.G. Slanger, B.J. Wood, G. Black, J. Geophys. Res. **76**, 8430 (1971)
55. E.A. Koroleva, L.E. Khvorostovskaya, Optika i Spectroscopy **33**, 627 (1972)
56. R.D. Penzhorn, Environ. Sci. Technol. (e.a.) **8**, 907 (1974)
57. P.J. Cvetanovic, Can. J. Chem. **52**, 1452 (1974)
58. E.C. Zipf, Bull. Am. Phys. Soc. **12**, 225 (1967)
59. S.V. Filseth, F. Stuhl, K.H. Welge, J. Chem. Phys. **52**, 239 (1970)
60. R.P. Wayne, Adv. Photochem. **7**, 311 (1969)
61. P. Borrell, P.M. Borrell, D. Pedley, Chem. Phys. Lett. **51**, 300 (1977)
62. A. Leiss et al., J. Photochem. **8**, 211 (1978)
63. T.S. Frankiewicz, R.S. Berry, J. Chem. Phys. **58**, 1787 (1973)
64. R.E. Huffman, J. Geophys. Res. (e.a.) **76**, 1028 (1971)
65. M.J. McEwan, L.N. Phillips, *Chemistry of the Atmosphere* (Edward Arnold, London, 1975)
66. J.F. Noxon, Space Sci. Rev. **8**, 92 (1968)
67. W.F.J. Evans, D.M. Hunten, E.J. Llewellyn et al., J. Geophys. Res. **73**, 285 (1968)
68. D.M. Hunten, in *The Radiating Atmosphere*, ed. by B.M. Cormac (Reidel Publications, Dodrecht, 1971)

Chapter 10
Infrared Atmospheric Emission

Abstract Emission of a flat layer, consisting of a gas with a weakly varied temperature in the perpendicular direction to the layer, is reduced to the case of the layer with a constant temperature. The average optical thickness of the Earth's atmosphere in an infrared spectrum range is $u \approx 2.7$ and is determined on the basis of the energetic balance of the Earth and its atmosphere within the framework of the standard atmosphere model due to emission and surviving of infrared photons. It is shown that infrared emission of the atmosphere is determined mostly by atmospheric water. One can separate the flux of outgoing infrared radiation of the atmosphere from that towards the Earth. The fluxes due to rotation-vibration transitions of atmospheric CO_2 molecules are evaluated. Doubling of the concentration of CO_2 molecules in the atmosphere that is expected over 130 years leads to an increase of the average Earth temperature by (0.4 ± 0.2) K mostly due to the flux towards the Earth if other atmospheric parameters are not varied. Various models with a water change give the temperature change (3.0 ± 1.5) K at doubling of the CO_2 molecule concentration. An observed temperature change (0.8 °C) during the industrial epoch may be realized if approximately 0.5 % of free water molecules become aerosols, and this testifies to an atmospheric instability.

10.1 Emission of a Flat Layer

The atmospheric emission is important for the Earth's energetic balance (see Fig. 2.13). Figure 10.1 also contains the average fluxes in the Earth's balance that are the ratio of the powers given in Fig. 2.13 to the area of the Earth's surface.Infrared emission of the atmosphere is determined by vibration-rotation transitions of some atmospheric molecules, mostly H_2O and CO_2, and also by radiation of aerosols and atmospheric particles. The character of this emission is called the greenhouse effect [1, 2] and corresponds to a large optical thickness of the atmosphere. This means that the atmosphere absorbs shortwave solar radiation and emits infrared radiation, and infrared photons are locked in the atmosphere. This allows one to separate outgoing from Earth fluxes of infrared atmospheric radiation. Therefore, the optical thickness u of the atmosphere is large, and below we determine this value which is defined as

B.M. Smirnov, *Microphysics of Atmospheric Phenomena*,
Springer Atmospheric Sciences, DOI 10.1007/978-3-319-30813-5_10

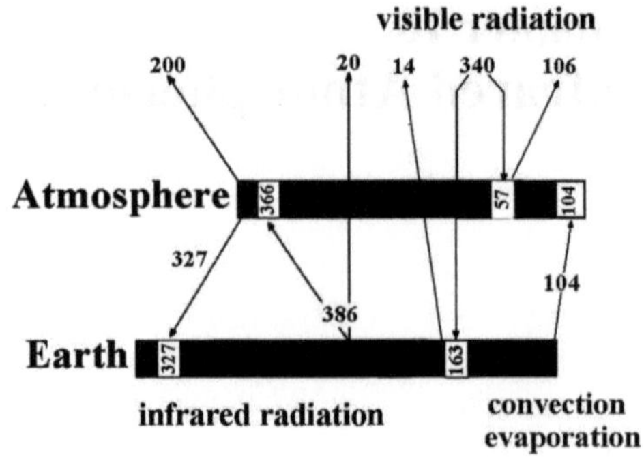

Fig. 10.1 Expressed in W/m^2 average energy fluxes in the atmosphere that determine the energetic balance of the Earth and are obtained on the basis of data of Fig. 2.13. Absorbed fluxes are given inside rectangles; scattering fluxes are indicated near *arrows*

$$u = \int_0^\infty k_\omega dh, \tag{10.1}$$

where k_ω is the absorption coefficient and h is an altitude over a sea level.

In this consideration we account for the atmosphere depth to be small compared to its radius, and the atmosphere is a flat layer. If the temperatures of layers at different altitudes are identical, the atmosphere emits as a Planck radiator of this temperature. Inasmuch as typical radiative times are large compared with collision times involving excited molecules in atmospheric air, radiative processes do not violate the Boltzmann distribution over excited vibration and rotation states that take place in the form:

$$N_* = \frac{g_*}{g_o} N_o \exp\left(-\frac{\Delta\varepsilon}{T}\right), \tag{10.2}$$

where N_o and N_* are the number densities of molecules in a lower o and excited $*$ transition states, g_o, g_* are the statistical weights for these states, and $\Delta\varepsilon$ is the molecule excitation energy. A frequency of this transition is equal approximately to $\hbar\omega \approx \Delta\varepsilon$.

In the case of a large optical thickness of the layer, the isotropic photon flux i_ω equals

$$i_\omega = \frac{\omega^2}{\pi^2 c^2}\left[\exp\left(\frac{\hbar\omega}{T}\right) - 1\right]^{-1} \tag{10.3}$$

This photon flux propagates inside the layer, and the thermodynamic equilibrium is supported by radiative processes with absorption and emission of these photons. Correspondingly, the photon flux through the boundary surface of this layer is equal to

$$j_\omega = \int_0^{\pi/2} i_\omega \cos\theta d(\cos\theta) \left(\int_{-\pi/2}^{\pi/2} d\cos\theta\right)^{-1} = \frac{i_\omega}{4} = \frac{\omega^2}{4\pi^2 c^2}\left[\exp\left(\frac{\hbar\omega}{T}\right) - 1\right]^{-1} \tag{10.4}$$

The characteristic of the transitions under consideration is the absorption coefficient k_ω at a given frequency ω that is defined as [3, 4]

$$k_\omega = N_o\sigma_{abs}(\omega) - N_*\sigma_{em}(\omega) \tag{10.5}$$

where $\sigma_{abs}(\omega)$ is the absorption cross section, and $\sigma_{em}(\omega)$ is the cross section of stimulated radiation. We consider the case of an optically dense gas, where

$$k_o L \gg 1 \tag{10.6}$$

Here $k_o \equiv k_{\omega_o}$, and the frequency ω_o correspond to the spectral line center, and L is the depth of the layer. Let us introduce the optical thickness u_ω of the layer

$$u_\omega = \int_0^L k_\omega dh, \tag{10.7}$$

where h is the layer altitude. In the case of an optically dense layer $u_\omega \gg 1$, thermodynamic equilibrium takes place between photons and air molecules. This thermodynamic equilibrium is valid if the mean free path of photons in air $\lambda_\omega \sim 1/k_\omega$ is small compared with a layer size L. Then the criterion (10.6) $u \gg 1$ holds true.

We now evaluate the photon flux emitted by atmospheric air within the framework of the standard atmosphere model where the air temperature varies slowly with altitude. In this case we represent Formula (10.4) in the form

$$j_\omega = \frac{\omega^2}{2\pi^2c^2}\int_0^1 d\,(\cos\theta)\int_0^\infty du_\omega \exp(-u_\omega/\cos\theta)F(u_\omega),$$
$$F(u_\omega) = \left[\exp(\left(\frac{\hbar\omega}{T}\right) - 1\right]^{-1} \tag{10.8}$$

Here we account for the altitude dependence for the atmosphere temperature by the dependence $F(u_\omega)$ which is weak. Expanding the latter function in a series

$$F(u_\omega) = F(u_o) + (u_\omega - u_o)F'(u_o) + \frac{1}{2}(u_\omega - u_o)^2F''(u_o), \tag{10.9}$$

one can choose the parameter u_o such that the second term is zero after integration. This yields [5, 6]

$$u_o(\omega) = 2/3, \tag{10.10}$$

and the radiation flux at a given frequency equals

$$j_\omega = j_{\omega_{ef}}(1-\alpha), \; u_o = \frac{2}{3}, \; \alpha = \frac{5F''(u_o)}{18F(u_o)} \tag{10.11}$$

where α is a small parameter. In particular, in the case $\hbar\omega \gg T$ a small parameter equals

$$\alpha = \frac{5}{18}\left(\frac{\hbar\omega}{T^2} \cdot \frac{dT}{du}\right)^2 \tag{10.12}$$

Let us determine this small parameter for the lower Earth atmosphere, taking the photon frequency at the maximum of the energy flux $\hbar\omega = 3T$ that gives according to the Wien law $\hbar\omega = 3T$. Next, within the framework of the standard atmosphere we have $T = 288K$, $dT/dh = 6.5$ K/km. Taking $du/dh \approx u/L$, and for radiation of the atmosphere $u \approx 2.7$, $L \approx 8$ km, one can obtain $\alpha \sim 0.01$.

Thus we reduce the emission flux of an optically thick layer with a weakly varied temperature to that of an identical temperature which is taken for a layer with the optical thickness $u_o = 2/3$ from its surface. It should be noted that for absorption of incident radiation that is directed perpendicular to the layer this optical thickness is close to $u = \ln 2$ according to Formula (9.8). As can be seen, these results for emission and absorption of infrared radiation are nearby.

10.2 Atmospheric Optical Thickness

In determining the optical thickness of the atmosphere in the infrared spectrum range, we are based on the Earth's energetic balance represented in Table 2.2, Figs 2.13, and 10.1. We find below the average optical thickness (10.1) of the atmosphere both from the atmosphere transparency in an infrared spectrum range and from atmospheric emission. The probability of the photon transfer through the atmosphere is $P = \exp(-u)f(u)$ and we take the distribution function $f(u)$ over optical thicknesses such that $f(u) = const$ in the range $u_1 \leq u \leq u_2$ and is zero outside this range. This distribution function accounts for the frequency dependence of the optical thickness. Correspondingly, the probability for an infrared photon to transfer through the atmosphere from the Earth's surface equals

$$P = \int_{u_1}^{u_2} f(u)du \int_0^1 \exp\left(-\frac{u}{\cos\theta}\right) d\cos\theta, \tag{10.13}$$

where θ is the angle between the photon motion direction and the Earth's surface. On the basis of this formula and the probability of the photon surviving $P = 1/20$ according to Fig. 10.1 one can determine the optical thickness as a result of photon propagation through the atmosphere. Taking $u_2/u_1 = 1.2, 1.5, 2, 3$ and the above probability of surviving, we obtain for the average optical thickness $\bar{u}$ of the atmosphere

$$\bar{u} = \int_{u_1}^{u_2} uf(u)du = 2.6 \pm 0.2 \tag{10.14}$$

This is a typical optical thickness of the atmosphere for thermal radiation.

Another method to find the average optical thickness of the atmosphere is based on parameters of atmospheric emission. On the basis of the powers of atmospheric infrared radiation towards the Earth's surface 1.67×10^{17} W and in a surrounding space 1.02×10^{17} W, one can determine the temperature of atmospheric layers $T_{\downarrow}$ and $T_{\uparrow}$ that are responsible for thermal radiation in the indicated directions in accordance with the data of Fig. 10.1. Modeling radiation of these layers by a blackbody and using the Stephan-Boltzmann law (2.32), we obtain for temperatures of the above layers

$$T_{\downarrow} = 276\,\mathrm{K}, \ \mathrm{T}_{\uparrow} = 244\,\mathrm{K} \tag{10.15}$$

In accordance with the temperature of the standard atmosphere, the radiation fluxes towards the Earth's surface and outside it are created by atmospheric layers at altitudes

$$h_{\downarrow} = 1.9\,\mathrm{km}, \ h_{\uparrow} = 6.8\,\mathrm{km} \tag{10.16}$$

In the following we use these data from the Earth's energetic balance for the analysis of atmospheric thermal radiation.

Let us take the altitude dependence of the absorption coefficient in the form

$$k_{\omega} \equiv \frac{du}{dh} = A \exp\left(-\frac{h}{\lambda}\right), \tag{10.17}$$

and the parameters A and λ follow from (10.10) for emission to the Earth's surface and outside. These equations have the form

$$\int\limits_{h=0}^{h=h_{\downarrow}} du = \frac{2}{3}; \ \int\limits_{h=h_{\uparrow}}^{\infty} du = \frac{2}{3} \tag{10.18}$$

From these equations it follows

$$A = 0.42\,\mathrm{km}^{-1}, \ \lambda = 5.5\,\mathrm{km}, \ u = 2.3 \tag{10.19}$$

Because of $\lambda < \Lambda$, atmospheric radiators are located in a narrower region compared with that occupied by air molecules and other molecules which take part in convection. This is indirect confirmation that atmospheric water creates emission of the atmosphere in a infrared spectrum range. Based on this, we consider one more model for atmospheric radiators representing the atmospheric absorption coefficient in the form

$$k_{\omega} = k_o, \ h < L; \ k_{\omega} = 0, \ h > L \tag{10.20}$$

On the basis of (10.18) one can find the parameters of this formula

$$k_o = 0.35\,\text{km}^{-1},\ L = 8.7\,\text{km};\ u = k_o L = 3.0 \qquad (10.21)$$

The statistical averaging of data of Formulas (10.14), (10.21) and (10.19) gives for the average optical thickness of the atmosphere:

$$\bar{u} = 2.7 \pm 0.3 \qquad (10.22)$$

We thus obtain nearby values of the average atmospheric optical thickness from different considerations. In addition, the average absorption coefficient drops with an increase of altitude faster than the number density of air molecules. This indicates indirectly that water is the basic atmospheric absorber. The concentration of free and bound water molecules in atmospheric air drops more strongly with increasing altitude than that for other molecules located in the atmosphere, and the average concentration of water molecules in the atmosphere is higher than that for other radiative atmospheric molecules. In particular, the average concentration of water molecules near the Earth's surface for the standard atmosphere is 1.0 % by weight [7], whereas the concentration of carbon dioxide is 0.04 %.

Emission of the Earth's atmosphere is determined by radiation of its components, that is, by atmospheric molecules and microparticles. The radiation flux of the atmosphere in the infrared spectrum range consists of two parts, directed to the Earth and outside it (see Fig. 10.1). Modeling the Earth's atmosphere by a blackbody, we have according to Wien's law [8, 9] that the radiation wavelength λ_m of the maximum radiative intensity is connected with the blackbody temperature T by the relation $\lambda_m T = 0.29\,\text{cm}\cdot\text{K}$. Hence, in the analysis of the atmospheric emission we are guided by infrared radiation in the wavelength range $\lambda = 10-15\,\mu\text{m}$. Evidently, water is the main atmospheric radiator. As we conclude from the analysis of the Earth's heat balance, basic atmospheric radiators are located at altitudes of a few kilometers, and water satisfies this requirement since it does not penetrate altitudes above 10 Km because of condensation at low temperatures. The water concentration in the atmosphere near the Earth's surface is 1.7 % by volume on average (Sect. 2.3) or 0.5 % as the molecule concentration. If total atmospheric water is transformed in the liquid state and this water is distributed over the globe surface, it would cover the globe with a layer of the depth 2.5 cm.

Figure 10.2 shows the absorption coefficient for bulk liquid water, and these data may be used for estimations of emission of water aerosols. Indeed, the atmospheric optical thickness $u = 2.7$ in the infrared spectrum range, estimated in the previous paragraph on the basis of the heat balance of the Earth, may be provided by the water layer of the depth $l = (8-30)\,\mu\text{m}$; that is, it includes less than 1 % of atmospheric water. Of course, it is an overstated estimation for an aerosol amount because the size of aerosols is comparable to the wavelength of radiation and they have weaker absorbance than a water layer. According to measurements in stratocumulus clouds, the absorption coefficient for the wavelength range 10–12 μm is $700-1000\,\text{cm}^2/\text{g}$ [11]. This leads to the depth of a formed water layer $30-40\,\mu\text{m}$ that corresponds

Fig. 10.2 Absorption coefficient for a bulk liquid water layer under normal conditions [10]. A cross gives the absorption coefficient for aerosols [11]

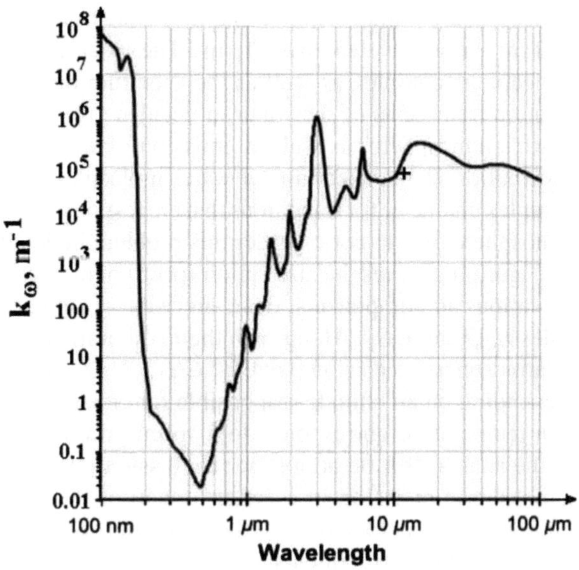

to the concentration of atmospheric water in aerosols of roughly (1–2) %. As can be seen, water aerosols may make a remarkable contribution to atmospheric emission. This result causes alarm with respect to climate change because transition of a small part of an atmospheric water vapor in aerosols may lead to a significant change of the atmospheric optical thickness. In particular, according to several studies [12–15] cosmic rays influence the formation of aerosols and clouds in the Earth's atmosphere.

Free water molecules include the basic water mass in the atmosphere because of a restricted humidity of the atmosphere which is 75 % on average at ground level. Therefore water molecules may make a contribution to emission of infrared radiation, especially for hot atmospheric regions. Let us consider properties of a free H_2O molecule [16, 17]. It has the C_{2v} symmetry, where two segments OH form the angle 104.5°, and the length of the bond OH is 0.957 Å for the ground molecule state. The H_2O molecule has three vibration degrees of freedom, and the symmetric vibration with motion of hydrogen atoms in identical directions with respect to the oxygen atom is characterized by the energy $\nu_1 = 3657\,\text{cm}^{-1}$ (the wavelength is $\lambda_1 = 2.734\,\mu\text{m}$). The energy of the antisymmetric vibration, where hydrogen atoms move in opposite directions with respect to the oxygen atom, is $\nu_2 = 3756\,\text{cm}^{-1}$ (the wavelength is $\lambda_2 = 2.662\,\mu\text{m}$). The bond lengths do not vary in the third vibration that takes place in the molecule plane and is accompanied by variation of the angle between segments. Its energy is equal to $\nu_3 = 1595\,\text{cm}^{-1}$ (the wavelength is $\lambda_3 = 6.269\,\mu\text{m}$). These vibrations are in an infrared spectrum range which is of importance at high temperatures.

It is essential for emission of a water vapor at room temperature that the dipole moment of the water molecule is nonzero and equals $1.85ea_o$ (e is the electron charge and a_o is the Bohr radius). Therefore transitions between rotation states of the water molecules are radiative active and they start from transition energies approximately 200 cm^{-1} (the wavelength is 50 μm). The spectrum of rotation transitions of the water molecule is complex as well as the spectrum of vibration-rotation transitions. The rotation transitions of the water molecules give the contribution to emission of the Earth's atmosphere, as well as emission of water aerosols. Because the frequency distribution in the spectrum of water molecules and aerosols has an irregular and complex character for transitions that are responsible for water emission in the Earth's atmosphere, we do not consider this problem in detail.

Figure 10.3 gives the absorption spectrum of a real atmosphere that is determined by various air components. Basic emission of atmospheric air proceeds due to molecules H_2O and CO_2. Along with them, other molecular components may contribute to atmospheric emission, such as CH_4, N_2O, freons, and SF_6, especially, locally. Depending on an altitude responsible for emission in a given spectrum range, a certain temperature corresponds to a spectrum range for a given component. Hence, in the analysis of the atmospheric emission [19–23], it is convenient to reduce the atmospheric emission in a certain range of the spectrum to the temperature of a blackbody that corresponds to the same spectral radiative flux. This may be obtained directly from measurements.

In particular, an example of such a measurement is represented in Fig. 10.4 where a measured radiative flux from the Earth's atmosphere as a function of a radiation frequency is compared with the flux of a thermal radiation at a given temperature. In this case, the equilibrium radiative flux is given by Formula (10.3), and the radiative flux in a solid angle $d\Theta$ is equal to $i_\omega d\Theta$, as well as a flux of outgoing equilibrium radiation in a solid angle $d\Theta$. If the temperature altitude profile of the atmosphere is known, measurements of the spectral flux of atmospheric radiation allow one to determine which altitudes are responsible for this radiation. One can identify also from this which molecular absorbers are responsible for emission in a given spectrum

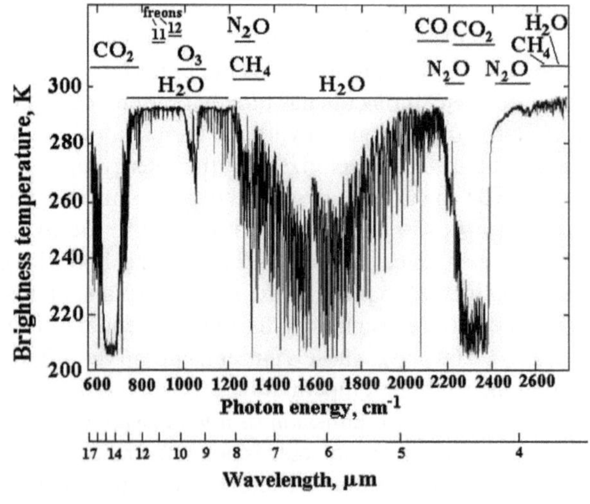

Fig. 10.3 Spectra of absorption of infrared radiation by atmospheric air owing to its molecular components [18]

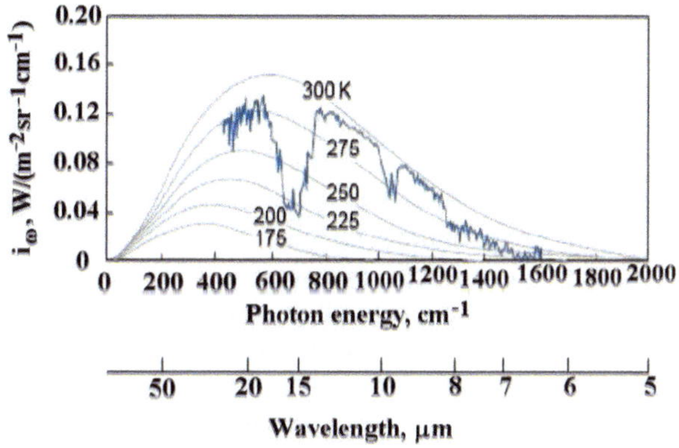

Fig. 10.4 Spectral flux of emission of atmospheric air in the infrared spectral range measured by a satellite over northern Africa (Niger Valley) at noon and those for equilibrium radiation at a given temperature [24]

range. One can add to this that according to the data of Fig. 10.4, the temperature of radiation created by CO_2 molecules is below the minimum temperature of the standard atmosphere at these altitudes. Hence, in this case the atmospheric temperature profile under conditions of this measurement differs from that for the standard atmosphere model.

Thus we have that the spectral radiative flux of the Earth's atmosphere may include some frequency bands created by vibration-rotation or rotation radiative transitions of certain components. One can introduce the effective temperature for a certain frequency band in accordance with Formula (10.10). The optical thickness of the atmosphere starting from this layer in the direction perpendicular to the Earth's surface equals 2/3 at a given frequency, and the temperature of this atmospheric layer is the effective temperature for emission at a given frequency. Below we use this concept in the analysis of emission of atmospheric CO_2 molecules.

Along with molecules, dust particles and aerosols may make a certain contribution to atmospheric emission. Emission of dust particles larger than 5 μm that rise from the Earth's surface does not go beyond the atmosphere because these particles are located close to the Earth's surface according to Formula (2.29). An exclusion relates to large solid particles of volcanic eruptions because the stream of erupted matter reaches high altitudes. In other cases nucleation processes, including formation of water aerosols from a water vapor, proceed in the troposphere, and the emission of liquid and solid particles contributes to emission of the atmosphere.

10.3 Atmospheric Absorption Coefficient Due to CO_2 Molecules

Atmospheric carbon dioxide molecules contribute to atmospheric emission because of a remarkable concentration of CO_2 molecules in the atmosphere and the absorption spectrum of these molecules corresponds to the maximum emission of a blackbody at room temperature. The role of this emission becomes stronger because an amount of CO_2 molecules in the Earth's atmosphere increases due to human activity, mostly because of combustion of fossil fuels and deforestation, and variation of this amount is detected reliably. Accumulation of CO_2 in the atmosphere causes the greenhouse effect in the atmosphere; that is, due to emission of atmospheric CO_2 molecules at higher altitudes, the radiative power decreases that leads to an increase of the average Earth temperature. Our goal is to evaluate the temperature change under real atmospheric conditions, and we do this below. It is important in this analysis that in determining the temperature change due to an increase of the concentration of atmospheric CO_2 molecules we assume other atmospheric parameters to be invariable.

Let us consider the spectrum of the CO_2 molecule. The CO_2 molecule is linear; that is, its three atoms are located in one line, and their displacements cause three types of vibrations, as shown in Fig. 10.5 [16, 25]. In this figure laser transitions of the wavelengths of 9.4 and 10.6 μm are present. In this context, the main vibration transition is 01 °0 → 00 °0 with the wavelength near 15 μm because it is important for atmosphere emission. According to the selection rules [25], there are two branches of vibration-rotation transitions, namely, the P-branch corresponds to transition from the rotation state with momentum j in the state with momentum $j' = j - 1$, and the transition energy is $\hbar\omega_o - 2Bj$, where $\hbar\omega_o = 667.4\,\mathrm{cm}^{-1}$ is the vibration transition energy, and $B = 0.39\,\mathrm{cm}^{-1}$ is the rotation constant. For R- branch the rotation momentum of the final state is $j' = j + 1$, and the transition energy equals $\hbar\omega_o + 2B(j+1)$.

Because of a long time of radiative transitions in molecules compared with collision ones in air, the number density of molecules N_j in a given rotation state j is determined by the Boltzmann formula

$$N_j = N_o(2j+1)\frac{B}{T}\exp\left[-\frac{Bj(j+1)}{T}\right], \tag{10.23}$$

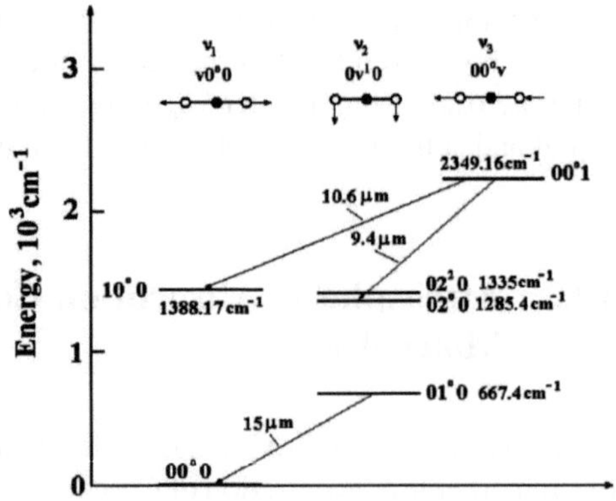

Fig. 10.5 Types of the lowest vibration states of the CO_2 molecules, energies of the lowest vibration levels, and parameters of radiative transitions

where N_o is the total number density of molecules for this vibration state, T is the gas temperature expressed in energetic units, and $Bj(j+1)$ is the excitation energy. Because usually $B \ll T$, we deal with a large number of rotation states $j \gg 1$.

Let us use a general formula for the absorption coefficient k_ω for a molecular gas for transitions from the ground vibrational state:

$$k_\omega = \frac{\pi^2 c^2}{\omega^2} \sum_j w_j a_\omega N_j \left[1 - \exp\left(-\frac{\hbar\omega_o}{T}\right)\right] \tag{10.24}$$

where $w_j = a_\omega/\tau$ is the rate of the radiative transition, τ is the radiative lifetime, and a_ω is the photon distribution function of an individual vibration-rotation transition. Returning to CO_2 molecules, we obtain that the absorption spectrum consists of many lines, and the frequency difference for neighboring transitions is $2B/\hbar$. Joining P and R branches, one can obtain from this that the absorption coefficient k_ω due to CO_2 molecules is symmetric with respect to the frequency ω_o. Due to the symmetry, a gas of CO_2 molecules is characterized by a simple spectrum in contrast to other radiative atmospheric molecules such as H_2O, CH_mF_{4-m}, CH_mCl_{4-m} (m is a whole number from 0 up to 4), which emission is the basis of the random model for the molecular spectrum (e.g., [26–28]). In the CO_2 case the absorption coefficient as a frequency function is an almost periodic function with the period $2B/\hbar$ that corresponds to the regular model of spectral lines or the Elsasser model [29].

If CO_2 molecules are located in air, and the number density of air molecules N_a is located in the range $10^{17}\,\mathrm{cm}^{-3} \ll N_a \ll 10^{20}\,\mathrm{cm}^{-3}$ [30], the impact theory holds true for broadening of the spectral lines, and the photon distribution function a_ω is

$$a_\omega = \frac{\nu}{\pi[(\omega-\omega_j)^2+\nu^2]}, \tag{10.25}$$

where ν is the width of the spectral line, and the transition energy is $\hbar\omega_j = \hbar\omega_o \pm 2Bj$ for $j \gg 1$. From this follows the expression of the absorption coefficient:

$$k_\omega = \frac{\pi^2 c^2}{\omega^2 \tau} N_o \frac{\hbar(\omega-\omega_o)}{T} \exp\left[-\frac{\hbar^2(\omega-\omega_o)^2}{4BT}\right]\left[1-\exp\left(-\frac{\hbar\omega}{T}\right)\right] \times \sum_j \frac{\nu}{\pi[(\omega-\omega_j)^2+\nu^2]}, \tag{10.26}$$

Summarizing this formula over rotation momenta j, use the Mittag-Leffler theorem, which takes the form [31]

$$\sum_{k=-\infty}^{\infty} \left[(x-k)^2+y^2\right]^{-1} = \frac{\pi \sinh 2\pi y}{y(\cosh 2\pi y - \cos 2\pi x)} \tag{10.27}$$

This leads to the oscillation form of the absorption coefficient k_ω as the frequency function that may be represented in the form (e.g., [32, 33])

$$k_\omega = S \sinh \frac{\pi \hbar \nu}{B} \left[\cosh \frac{\pi \hbar \nu}{B} - \cos \frac{\pi \hbar (\omega - \omega_o)}{B} \right]^{-1}, \tag{10.28}$$

with the transition intensity S given by formula

$$S = \frac{\pi^2 \omega^2}{c^2} \frac{\hbar N_a c_m}{2B\tau} \frac{\hbar(\omega - \omega_o)}{T} \exp\left[-\frac{\hbar^2 (\omega - \omega_o)^2}{4BT} \right] \left[1 - \exp\left(-\frac{\hbar \omega}{T} \right) \right] \tag{10.29}$$

Here $N_o = N_a c_m$, so that c_m is the concentration of CO_2 molecules, and N_a is the number density of air molecules.

In evaluating the absorption coefficient k_ω due to atmospheric CO_2 molecules, we use the radiative lifetime for the vibration transition at the wavelength 15 μm as $\tau = 0.66$ s, and take the spectral line width at atmospheric pressure and room temperature to be $\nu = 0.08\,\mathrm{cm}^{-1}$ [30]. For definiteness, we determine the absorption coefficient at the altitude 15 km, where the temperature is $T = 217$ K for the standard atmosphere, the pressure is $p = 0.12$ atm, and the number density of air molecules is $N_a = 4.0 \times 10^{18}\,\mathrm{cm}^{-3}$. This leads to the value of the reduced width of a spectral line $y = \pi \hbar \nu / B = 0.10$. From this one can obtain the following relation for the ratio of neighboring maximum and minimum of the absorption coefficient

$$\frac{k_{\max}}{k_{\min}} = \frac{\cosh \frac{\pi \nu}{B} + 1}{\cosh \frac{\pi \nu}{B} - 1} \approx \frac{2B}{\pi \nu} = 380 \tag{10.30}$$

Figure 10.6 contains the frequency dependence for the absorption coefficient due to absorption of CO_2 molecules at the altitude of 15 km which corresponds to Formulas (10.28) and (10.29) and the above parameters. We assume that carbon dioxide is mixed with air at this altitude and use the contemporary value of its concentration $c_m = 4.0 \times 10^{-4}$. Figure 10.6b contains the same dependence on another scale. As follows from this figure, the averaging absorption coefficient over frequency oscillations is closer to the minimum values of k_ω rather than to its maximum values.

Let us fulfill an average over oscillations of the optical depth of the atmosphere for a given altitude h defined in accordance with formula

$$u_\omega(h) = \int_h^\infty k_\omega(z) dz, \tag{10.31}$$

where z is a current atmosphere altitude. The number density of air molecules is given by Formula (2.3) that has the form

$$N_a(h) = N_o \exp\left(-\frac{z}{\Lambda} \right), \tag{10.32}$$

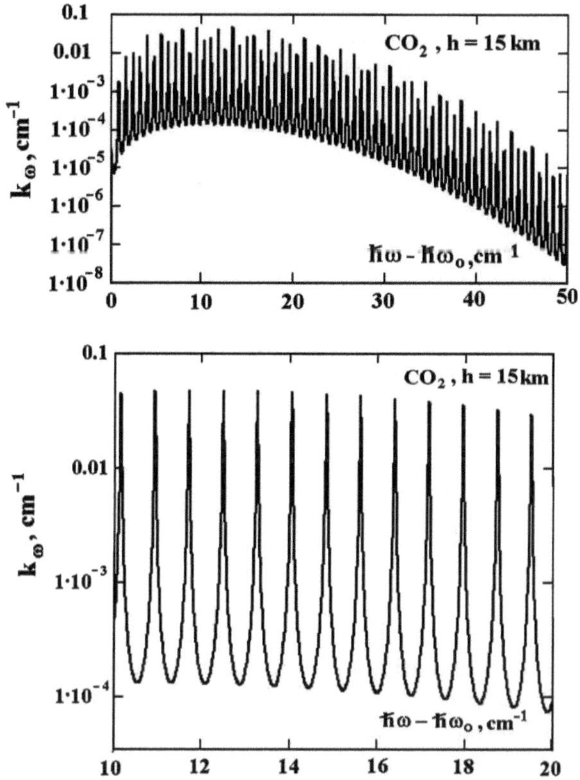

Fig. 10.6 The absorption coefficient k_ω of atmospheric carbon dioxide at the altitude of 15 Km in wide **a** and narrow **b** frequency ranges

and in the altitude range 10–30 km we have for the standard atmosphere $\Lambda = (6.4 \pm 0.2)$ km. Reducing the atmosphere with a weakly varied temperature to that with unvaried temperature, one can find the effective temperature T_ω for emission at a given radiation frequency ω on the basis of Formula (10.10)

$$T_\omega = T(z_\omega), \int_{k_\omega}^{\infty} k_\omega \omega = 2/3 \tag{10.33}$$

In averaging the logarithm of the optical depth, we obtain for the average absorption coefficient $k_{ef}(\omega)$ at a given frequency ω the relation:

$$k_{ef}(\omega) = k_{\min}(\omega)\exp(I), \tag{10.34}$$

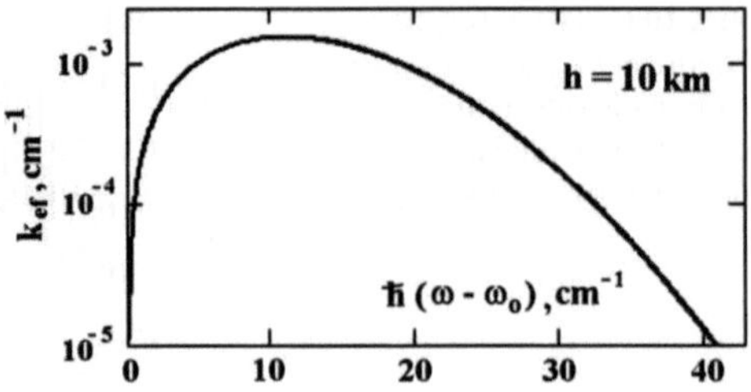

Fig. 10.7 Effective absorption coefficient due to atmospheric CO_2 molecules at the altitude of 10 km

and

$$I = \frac{1}{2\pi}\int_0^{2\pi} \ln\left(\frac{k_\omega}{k_{min}(\omega)}\right) dt =$$
$$\frac{1}{2\pi}\int_0^{2\pi} \ln\left[\cosh(y) + 1\right] - \ln\left[\cosh(y) - cos(t)\right] dt, \; y = \frac{\pi\hbar\omega}{B}, \; t = \frac{\pi\hbar\omega}{B} \tag{10.35}$$

where t is the reduced frequency. Guided by altitudes where $y \ll 1$, we obtain $I = 1.18$, and the ratio of the effective absorption coefficient $k_{ef}(\omega)$ to its minimum value $k_{min}(\omega)$ is $\exp(I) \approx 3.6$. This is lower than the ratio of the maximum value of the absorption coefficient $k_{max}(\omega)$ to its effective value $k_{ef}(\omega)$, that is, 88. Figure 10.7 gives the frequency dependence for the absorption coefficient at the altitude of 10 km.

We now determine the effective atmospheric temperature that corresponds to emission at a given frequency. According to satellite measurements [21] for the spectrum band near 15 μm, this temperature is close to 220 K which corresponds to the minimum atmospheric temperature in the region between the troposphere and stratosphere. On the basis of Formulas (10.31) and (10.32) using the dependence $k_{min}(\omega) \sim N_a^2$ we have at these altitudes

$$u_\omega(h) = \int_h^\infty k_{ef}(\omega, z)dz = \frac{k_{ef}(\omega, h)\Lambda}{2} \tag{10.36}$$

The effective temperature for radiation at a given frequency is equal to the temperature of an atmospheric layer at an altitude h, so that the optical depth for this layer is $u_\omega(h) = 2/3$.

10.4 Greenhouse Effect Due to Atmospheric CO_2

Spectra of infrared radiation responsible for atmospheric emission are created mostly both by water (molecules and aerosols) and by CO_2 molecules. These spectra are entangled, but it is not important for outgoing radiation because regions of location of these components are separated. Indeed, water is located and emits radiation from altitudes below 10 km, whereas the emission of CO_2 molecules that contributes to atmospheric radiation occurs mostly at higher altitudes. Therefore in the analysis of the greenhouse effect which arises due to emission of CO_2 molecules, we ignore emission of other atmospheric components because outgoing radiation is created by CO_2 molecules in this case. Correspondingly, emission of CO_2 molecules leads to the greenhouse effect at frequencies where this radiation is created by atmospheric layers at altitudes above approximately 10 km.

We now determine the contribution of emission of CO_2 molecules in the band of 15 μm to the heat balance of the atmosphere. We base this on the above result that the temperature of an atmospheric layer, which is responsible for atmospheric emission, is equal to $T_a = 244$ K, and these layers are located at an altitude below about 10 km. Introducing the spectral band width $\Delta\omega$ created by CO_2 molecules at altitudes above 10 km, we have for the radiative energy flux of this band in accordance with Formula (10.4)

$$\Delta J = \hbar\omega j_\omega d\omega = \frac{1}{4\pi^2 c^2} \frac{\hbar\omega^3 \Delta\omega}{\exp\left(\frac{\hbar\omega}{T}\right) - 1} \tag{10.37}$$

We use the following algorithm for the analysis of the greenhouse effect due to emission of CO_2 molecules. Assuming that the emission of CO_2 molecules does not influence the atmospheric temperature, within the framework of the standard atmosphere we use the altitude dependence for the temperature of atmospheric air $T(h)$ in accordance with Fig. 2.1. Assuming $y \ll 1$, where the parameter y is defined by Formula (10.35), one can obtain that radiation at a frequency ω is created at altitude h and corresponds to the temperature $T(h)$. On the basis of Formulas (10.34) and (10.36) we obtain the connection between the above parameters from the relation

$$u_\omega(h) = k_{\min}(\omega, h) \exp(I) \frac{\Lambda}{2} = S(\omega, h) \Lambda \exp(I) \frac{y}{4}, \tag{10.38}$$

where the parameter S is given by Formula (10.29). From this relation one can determine the radiative temperature $T(\omega)$ in the range of emission created by altitudes $h > 10$ km. In this case emission at a given frequency is created at an altitude h that follows from the relation (10.10) $u_\omega(h) = 2/3$, and the atmosphere temperature at a given frequency ω is the temperature $T(h)$ for this altitude according to Fig. 2.1.

Let us use Formula (10.38) for determination of the parameters of outgoing atmospheric radiation emitted by CO_2 molecules. Note that according to Formula (10.23) the maximum number density of vibrationally excited molecules corresponds to the rotation number $j_{\max} = 2\sqrt{T/B}$, and in reality $j_{\max} = 10 - 14$. Figure 10.8 gives the altitudes where the atmospheric emission is created owing to transitions from the first vibrationally excited state of the CO_2 molecule with the rotation number $j_{\max}$. This means that higher altitudes of the atmosphere do not contribute to emission of CO_2 molecules, and outgoing radiation due to CO_2 molecules is created in a low stratosphere.

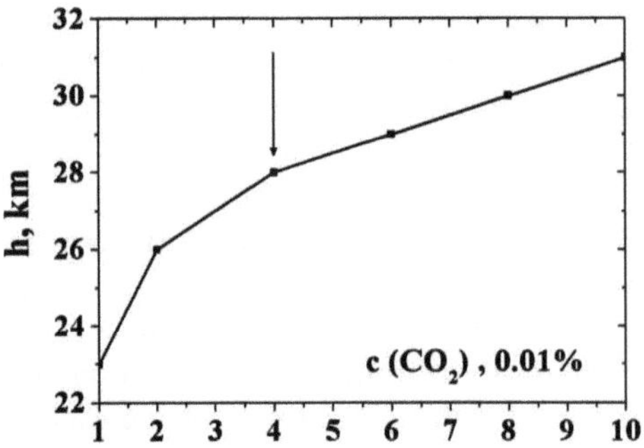

Fig. 10.8 Dependence on the concentration of carbon dioxide molecules for the altitude h where the outgoing atmospheric emission is formed due to radiative transitions in CO_2 molecules in the band of 15 μm. The arrow indicates the contemporary content of atmospheric CO_2

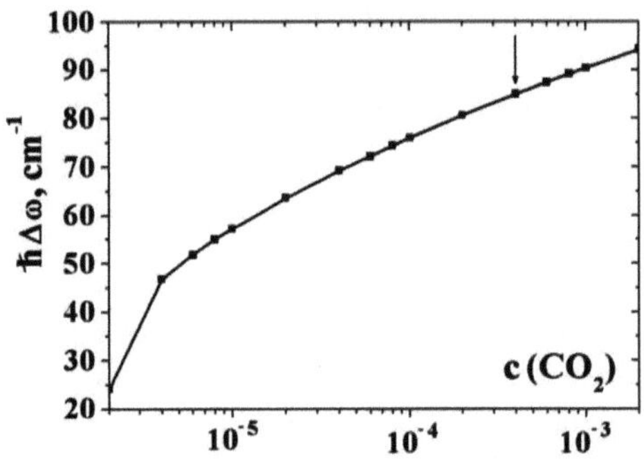

Fig. 10.9 Width of the spectral band $\Delta\omega$ for outgoing radiation created by CO_2 molecules at altitudes above 10 km. The arrow indicates the contemporary content of atmospheric CO_2

Figure 10.9 contains the width of a spectral band for radiation of CO_2 molecules where this radiation is formed at altitudes above 10 km. This radiation originates at higher altitudes of the atmosphere compared with that emitted by water molecules and aerosols, and therefore it determines an outgoing atmospheric radiation at given frequencies which is not entangled with water emission from lower altitudes. From this one can find the part of the atmospheric radiation flux δ that is created by CO_2 molecules and is determined by formula

$$\delta = \frac{\Delta J}{J_o}, \tag{10.39}$$

where ΔJ is the radiative flux created by atmospheric CO_2 molecules and is given by Formula (10.37), and $J_o = \sigma T_a^4$ is the total radiative flux of outgoing atmospheric radiation.

Thus, CO_2 molecules that provide atmospheric radiation in a frequency range under consideration are located at altitudes above layers where outgoing thermal radiation is created due to water molecules and aerosols. Let us connect the flux of outgoing radiation due to CO_2 molecules on the basis of Formula (10.10) with the optical thickness of the atmosphere $u_\omega(h)$ for a given radiation frequency ω at the altitude above a given one h. Assuming $y \ll 1$, where y is defined by Formula

(10.36), we have on the basis of Formulas (10.31) and (10.32) for the atmospheric optical thickness at altitudes above a given h

$$u_\omega(h) = k_{\min}(\omega, h)\exp(I)\frac{\Lambda}{2} = S(\omega, h)\Lambda\exp(I)\frac{y}{4}, \tag{10.40}$$

where the parameter S is determined by Formula (10.29). From this on the basis of Formula (10.10) one can find an altitude responsible for thermal radiation at a given frequency.

As shown, the absorption coefficient as a function of a frequency decreases as we move from the transition center. Let us introduce the width of a radiation band as $\hbar\Delta\omega = 2\hbar(\omega_b - \omega_o)$, where the boundary frequency ω_b is given by

$$u(\omega_b, h_o) = 2/3 \tag{10.41}$$

Here $h_o \approx 10$ km is the boundary of the lower atmosphere whose thermal radiation is created by water molecules and aerosols. Figure 10.9 gives the dependence of a radiation band $\Delta\omega$ on the concentration of CO_2 molecules. Because the band boundary is determined by the tail of the distribution function over rotation momenta, Formula (10.41) gives for the boundary of a radiation band due to atmospheric carbon dioxide

$$\frac{d\hbar\omega_b}{d\ln c} = \frac{2BT}{\hbar(\omega_b - \omega_o)} \tag{10.42}$$

In particular, Formula (10.42) gives for the right-hand side of (10.42) $d\hbar\omega_b/d\ln c = 3.4\,\mathrm{cm}^{-1}$ at the contemporary concentration of carbon dioxide in air. From this and Fig. 10.9 it follows that the width of a radiation band increases slightly with an increasing concentration of carbon dioxide at its contemporary values. In particular, doubling of the contemporary concentration of CO_2 molecules leads to an increase of the width of a radiation band at an altitude of 10 km from 85 up to 89 cm^{-1}.

The above analysis shows that the greenhouse effect due to carbon dioxide acts as follows. Basic thermal radiation of the Earth's atmosphere is created by water molecules and aerosols in lower atmospheric layers. But thermal radiation of the atmosphere due to carbon dioxide inside the width of a radiation band $\hbar\Delta\omega = 2\hbar(\omega_b - \omega_o)$ is created by CO_2 molecules located in a space above the boundary of a lower atmosphere $h_o \approx 10$ km. An altitude $h(\omega)$ that is responsible for radiation at a frequency ω is given by Formula (10.10) according to which $u_\omega(h[\omega)] = 2/3$. Figure 10.10 gives this dependence at different concentrations of CO_2 molecules in atmospheric air.

The effective radiation temperature is equal to the temperature of a layer at an altitude h such that the atmospheric optical thickness above this layer is $u_\omega(h) = 2/3$. On the basis of Formulas (10.31) and (10.32) using the dependence $k_{\min}(\omega) \sim N_a^2$, one can reduce this dependence to the form

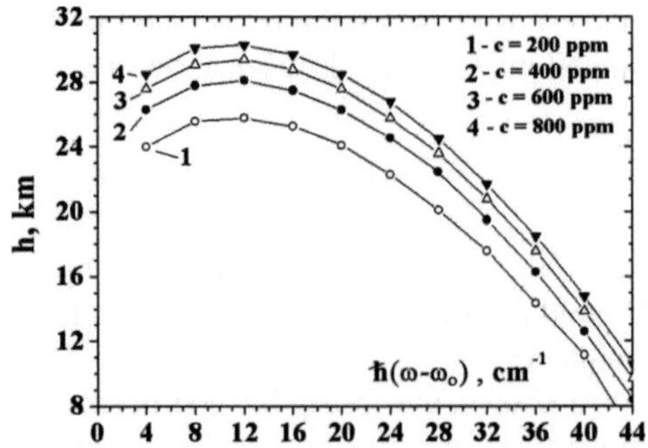

Fig. 10.10 Atmospheric altitudes responsible for emission of atmospheric CO_2 for a given radiation frequency at various concentrations of atmospheric CO_2

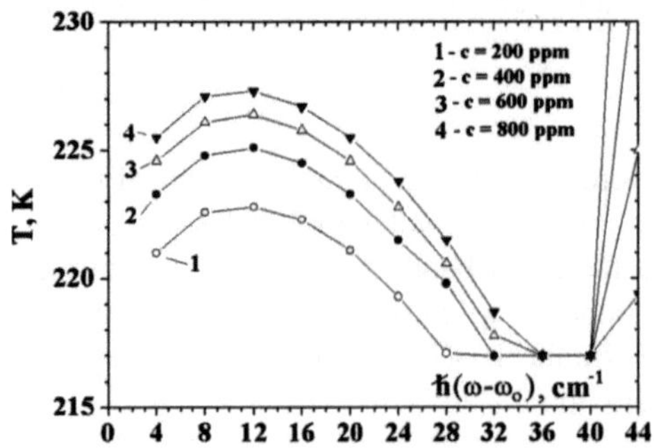

Fig. 10.11 Radiative temperature as a frequency function at indicated concentrations of atmospheric CO_2 molecules

$$u_\omega(h) = \int_h^\infty k_{ef}(\omega, z)dz = \frac{k_{ef}(\omega, h)\Lambda}{2} = \frac{2}{3} \tag{10.43}$$

Figure 10.11 gives the dependence of the radiative temperature on the radiation frequency in accordance with Formula (10.43).

Note that the effective radiation temperature for a given frequency increases with the increasing concentration of carbon dioxide because it leads to an altitude increase for a layer responsible for emission at a given frequency, but the stratosphere ($h >$ 10 km) temperature increases with increasing altitude, in contrast to the troposphere ($h <$ 10 km). One can add to this that according to satellite measurements [21] the radiative atmospheric temperature at wavelengths near 15 μm is close to 220 K which corresponds to the minimal temperature in a region between the troposphere and stratosphere.

Thus, the radiation flux at a given frequency ω is the equilibrium radiation flux at the radiative temperature which coincides with that of a layer with the optical thickness $u_\omega(h) = 2/3$, and the optical thickness is taken from an indicated altitude up to infinity in accordance with Formula (10.43). From this it follows that an injection of additional carbon dioxide leads to emission of radiation at higher altitudes. The change of the radiation flux is given by

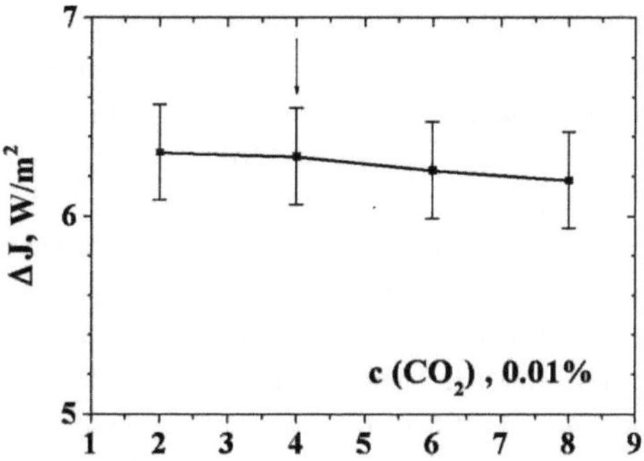

Fig. 10.12 Change of the flux of outgoing atmospheric radiation ΔJ as a result of injection of carbon dioxide in the air atmosphere if it was absent there at the beginning. The *arrow* indicates the contemporary content of carbon dioxide

$$\Delta J = \int\limits_{\omega-\Delta\omega/2}^{\omega+\Delta\omega/2} \frac{\hbar\omega^3 d\omega}{4\pi^2 c^2} \left\{ \left[\exp\left(\frac{\hbar\omega}{T_\uparrow}\right) - 1 \right]^{-1} - \left[\exp\left(\frac{\hbar\omega}{T}\right) - 1 \right]^{-1} \right\}, \quad (10.44)$$

and in the first approximation we assume that atmospheric carbon dioxide does not influence atmospheric emission. Figure 10.12 gives the change in the flux of outgoing thermal radiation as a result of carbon dioxide injection in the atmosphere at a given CO_2 concentration. If carbon dioxide is absent in the atmosphere, within the framework of the blackbody model, the equilibrium radiation is characterized by a temperature $T_\uparrow = 244\,\mathrm{K}$, as follows from Formula (10.15). This corresponds to the flux $J_o = \sigma T_\uparrow^4 = 200\,\mathrm{W/m^2}$ of outgoing radiation (see Fig. 10.1). A decrease of the radiation flux as a result of injection of carbon dioxide in the atmosphere is given in Fig. 10.12. The presence of CO_2 in the atmosphere leads to a decrease of the Earth's temperature compared with the case where carbon dioxide is absent in atmospheric air. As can be seen, the change of the flux of outgoing radiation is relatively small, and the corresponding increase of the Earth's temperature ΔT as a result of carbon dioxide insertion in the atmosphere is given by

$$\Delta T = \frac{T}{4}\frac{\Delta J}{J_o} \approx -1.8K \quad (10.45)$$

As follows from Fig. 10.12, the Earth's temperature decreases (not increases!) as carbon dioxide is inserted in the atmosphere. But a temperature change is small in variation of the CO_2 concentration from its contemporary value. In particular, the variation of the Earth's temperature as a result of doubling of the CO_2 concentration from its contemporary value does not exceed the accuracy of evaluations ($\sim$0.1 °C). Hence, one can conclude that contemporary variation of the carbon dioxide content does not lead to a change of the outgoing radiation flux and therefore does not act on the Earth's temperature change due to outgoing radiation.

Thus, the presence of carbon dioxide in the atmosphere decreases the outgoing atmospheric radiative flux that leads to a decrease of the Earth's temperature by approximately 1.8 K. Change of the average temperature at doubling of the concentration of atmospheric CO_2 molecules does not exceed 0.1 K which is the accuracy of this estimation. The favorable conditions should be noted that allow us to provide a high accuracy and reliability of the above results. A long lifetime of CO_2 molecules in the atmosphere leads to a strong mixing between carbon dioxide and air molecules, so that at high altitudes far from sources and absorbers of carbon dioxide the concentration of CO_2 molecules is constant both in space and in time. Outgoing thermal radiation due to carbon dioxide is formed at high altitudes, so that thermal radiation due to carbon dioxide is regular. Therefore one can evaluate with a high reliability the local vertical profile of the atmospheric temperature on the basis of measured current fluxes as a frequency function. In addition, the contribution of carbon dioxide to the total radiative flux is small that allows one to use the perturbation theory.

In analyzing the outgoing atmosphere radiation due to CO_2 molecules, we are restricted ourselves by the strongest vibration transition. We now estimate the contribution of other transitions, the parameters for which the zero rotation momentum $j = 0$ of the upper state by energy is given in Table 10.1 [34]. Because the number density of vibrationally excited CO_2 molecules is relatively small at room temperature, transition 1 of Table 10.1 gives the main contribution to the flux of outgoing radiation due to CO_2 molecules. This transition creates a radiation band that lies between 625 and 710 cm^{-1} at the contemporary concentration of CO_2 molecules. Below we neglect the transition 4 of Table 10.1 because it is screened by the main transition, and also transitions 5–7, because they proceed from more excited states with smaller populations. Let us take into account the contribution to outgoing radiation due to the transition 2 with the center at 721 cm^{-1}, and also due to the transition 3 with the center at 618 cm^{-1}. The energies of these transitions are located at the tail of the atmospheric absorption coefficient for the main transition outside its radiation band. Therefore, one can assume these transitions to be not overlapped with others. One can calculate a radiation band for the transition with the center at 721 cm^{-1} that is approximately 0.8 cm^{-1} at the contemporary concentration of CO_2 molecules,

Table 10.1 Parameters of vibration-vibration transitions for CO_2 molecules [34] which are of interest for thermal atmosphere radiation

Number	Vibration transition	ΔE, cm^{-1}	E_o, cm^{-1}	λ, μm	τ, c
1	$01^10 \to 00^00$	667.38	667.38	14.98	0.64
2	$10^00 \to 01^10$	720.81	1388.19	13.87	2.1
3	$02^00 \to 01^10$	618.03	1285.41	16.18	0.83
4	$02^20 \to 01^10$	667.75	1335.13	14.98	3.2
5	$00^01 \to 10^00$	960.96	2349.15	10.41	2.9
6	$00^01 \to 02^00$	1063.74	2349.15	9.401	5.0
7	$00^01 \to 00^00$	2349.15	2349.15	4.257	2.2×10^{-3}

Here ΔE is the energy of a vibration transition, E_o is the excitation energy for the upper transition state, λ is the transition wavelength, and τ is the radiative lifetime for the upper vibration state of this transition whereas the maximum optical thickness of the layer at the altitude of 10 km for transition with the center at 618 cm^{-1} does not exceed 0.5 at the altitude of 10 km. Thus, one can conclude that the part of outgoing infrared radiation determined by CO_2 molecules is created by the main vibration transition with the center at 667 cm^{-1}, whereas the contribution of other vibration transitions does not exceed a few percent. Within the limits of accuracy, the average flux of outgoing thermal radiation of the atmosphere is not changed as a result of doubling of the contemporary concentration of CO_2 molecules within the accuracy of evaluations.

10.5 Thermal Radiation of Atmospheric CO_2 Molecules Towards the Earth

The greenhouse effect in the Earth's atmosphere determines the Earth's energetic balance, and we consider this effect due to a change of the concentration of atmospheric CO_2 molecules. Because this change is relatively small, one can analyze this effect under stable values of other atmospheric parameters. The contribution of carbon dioxide to flux of outgoing infrared radiation of the atmosphere as a function of the CO_2 concentration is given in Fig. 10.12. We now evaluate the radiation flux towards the Earth's surface due to atmospheric CO_2 molecules.

We use the above assumption that for the standard atmosphere model and a contemporary content of carbon dioxide the absorption coefficient k_ω in the infrared spectrum range is independent of a frequency and according to Formula (10.19) is equal to $\kappa_o \approx 0.42\,\text{km}^{-1}$ at the Earth's surface. We assume atmospheric absorption to be determined by water vapor and aerosols, and additional carbon dioxide absorbs independently of them. Therefore the total absorption coefficient $k_t(\omega)$ is a sum of those due to water κ_o and owing to carbon dioxide k_ω determined by Formula (10.28). We have under these conditions

$$k_t(\omega) = \kappa_o + k_\omega \tag{10.46}$$

Carbon dioxide absorbs the following part of Earth's thermal radiation at a frequency ω

$$\alpha_\omega = \frac{k_\omega}{\kappa_o + k_\omega} \tag{10.47}$$

Correspondingly, a distance from the Earth's surface for a layer that determines the radiation temperature according to Formula (10.10), is equal to

$$l = h_\downarrow(1 - \alpha_\omega), \tag{10.48}$$

and the temperature T_l of a radiating layer is given by

$$T_l = T_o - l\frac{dT}{dh} \tag{10.49}$$

Here $T_o = 288\,\mathrm{K}$ is the average Earth's temperature and $dT/dh = 6.5$ K/km is the temperature gradient. On the basis of these formulas we have by analogy with Formula (10.41) for an additional radiation flux $\Delta J_\downarrow$ owing to presence of carbon dioxide in the atmosphere:

$$\Delta J_\downarrow = \int\limits_{\omega-\Delta\omega/2}^{\omega+\Delta\omega/2} \frac{\hbar\omega^3 d\omega}{4\pi^2 c^2}\left\{\left[\exp\left(\frac{\hbar\omega}{T_\downarrow}\right)-1\right]^{-1} - \left[\exp\left(\frac{\hbar\omega}{T_l}\right)-1\right]^{-1}\right\} \tag{10.50}$$

Here $\Delta\omega$ is the width of the radiation band, $T_\downarrow = 276\,\mathrm{K}$ is the temperature of an atmosphere layer that is responsible for radiation at a given frequency at the contemporary concentration of atmospheric carbon dioxide, and $T_l(\omega)$ is the same temperature at a given concentration of carbon dioxide.

We assume an additional radiation flux $\Delta J_\downarrow$ towards the Earth's surface due to atmospheric carbon dioxide to be small compared to the total radiation flux towards the Earth's surface that is equal to $J_o = 327\,\mathrm{W/m^2}$ according to Fig. 10.1. We then have according to Formula (10.42) for the temperature change of the Earth's surface that compensates an additional flux $\Delta J_\downarrow$ to the Earth's surface:

$$\Delta T = \frac{T_\downarrow}{4}\frac{\Delta J_\downarrow}{J_o} = \lambda \Delta J_\downarrow, \tag{10.51}$$

where $T_\downarrow$ is the temperature of a radiating layer (10.15) that determines the energetic balance of the Earth, $\lambda = 0.37\,\mathrm{K \cdot m^2/W}$.

In contrast to outgoing radiation which flux is determined by vibration transitions with the center at the transition energy of $667\,\mathrm{cm}^{-1}$, three vibration transitions of Table 10.2 give the contribution to the radiation flux towards the Earth's surface, and one can ignore transitions other than those of Table 10.2. The latter means that the optical thickness with respect to these transitions $u = k_\omega h_\downarrow < 2/3$ at any frequency of these transitions. It should be noted that a radiation band for a given transition is not overlapped with other transitions. But, as follows from Table 10.2, the transition with the center at the energy of $667\,\mathrm{cm}^{-1}$ is overlapped with neighboring transitions. We use the random model for radiative transitions [26–28, 32] to determine the spectral radiative flux in a frequency range where neighboring transitions are overlapped. Within the framework of the used method where we operate with the absorption coefficient averaged over oscillations inside a given vibration-rotation transition, the random model represents the absorption coefficient as a sum of those due to neighboring transitions. Figure 10.13 contains the average temperature change as a function of the concentration of CO_2 molecules under the assumption that other atmospheric parameters are conserved. In particular, doubling of the concentration of CO_2 molecules compared to the contemporary content of an increase of the global Earth temperature is

Table 10.2 Parameters of radiation of atmospheric carbon dioxide for indicated transitions obtained under the assumption that spectra of a given and a neighboring transition do not overlap

Parameter	$c = 0.02\,\%$	$c = 0.04\,\%$	$c = 0.06\,\%$	$c = 0.08\,\%$
$\Delta J_{\downarrow}(667)$, $\mathrm{W/m^2}$	–1.0	0	0.55	0.93
$\Delta\omega_{\downarrow}(667)$, $\mathrm{cm^{-1}}$	110.3	114.3	116.7	118.3
$\Delta T_{\downarrow}(667)$, K	–0.19	0	0.11	0.18
$\Delta J_{\downarrow}(721)$, $\mathrm{W/m^2}$	–2.22	0	1.14	1.87
$\Delta\omega_{\downarrow}(721)$, $\mathrm{cm^{-1}}$	77.0	83.1	86.3	88.5
$\Delta T_{\downarrow}(721)$, K	–0.43	0	0.22	0.36
$\Delta J_{\downarrow}(618)$, $\mathrm{W/m^2}$	–1.56	0	0.80	1.34
$\Delta\omega_{\downarrow}(618)$, $\mathrm{cm^{-1}}$	87.3	92.8	95.4	97.4
$\Delta T_{\downarrow}(618)$, K	–0.30	0	0.16	0.26

Here $\Delta J_{\downarrow}$ is the change of the radiative flux towards the Earth's surface due to a given transition of atmospheric CO_2 molecules according to Formula (10.50), $\Delta T_{\downarrow}$ is the variation of the average Earth temperature owing to an indicated vibration transition of atmospheric CO_2 molecules, and $\Delta\omega$ is the width of a radiation band for this transition. The argument indicates the energy (in $\mathrm{cm^{-1}}$) of the vibration-vibration transition at zero rotation number

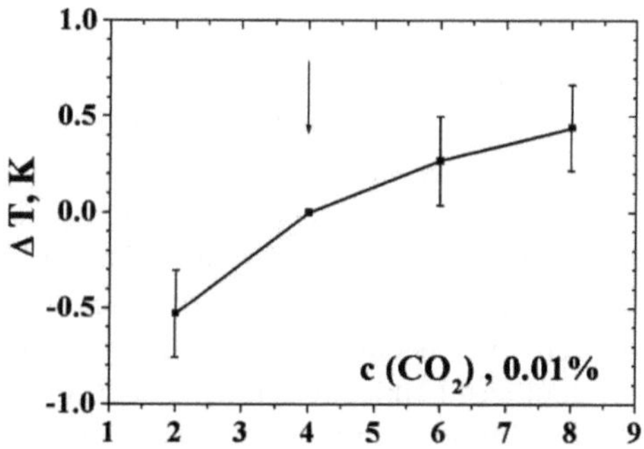

Fig. 10.13 Variation of the Earth's surface temperature depending on the concentration of atmospheric CO_2 molecules. Data of Table 10.2 are used together with the random model in a spectrum range where spectra of neighboring transitions overlap

$$\Delta T = 0.4 \pm 0.2K \tag{10.52}$$

Thus, we analyze above the character of atmospheric radiation due to atmospheric carbon dioxide. The basis of this analysis is a uniform distribution of CO_2 molecules in a space, a simple spectrum of linear CO_2 molecules, and a large optical thickness of the atmosphere that allows us to reduce the emission of a layer with a varied temperature over altitudes to the emission of a layer with a constant temperature. The accuracy of this method as a result of expansion over a small parameter is determined mostly by errors of the standard atmosphere model. As follows from the above analysis, the flux of outgoing radiation is practically independent of the carbon dioxide concentration if the latter is of the order of the contemporary one. The above analysis gives a simple algorithm to determine a current profile of the temperature of upper tropospheric layers on the basis of satellite measurements of the spectral fluxes due to carbon dioxide.

The accuracy of determination of the spectral radiative flux towards the Earth is worse than that for outgoing radiation for the following reasons. First, there are sources and absorbers of carbon dioxide molecules near the Earth's surface that lead to a nonuniform distribution of the number density of carbon dioxide molecules in a space near the Earth's surface. Second, radiation of CO_2 molecules is screened partially by radiation of other molecules, aerosols, and particles. Third, in contrast to outgoing radiation which is determined by one vibration transition, several transitions give the contribution to the radiation flux due to CO_2 molecules, and spectra of these transitions are overlapped. These factors decrease the accuracy of determination of radiative fluxes towards the Earth's surface, estimated as 50 %.

10.6 Climate Change Due to Greenhouse Effect

The Earth's climate varies now [35–38]. We now consider the climate change from the standpoint of accumulation of carbon dioxide in the Earth's atmosphere and the greenhouse effect that is supported by this factor, although the preceding analysis proves that atmospheric water is the main component determining the atmospheric greenhouse effect, and therefore the change of an amount of atmospheric water influences climate change in the first place [39, 40]. It should be noted that a water vapor in the form of free water molecules is a radiator of infrared radiation, whereas aerosols influence the Earth's energetic balance both as radiators of infrared radiation and also scatterers or absorbers of solar radiation [41–44]. Processes involving atmospheric aerosols are determined by processes over land and by human activity. Therefore the aerosol concentration in the northern hemisphere is three times more than that in the southern hemisphere [45]. This testifies to the qualitative analysis based on the standard atmosphere model which is used here.

In analyzing climate change, we use as a basis numerical factors that characterize this change. According to NASA information [46] the average Earth temperature has increased by

$$\Delta T = (0.8 \pm 0.1)\,^{\circ}\mathrm{C} \tag{10.53}$$

from 1880 up to now. Unfortunately, this value is not grounded. Moreover, an increase of the average Earth temperature is a nonmonotonic function of time [47], and this time dependence is not correlated with accumulation of atmospheric carbon dioxide [47].

In addition (10.53), the volume of glaciers diminishes by 13 % per decade, that is, a decrease of glacier volume twice proceeds approximately through 50 years [46]. The latter is determined by atmospheric dust accumulated in glaciers. An essential contribution to its amount proceeds from volcanic eruptions [48] because its material contains micron and submicron particles. In particular, the eruption of the Krakatau volcano in Indonesia changed the Earth's energetic balance for several months. Let us estimate the total mass of atmospheric micron-size aerosols that can screen solar radiation. Evidently, solar radiation is scattered effectively on particles whose size is compared to the wavelength of radiation, and we take the particle radius r to be $r = 1\,\mu$m. For total screening the total area of particle projections must cover the Earth's surface, and in this case the aerosol number per unit Earth's area is $n = 1/(\pi r^2) \sim 3 \times 10^7\,\text{cm}^{-2}$ which corresponds to the aerosol mass per unit Earth's area $\rho \sim 1 \times 10^{-4}\,\text{g/cm}^2$ compared to the air mass per unit Earth's area $1 \times 10^3\,\text{g/cm}^2$. As can be seen, a relatively small dust amount is able to screen solar radiation. Note that the fall velocity for a micron-size aerosol according to Stokes law is $\sim$0.01 cm/s, so that being thrown at altitudes of $\sim$10 km, a particle may be located in the atmosphere for several months. The same situation is realized if dust particles occur in the atmosphere as a result of nuclear explosion [49, 50]. This atmospheric state is called nuclear winter or atomic winter [49]. Note that the above estimations relate to optically dark materials that absorb visible solar radiation, and as follows from Fig. 10.2, water aerosols do not relate to these materials.

Let us consider briefly various mechanisms of climate change. Solar radiation plays a key role in the Earth's energetic balance, and variations of solar irradiance may lead to a variation of the Earth's temperature. Although solar irradiance is often called a solar constant, its small fluctuations may be essential. From the above consideration it follows that yearly fluctuations of this value may be 5 W/m^2, and this is on the order of 1 W/m^2 for the average solar flux on the Earth's surface. According to Formula (10.44) this leads to fluctuations of $\sim$1 K in the Earth's temperature and must be taken into consideration. In addition, note that the area of the Earth's surface covered by clouds correlates with the flux of cosmic radiation [12, 51]. Inasmuch as cloud aerosols contribute to thermal radiation of the Earth, cosmic radiation may influence Earth's climate [51].

Greenhouse gases other than carbon dioxide may be taken into account in the energetic balance of the Earth. Moreover, starting in 1750 the methane concentration in the atmosphere has increased by more than 150 % [52]. In the same manner stratospheric ozone as a greenhouse atmospheric gas may influence thermal emission [53]. The same action may be followed from a stratospheric water vapor [54, 55]. But the total contribution of these greenhouse gases to thermal emission flux of the atmosphere does not exceed 10 %, and we do not account for them.

In the above analysis, on the basis of the standard atmosphere model we operate with average atmospheric parameters that are identical for the total globe. This may give an error if nonuniformities arise in a stepwise manner. As an example of this one can consider oceanic streams whose existence determines the climate of neighboring regions. For instance, the Gulf Stream transfers warm water over large distances and in this manner influences the climate of large regions. Variations of atmospheric parameters may cause a change in variations of conditions for existence of these streams. In other words, this ocean-atmosphere system may be found in two states, with these streams and without them [56–60]. In contrast to the above examples where variations of the atmospheric parameters proceeded in a continuous way, in this case such a transition may have a stepwise character as a result of the continuous variation of atmospheric parameters.

Thus, there are many factors that influence the climate including industrial and practical human activity. In this analysis we include in the climate concept the energetic balance of the Earth, although this concept is wider. Moreover, often climate change is assumed as a result of the greenhouse atmospheric effect due to atmospheric carbon dioxide. The problem is that under the action of contemporary propaganda most people are convinced of carbon dioxide as the main reason for climate change. We above ascertain the basis of this belief and show its falsehood.

Following the Arrhenius concept [61] of 1896, it is convenient to use the average Earth temperature as the climate characteristic. Of course, other parameters may be used as the climate indicator. In particular, the rising sea level that is now (2.9 ± 0.4) mm/year [62] may be used as this characteristic. Note there are severe difficulties in determination of the global temperature change because its local, seasonal, and daily fluctuations are on the order of tens of degrees, whereas an accuracy of its average value must not exceed $0.1°$ K. But such an accuracy of measurements requires monitoring from thousands of meteorological stations; that is, the result (10.53) may be obtained at the cost of large efforts. Moreover, this accuracy in 1880 causes doubts. Another characteristic to describe the atmospheric state is the parameter related to varying atmospheric components. According to the Arrhenius concept [61], the most convenient component of this type is the concentration of CO_2 molecules. As can be seen, in this consideration the concentration of CO_2 molecules is an indicator of the atmospheric state, but not a reason for climate change. The Arrhenius concept allows one to introduce some parameters describing climate change. One of them is equilibrium climate sensitivity (ECS) [63], which is the variation of the global Earth temperature δ at the doubling of the atmospheric carbon dioxide concentration.

In particular, let us determine ECS under the assumption that other atmospheric parameters influencing the value (10.53) vary subsequently proportionally to the change of the concentration of CO_2 molecules. Then we obtain for ECS:

$$\delta = \Delta T \frac{\ln 2}{\ln(c_2/c_1)}, \tag{10.54}$$

where c_1 and c_2 are the concentrations of CO_2 molecules at the beginning and end of a period with a change ΔT for the average Earth temperature. On the basis of measurements at the Mauna Loa Observatory [64, 65] (Fig. 2.11) we have that the concentration of CO_2 molecules varied from 316 in 1959 to 402 ppm in 2015. In 1880 the concentration of CO_2 molecules was (280–290) ppm. Then Formula (10.54) gives

$$\delta = \Delta T \frac{\ln 2}{\ln(c_2/c_1)} = (1.6 \pm 0.3)K \tag{10.55}$$

One can compare this with values under various assumptions for variation of other atmospheric parameters. We join these evaluations and then on the statistical average of the results [66–73]) we have

$$\delta = (3.0 \pm 1.5)\,°\mathrm{C} \tag{10.56}$$

It should be noted that under the assumptions that other atmospheric parameters do not vary at doubling of the concentration of CO_2 molecules, according to Formula (10.52) ESC equals $\delta T = (0.4 \pm 0.2)\,\mathrm{K}$.

One can see a large difference between Formulas (10.52) and (10.55) or (10.56) for equilibrium climate sensitivity. These formulas correspond to different conditions of climate change, so that in the first case the carbon dioxide concentration varies at the conservation of other atmospheric parameters, whereas in the second case other atmospheric parameters vary simultaneously with the CO_2 content as it takes place in reality. Comparison of these formulas exhibits that an increase of the carbon dioxide concentration makes a small contribution to the atmospheric greenhouse effect. We also note that if the concentration of atmospheric CO_2 molecules varies from 290 to 400 ppm, as estimated from 1880 to 2015, Earth's temperature varies by approximately 0.2 K which follows from the data of Table 10.2. This value corresponds to conditions where other atmospheric parameters do not vary and is less than the value under real conditions (Formula (10.53)).

We now evaluate the change of the average Earth temperature with an increasing amount of atmospheric water within the framework of the model (10.20) for the absorption coefficient. We assume the proportional change of the amount of free water molecules and aerosols, and the size distribution function of aerosols is conserved in this change. Then the absorption coefficient K is equal to

$$k = k_o + \Delta k, \;\; \Delta k = \frac{\Delta \rho}{\rho}, \tag{10.57}$$

where ρ is the water density in the atmosphere and $\Delta\rho$ is its change. From this it follows for the altitude change

$$\Delta h = h_{\downarrow}\frac{\Delta k}{k_o}, \tag{10.58}$$

so that the altitudes $h_{\downarrow} - \Delta h$ and $h_{\uparrow} - \Delta h$ are responsible for emission towards the Earth and outside, respectively. Repeating the operations used in the temperature change as a result of variation of carbon dioxide, we obtain for the variation of the Earth's temperature ΔT

$$\Delta T = \frac{T_E}{4}\frac{\Delta J}{J_E}, \;\; \Delta J = 4\Delta h \frac{dT}{dh}\left(\frac{J_{\downarrow}}{T_{\downarrow}} - \frac{J_{\uparrow}}{T_{\uparrow}}\right) \tag{10.59}$$

Here $J_{\downarrow} = 327\,\mathrm{W/m^2}$, $J_{\downarrow} = 200\,\mathrm{W/m^2}$, $J_E = 386\,\mathrm{W/m^2}$ are the radiation fluxes given in Fig. 10.1, $T_E = 288\,\mathrm{K}$ is the Earth's temperature, $T_{\downarrow} = 276\,\mathrm{K}$, $T_{\uparrow} = 244\,\mathrm{K}$ are the temperatures responsible for atmospheric emission towards the Earth and outside, Δh is given by Formulas (10.57) and (10.58), and the temperature gradient in the atmosphere is equal to $dT/dh = 6.5\,\mathrm{km}$. Figure 10.14 contains the temperature change in accordance with Formula (10.59). Note that the absorption coefficient k is proportional to the atmospheric water concentration, and we assume proportional variations of free and bound (aerosol) water molecules. In particular, according to the

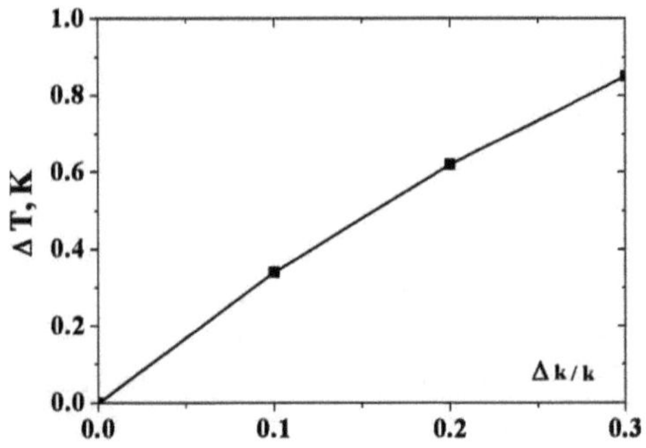

Fig. 10.14 Variation of the temperature of the Earth's surface depending on the total concentration of atmospheric water molecules in the free state and in the bound state in aerosols. Data of Table 10.2 are used together with the random model in a spectrum range where spectra of neighboring transitions overlap

data of Fig. 10.14, the NASA value for the temperature variation ($\Delta T = 0.8$ K) may be attained if approximately 0.5 % of atmospheric water molecules are converted into aerosols. This causes alarm because such conversion may be realized by a low energy action, that testifies to the instability of the real atmosphere. Note that the real optical thickness of the Earth's atmosphere over oceans is near 0.14 [74, 75] and approximately 20 % of this value has an anthropogenic nature.

Note that the problem of climate change exceeds the bounds of our analysis. Indeed, as a result of the above analysis we give reliable estimations that are valid under indicated conditions. In reality, although the NASA value for the Earth's temperature variation during the last century $\Delta = 0.8$ K [46] corresponds more or less to other estimations, this value does not vary monotonically during this time and does not correlate with the variation of the CO_2 concentration [47]. Moreover, some scientists, including L. Pauling, consider accumulation of carbon dioxide in the atmosphere as a result of ocean warming, because the mass of atmospheric CO_2 is 1.5 % of the oceanic one [47]. Therefore the problem of the Earth's warming is complex and above we analyze only some elements of this problem. It should be added to this that the atmospheric sciences have some methods and mathematical codes that allow one to analyze complex atmospheric situations. In the preceding we consider only simple elements of atmospheric physics, but usage of physical information and methods leads to reliable results of such an analysis.

References

1. J.B.J. Fourier, Annal. Chem. Phys. **27**, 136 (1824)
2. J.B.J. Fourier, Mem. Acad. R. Sci. **7**, 569 (1827)
3. I.I. Sobelman, *Atomic Spectra and Radiative Transitions* (Springer, Berlin, 1979)
4. V.P. Krainov, H.R. Reiss, B.M. Smirnov, *Radiative Processes in Atomic Physics* (Wiley, New York, 1997)

5. B.M. Smirnov, *Physics of Weakly Ionized Gases* (Mir, Moscow, 1980)
6. B.M. Smirnov, *Physics of Ionized Gases* (Wiley, New York, 2001)
7. *U.S. Standard Atmosphere.* (U.S. Government Printing Office, Washington, 1976)
8. W. Wien, Wied. Ann. Phys. Chem. **58**, 662 (1896)
9. F. Reif, *Statistical and Thermal Physics* (McGrow Hill, Boston, 1965)
10. https://en.wikipedia.org/wiki/Electromagnetic-absorption-by-water
11. C.M.R. Platt, Quart. J. R. Meteorolog. Soc. **102**, 553 (2006)
12. H. Svensmark, E. Frils-Christensen, J. Atmos. Terr. Phys. **59**, 1225 (1997)
13. H. Svensmark et al., Proc. R. Soc. **A463**, 385 (2007)
14. H. Svensmark, T. Bondo, J. Svensmark, Geophys. Res. Lett. **36**, L151001 (2009)
15. H. Svensmark, M.B. Enghoff, J.O.P. Pedersen, Phys. Lett. A **377**, 2343 (2013)
16. G. Herzberg, *Molecular Spectra and Molecular Structure* (Princeton, Van Nostrand Reinhold, 1945)
17. http://www1.lsbu.ac.uk/water-vibration-spectrum
18. W.K. Toshiba, T. Woods, F. Eparvier et al., J. Atmos. Sol-Terr. Phys. **62**, 1233 (2000)
19. K. Ya, *Kondratyev, Radiation in the Atmosphere* (Academic Press, New York, 1969)
20. E.J. McCartney, *Absorption and Emission by Atmospheric Gases* (Wiley, New York, 1983)
21. K.N. Liou, *An Introduction to Atmospheric Radiation* (Academic Press, Amsterdam, 2002)
22. G.W. Petry, *A First Course in Atmospheric Radiation* (Sunlog Publishing, Madison, 2006)
23. W. Zdunkowski, T. Trautmann, A. Bott, *Radiation in the Atmosphere* (Cambridge University Press, Cambridge, 2007)
24. R.A. Hanel et al., J. Geophys. Res. **77**, 2629 (1972)
25. M.A. El'yashevich, Molecular Spectroscopy (Moscow, Fizmatgiz, 1963; in Russian)
26. R.M. Goody, Y.L. Yung, *Atmospheric Radiation: Theoretical Basis* (Oxford University Press, New York, 1989)
27. R.M. Goody, *Atmospheric Radiation: Theoretical Basis* (Oxford University Press, London, 1964)
28. R.M. Goody, *Principles of Atmospheric Physics and Chemistry* (Oxford University Press, New York, 1995)
29. W.M. Elsasser, Phys. Rev. **54**, 126 (1938)
30. B.M. Smirnov, G.V. Schlyapnikov, Sov. Phys. Uspekhi **130**, 377 (1980)
31. E.T. Whittaker, G.N. Watson, *Modern Analysis* (Cambridge University Press, London, 1940)
32. B.M. Smirnov, *Plasma Processes and Plasma Kinetics* (Wiley, Berlin, 2007)
33. M.L. Salby, *Physics of the Atmosphere and Climate* (Cambridge University Press, Cambridge, 2012)
34. A.A. Radzig, B.M. Smirnov, *Reference Data on Atoms, Molecules and Ions* (Springer, Berlin, 1985)
35. D. Martyn, *Climates of the World* (Elsevier, Amsterdam, 1992)
36. T.R. Karl, K.E. Trenberth, Science **302**, 1719 (2003)
37. *State of the climate in 2003* ed. by D.H.Levinson, A.M. Waple, Bull. Am. Meteorol. Soc. **85**(6), S1 (2004)
38. *State of the climate in 2004* ed. by D.H.Levinson, Bull. Am. Meteorol. Soc. **86**, S1 (2005)
39. A. Dai, K.E. Trenberth, T.R. Karl, J. Clim. **12**, 2451 (1999)
40. C.A. Smith, J.D. Haigh, R. Toumi, Geophys. Res. Lett. **28**, 179 (2001)
41. V. Ramanathan et al., Science **294**, 2119 (2001)
42. Y.J. Kaufman, D. Tanre, O. Boucher, Nature **419**, 215 (2002)
43. D.P. Wylie et al., J. Climate **18**, 3021 (2005)
44. N. Bellouin et al., Nature **438**, 1138 (2005)
45. https://www.wmo.int/pages/themes/climate/causes-of-climate-change.php
46. http://climate.nasa.gov
47. J.M. Kauffman, J. Sci. Explor. **4**, 723 (2007)
48. M.M. Joshi, K.P. Shine, J. Clim. **16**, 3525 (2003)
49. P.J. Crutzen, J.W. Birks, Ambio **11**, 114 (1982)
50. R.P. Turco, O.B. Toon, T.P. Ackerman, J.B. Pollack, C. Sagan, Science **222**, 1283 (1983)

51. N.J. Shaviv, Phys. Rev. Lett. **89**, 051102 (2002)
52. http://www.physicalgeography.net/fundamentals/7h.html
53. D.J. Karol, Science **302**, 236 (2003)
54. S. Solomon, K.H. Rosenlof, R.W. Portmann et al., Science **327**, 1219 (2010)
55. S. Fueglistaler et al., The relation between atmospheric humidity and temperature trends for stratospheric water. J. Geophys. Res. **118**, 1052 (2013)
56. S. Manabe, Mon. Weat. Rev. **97**, 739 (1969)
57. S. Manabe, R.J. Stouffer, J. Climate **1**, 841 (1988)
58. W.M. Washington, G.A. Meehl, Clim. Dynam. **4**, 1 (1989)
59. S. Manabe, R.J. Stouffer, J. Clim. **12**, 2224 (1999)
60. R.J. Stouffer, S. Manabe, Clim. Dynam. **20**, 759 (2003)
61. S. Arrhenius, Phil. Mag. **41**, 237 (1896)
62. https://en.wikipedia.org/wiki/Global-warming
63. https://en.wikipedia.org/wiki/Climate-sensitivity
64. https://en.wikipedia.org/wiki/Mauna-Loa-Observatory
65. http://www.esrl.noaa.gov/gmd/ccgg/trends
66. G.N. Plass, Tellus **8**, 140 (1956)
67. N. Andronova, M.E. Schlesinger, J. Geophys. Res. **106**, D22605 (2001)
68. M.A. Snyder, J.L. Bell, L.C. Sloan, Geophys. Res. Lett. **29**, 014431 (2002)
69. J.D. Annan, J.C. Hargreaves, Geophys. Res. Lett. **33**, L06704 (2006)
70. A. Ganopolski, T. Schneider von Deimling, Geophys. Res Lett. **35**, L23703 (2008)
71. M.E. Walter, Not. Am. Mat. Soc. **57**, 1278 (2010)
72. A. Schmittner, N.M. Urban, J.D. Shakun et al., Science **334**, 1385 (2011)
73. J.T. Fasullo, K.E. Trenberth, Science **338**, 792 (2012)
74. L.A. Remer et al., J. Atmos. Sci. **62**, 947 (2005)
75. Y.I. Kaufman et al., Geophys. Res. Lett. **32**, L17804 (2007)

Chapter 11
Local Atmospheric Photoprocesses

Abstract Recombination radiation and bremsstrahlung are evaluated for a conductive lightning channel contained an air plasma. Reflection of radio waves and their absorption as a result of interaction with ionospheric layers are analyzed. Photoionization processes are considered which are responsible for propagation of the ionization wave. Luminous phenomena are represented in an upper atmosphere at altitudes below 100 km which develop under the action of the Earth electric field.

11.1 Radiation of Lightning Channel

Along with the above global atmospheric photoprocesses in visible and infrared spectrum ranges, some local photoprocesses may accompany some phenomena in the Earth atmosphere. We below consider such photoprocesses which are of importance for atmospheric phenomena starting from emission of the lightning channel. Due to a high temperature, this channel contains dissociated and partially ionized air. Assuming for simplicity this channel to be contained in a uniform volume of the cylinder shape with thermodynamic equilibrium in the plasma due to a high air pressure, we calculate the radiation power of this conductive channel. Basic ways of radiation for a hot plasma of atmospheric air include bremsstrahlung at electron-ion scattering and photorecombination radiation resulted from electron capture by an ion with formation of a bound electron state. If this bound state corresponds to a highly excited atom state, one can consider these processes as a result of scattering of a classical electron on a changed nucleus with photon formation according to the scheme

$$e + A^{+} \to e' + A, \quad e + A^{+} \to A(n) \tag{11.1}$$

Here e' is a scattering electron, and n is the principal quantum number of a forming atom. Since an electron is considered as a classical particle, one can combine the processes of bremsstrahlung and photorecombination and use Kramers formulas [1] for the cross section of these processes. Kramers formulas lead to the following

B.M. Smirnov, *Microphysics of Atmospheric Phenomena*,
Springer Atmospheric Sciences, DOI 10.1007/978-3-319-30813-5_11

differential cross section $d\sigma$ for scattering of a classical electron of a velocity v on a motionless ion with emission of radiation with frequencies in a range from ω up $\omega + d\omega$ [2, 3]

$$d\sigma = \frac{16\pi}{3\sqrt{3}} \frac{e^6}{m_e^2 c^3 v^2} \frac{d\omega}{\hbar\omega} \tag{11.2}$$

One can separate from this photorecombination radiation, where a highly excited state with the principal quantum number n, is formed, and the energy conservation law has the form

$$\hbar\omega = \frac{m_e v^2}{2} + \frac{m_e e^4}{2\hbar^2 n^2},$$

and according to Kramers formulas $n \gg 1$. This allows one to obtain from formula (11.2) the photorecombination cross section as [4]

$$\sigma_{rec} = \frac{16\pi}{3\sqrt{3}} \frac{e^{10}}{m_e c^3 v^2 n^3 \hbar^4 \omega}, \tag{11.3}$$

Though this formula holds true for large principal quantum numbers n, where a classical description of this process is valid, we use it also for the ground atom state using that the Kramers formula leads to an error of 25 %, if we use it for photoionization of the hydrogen atom in the ground state [2, 5].

The radiation power P_{rad} per unit volume for a uniform column of ionized air is equal

$$P_{rad} = \langle N_e N_i \hbar\omega v \sigma_{rec} \rangle = \frac{16\pi}{3\sqrt{3}} \left(\frac{e^2}{\hbar c}\right)^3 \frac{e^4}{m_e n^3} \left\langle \frac{1}{v} \right\rangle, \tag{11.4}$$

Here an air plasma is quasineutral, i.e. the electron and ion number density are identical $N_e = N_i$, and an average in the form of triangle parentheses corresponds to the Maxwell distribution function of electrons, and this gives for the specific power of photorecombination radiation as a result of electron-ion scattering with formation of an excited atom with the principal quantum number n [4]

$$\frac{P_{rad}}{N_e^2} = \frac{16\sqrt{2\pi}}{3\sqrt{3}} \left(\frac{e^2}{\hbar c}\right)^3 \frac{e^4}{(m_e T_e)^{1/2} n^3}, \tag{11.5}$$

where T_e is the electron temperature. As is seen, the main contribution to the radiation power follows from small n where the classical description of this process is not valid. If we introduce formally the atom ionization potential J as

$$J = \frac{m_e e^4}{2\hbar^2 n^2},$$

one can obtain on the basis of formula (11.5) for the specific power of radiation of a hot air plasma [6]

$$\frac{P_{rad}}{N_e^2} = \frac{\kappa_o J^{3/2}}{T_e^{1/2}}, \tag{11.6}$$

If the atom ionization potential J and the electron temperature T_e are expressed in eV, the proportionality coefficient in formula (11.6) is equal $\kappa_o = 8 \times 10^{-33}\,\mathrm{W \cdot cm^3}$. Since the classical approach which is used in derivation of this formula is not valid, one can use this formula as an estimation.

Taking a typical temperature of the lightning conductive channel during propagation of an electric current through it $T_e = 2 * 10^4\,\mathrm{K}$ and the ionization potential of the nitrogen atom to be $J = 14.5\,\mathrm{eV}$, we obtain from formula (11.6) $P_{rad}/N_e^2 \sim 10^{-30}\,\mathrm{W/cm^3}$. Taking for simplicity the number density of electrons and atoms to be identical, one can obtain at the temperature $T = 2 \cdot 10^4\,\mathrm{K}$ and atmospheric pressure of air $p = 1\,\mathrm{atm}$, that gives $N_e \sim 2 \times 10^{17}\,\mathrm{cm^{-3}}$ and the specific power of radiation $P_{rad} \sim 4 \cdot 10^4\,\mathrm{W/cm^3}$. Taking a typical channel radius during passing of the electric current $r \sim 0.3\,\mathrm{cm}$ and the pulse duration $\tau \sim 70\,\mu\mathrm{s}$, we have for the specific radiation energy $\varepsilon_{rad} = P_{rad} \cdot \pi r^2 \cdot \tau \sim 1\,\mathrm{J/cm}$. Taking the charge per pulse of middle lightning $Q = 2\,\mathrm{C}$ and the cloud-earth electric field strength $E = 200\,\mathrm{V/cm}$, one can find the electric energy per lightning pulse $QE = 400\,\mathrm{J/cm}$. As is seen, the radiative energy is a part of percent of the lightning electric energy.

11.2 Reflection of Electromagnetic Waves from Ionosphere

Another problem of the local radiative processes in the atmosphere is interaction of radio waves with an ionospheric plasma. In contrast to the previous problem where the radiative process results from electron collisions with atoms and ions, in this case scattering and absorption of electromagnetic waves results from collective properties of the ionospheric plasma. In considering the character of propagation of an electromagnetic wave through a quasi-uniform plasma, we use the standard method (for example, [4, 7–13]) to analyze the interaction between an electromagnetic wave and plasma. We are based on Maxwell equations for electromagnetic wave which has the form

$$curl\mathbf{E} = -\frac{1}{c}\frac{\partial \mathbf{H}}{\partial t}, \quad curl\mathbf{H} = \frac{4\pi}{c}\mathbf{j} - \frac{1}{c}\frac{\partial \mathbf{E}}{\partial t}, \tag{11.7}$$

where $\mathbf{E}$ and $\mathbf{H}$ are the electric and magnetic field strengths of an electromagnetic wave, $\mathbf{j}$ is the electric current density induced by the electromagnetic wave, and c is light velocity. Applying the curl operator to the first equation of this set and the operator $-(1/c)\,(\partial/\partial t)$ to the second equation of the set, and then eliminating the magnetic field from the resulting equations, we obtain

$$\nabla div\mathbf{E} - \Delta\mathbf{E} + \frac{4\pi}{c^2}\frac{\partial\mathbf{j}}{\partial t} - \frac{1}{c^2}\frac{\partial^2\mathbf{E}}{\partial t^2} = 0 \quad (11.8)$$

Because of the quasineutrality of the plasma, according to the Poisson equation we have $div\mathbf{E} = 0$. Next, the electric current due to motion of the electrons is $\mathbf{j} = -eN_o\mathbf{w}$, where N_o is the average number density of plasma electrons in the absence of an electromagnetic wave, and $\mathbf{w}_e$ is the electron velocity under the action of the electromagnetic field. The equation of motion for the electrons is $m_e d\mathbf{w}_e/dt = -e\mathbf{E}$, that gives

$$\frac{\partial\mathbf{j}}{\partial t} = -eN_o\frac{d\mathbf{w}_e}{dt} = \frac{e^2N_o}{m_e}\mathbf{E} \quad (11.9)$$

From this we obtain the equation for the electric field of the electromagnetic wave

$$\Delta\mathbf{E} - \frac{\omega_p^2}{c^2}\mathbf{E} + \frac{1}{c^2}\frac{\partial^2\mathbf{E}}{\partial t^2} = 0, \quad (11.10)$$

where ω_p is the plasma frequency

$$\omega_p = \sqrt{\frac{4\pi N_o e^2}{m_e}} \quad (11.11)$$

Taking the wave parameters in the standard form as

$$E = E'\exp[i(kx - \omega t)],\ N = N_o + N'\exp[i(kx - \omega t)], \quad (11.12)$$

where x is the coordinate in the direction of the wave propagation and substituting it in the above equation, we obtain the following dispersion relation for propagation of an electromagnetic wave in a plasma

$$\omega^2 = \omega_p^2 + c^2k^2 \quad (11.13)$$

In the limit of low electron number density ($\omega_p \to 0$), the dispersion relation corresponds to an electromagnetic wave propagating in a vacuum, $\omega = kc$. The second term in the right hand side of equation (11.13) takes in consideration the interaction between the electromagnetic wave and plasma.

As it follows from the dispersion relation (11.13), an electromagnetic wave can propagate in a plasma if its frequency exceeds the plasma frequency ω_p. Let us introduce the boundary number density of electrons N_o for a given frequency ω of an electromagnetic wave such that $\omega = \omega_o \equiv \omega_p(N_o)$ and assume the linear dependence of the electron number density near the boundary, so that

$$\omega_p^2 = \omega_o^2\left[1 + \frac{d\ln N_e}{dh}(h - h_o)\right], \quad (11.14)$$

where h_o is the boundary altitude. From this and the dispersion relation (11.13) it follows for the depth l of penetration of the electromagnetic wave inside a plasma

$$l = \left(\frac{c^2 dh}{\omega_p^2 d\ln N_e}\right)^{1/3}, \tag{11.15}$$

and the value l is determined with the accuracy up to a numerical factor.

Absorption takes place as a result of electron collisions with molecules. In particular, in tau-approximation which is used in the plasma theory usually, a typical time τ of these collisions is determined as

$$\frac{1}{\tau} = \sum_j \sqrt{\frac{8T}{\pi m_e}} \sigma_j N_j, \tag{11.16}$$

where the parameters N_j, σ_j are the number density, and the cross section for electron scattering on molecules of $j-$ component correspondingly. The reduced mass of colliding electron and molecule is equal practically to the electron mass m_e. From this instead of the dispersion relation (11.13) we have

$$\omega^2 = c^2 k^2 + \omega_p^2 \left(1 - \frac{i}{\omega\tau}\right) \tag{11.17}$$

The last term in the dispersion equation is responsible for absorption of an electromagnetic wave in the plasma. One can see that the wave is reflected from a plasma, i.e. wave absorption is negligible if the following criterion holds true

$$\tau \gg \tau_o = \left[\omega_p c^2 \left(\frac{d\ln N_e}{dh}\right)^2\right]^{-2/3} \tag{11.18}$$

Returning to the Earth atmosphere, we have that reflection of electromagnetic waves takes place from the ionosphere where electrons are present. Figure 6.3 contains a typical space distribution of electrons and positive ions in lower layers of the ionosphere [14], and electrons are responsible for reflection of radio waves. For simplicity, we approximate the measured electron number density of this figure by two straight lines and treat this space distribution from the standpoint of reflection of radio-waves. The results of this treatment are represented in Table 11.1. This Table contains the parameters of the standard atmosphere at indicated altitudes, namely, the temperature T, and the number density of air molecules N_{air}, and the approximated number density of electrons N_e. The frequency ω_o and the corresponding wavelength λ_o of electromagnetic waves are given in Table 11.1.

In addition, the depth of penetration l of the electromagnetic wave of a given frequency ω_o deep into the ionospheric layer in accordance with formula (11.15)

Table 11.1 Parameters of ionospheric layers and the character of reflection of radio waves from them

h, km	80	120	200
T, K	195	334	859
N_{air}, cm^{-3}	4.0×10^{14}	4.8×10^{11}	1.0×10^{10}
N_e, cm^{-3}	1.4×10^{4}	1.3×10^{5}	2.6×10^{5}
ω_o, MHz	2.6	23	47
λ_o, m	740	82	40
l, km	0.62	1.2	0.71
τ_o, μs	11	14	18
v_T, cm^{-3}	3.8×10^{4}	4.9×10^{4}	7.9×10^{4}
N_{cr}, cm^{-3}	7.8×10^{15}	3.7×10^{15}	2.9×10^{15}

and a typical time τ_o given by formula (11.18). We give also the critical number density N_{cr} of atmospheric atoms and molecules in accordance with formula

$$N_{cr} = \frac{1}{v_T \sigma \tau_o}, \tag{11.19}$$

where v_T is a thermal velocity of electrons, σ is the cross section of electron-molecule collisions which is assumed to be independent of the electron velocity and in accordance with measurements we take for estimations $\sigma = 3 \times 10^{-16}\,cm^2$. Table 11.1 contains the values of the critical number density N_{cr} of atmospheric atoms and molecules which characterizes the character of absorption of corresponding electromagnetic waves. Namely, if the criterion

$$N_{cr} \gg N_{air} \tag{11.20}$$

is fulfilled, the corresponding electromagnetic wave is reflected from an ionospheric layer and absorption is weak. In the opposite case an electromagnetic wave is absorbed at the ionosphere boundary for waves of a given frequency and does not scatter from it. As it follows from the Table 5.3 data, absorption in the ionospheric layers under real conditions is weak.

The maximum electron number density of the order of $10^6\,cm^{-3}$ may be reached at an altitude of about 200 km. Therefore, the best reflection conditions are fulfilled for short waves within the range of 30–100 m. Because the electron number density at any particular altitude depends on season, time of day and other factors, the quality of radio communications at a given wave length varies continually. Long waves are reflected at low altitudes where the atmospheric density is relatively high, and reflection of the waves is accompanied by damping.

11.3 Photoprocesses in Ionization Wave

Let us return to the problem of propagation of the ionization wave in the form of a streamer or leader which is considered in Sect. 5.5. Our task is to analyze the character of photoionization processes in the ionization wave. Indeed, in this case photons formed in a head of the wave as a result electron-molecule collisions, create seed electrons in front of the wave, and these electrons are multiplied and become the head of the ionization wave. We now consider possible mechanisms of these photoprocesses.

The most likely photoionization process in air [15, 16] proceeds in the wavelength range 980 Å $< \lambda <$ 1025 Å, since the threshold wavelength for ionization of oxygen molecules 1025 Å [17] exceeds the threshold wavelength 980 Å [18] for ionization of nitrogen molecules. As a result, resonant radiation due to radiative transitions involving discrete states of the nitrogen molecule leads to photoionization of the oxygen molecule.

The photoionization process is of importance for propagation of the ionization wave in atmospheric air. Therefore radiation of stable nitrogen molecules in the wavelength range 980 Å$< \lambda <$ 1025 Å [15] leads to photoionization of oxygen molecules. The Birge-Hopfield I radiation transitions $b^1\Pi_u \to X^1\Sigma_g$ for the nitrogen molecules, the Birge-Hopfield II radiation transitions $b'\,{}^1\Sigma_u^+ \to X^1\Sigma_g$, and Carroll-Yoshino Rydberg transitions $c_4'\,{}^1\Sigma_u^+ \to X^1\Sigma_g$ for the nitrogen molecule can lead to photoionization of oxygen molecules. In all these cases, the transitions proceed in the ground electron state $X^1\Sigma_g$ of the nitrogen molecule, and excited vibrational states of the ground electron state of the nitrogen molecules partake in these transitions. Therefore emitted photons are not absorbed by nitrogen molecules in cold air [15]. The radiative lifetime for upper states of these transitions ($b^1\Pi_u$, $b'\,{}^1\Sigma_u^+$, $c_4'\,{}^1\Sigma_u^+$) lies in the range $\tau = 1$–$10\,$ns [19, 20], and the average cross section of these photons by oxygen molecules as a result of their ionization is equal $\sigma_{ph} = 0.035\,\mathrm{cm}^{-1}\mathrm{Torr}^{-1} \approx 1 \times 10^{-18}\,\mathrm{cm}^2$ [19]. In particular, from this it follows that if the ionization wave propagates in atmospheric air, the mean free path of photons emitted near the wave front is equal $\lambda_{ph} = 1/([O_2]\sigma_{ph}) \approx 0.3\,\mathrm{cm}$, where $[O_2]$ is the number density of oxygen molecules. Evidently, a transverse size of the ionization wave exceeds this mean free path.

Argon is one of air components. Though its concentration in air is small (1 %), argon may partake in processes which proceed in ionized air. In particular, radiation transitions between excited states of argon atoms are used for determination of the electron temperature of the air plasma [21, 22]. Excitation of lower excited states of the argon atom with transitions of a valence electron between $3p$ and $4s$ states is characterized by excitation energies in a range 11.5–11.8 eV that is below the ionization potential for the oxygen molecule 12.07 eV [23]. Hence propagation of the ionization wave in air may proceed as a result of the Firsov mechanism [24–26] according to which emission of a photon near the front of the ionization wave leads to formation of an excited atom on some distance from the front, and then excited atom partakes in the process of associative ionization that leads to electron and ion formation at some distance from the ionization wave front. In dry air this process may proceeds according to the scheme

$$e + Ar(^1S_0, 3p^6) \to e + Ar(^1P_1, 3p^54s),\ Ar(^1P_1, 3p^54s) \to Ar(^1S_0, 3p^6) + \hbar\omega,$$
$$Ar(^1P_1, 3p^54s) + O_2 \to Ar^+ - O_2 + e \qquad (11.21)$$

The excitation energy of the resonant state of the argon atom $Ar(^1P_1, 3p^54s)$ is 11.82 eV and corresponds to the wavelength 104.8 nm of an emitted photon. Since this excitation energy is less than the ionization potential of the oxygen molecule that is 12.07 eV, for the process (11.21) it is necessary the binding energy of the forming ion $Ar^+ - O_2$ to be higher 0.25 eV. Evidently, it holds true because the dissociation energy of a similar complex ion $Kr^+ - O_2$ is equal to 0.33 eV [27]. Moreover, according to a contribution of thermal excitation of molecular particles the threshold of associative ionization decreases.

In addition, the rate constant of the quenching process opposited to the first process of the scheme (11.21), is equal approximately $k_q = 3.9 \times 10^{-9}\,\mathrm{cm}^3/\mathrm{s}$ [28]. If we take for definiteness a typical electron energy or the electron temperature at the ionization wave front to be $T_e = (4 - 6)\,\mathrm{eV}$ that corresponds to the electric field strength $E = 40\,\mathrm{kV/cm}$, one can obtain for the rate constant of the first process of the scheme (11.21) to be found in a range $k_{ex} = (2 - 6) \times 10^{-10}\,\mathrm{cm}^3/\mathrm{s}$. Since the number density of argon atoms in atmospheric air at room temperature is $N_{Ar} \sim 3 \times 10^{17}\,\mathrm{cm}^{-3}$, the excitation rate with transition in the resonantly excited state is of the order of $10^8\mathrm{s}^{-1}$ at the ionization wave front. Next, the absorption coefficient of a resonant photon at the spectral line center is $3 \times 10^5\,\mathrm{cm}^{-1}$ [29], and reabsorption of resonant photons starts from distances of the order of $1/k_o \sim 0.03\,\mu\mathrm{m}$. These processes lead to formation of free electrons as a result of the last process of the scheme (11.21) and correspond to an effective time $\tau_{ef}(L)$ for photon propagation on a distance L according to equation

$$\tau_{ef}(L)N(O_2)k_{as} \sim 1 \qquad (11.22)$$

Here $N(O_2) \approx 6 \times 10^{18}\,\mathrm{cm}^{-3}$ is the number density of oxygen molecules, k_{as} is the rate constant of the process of associative ionization. On the basis of the Veklenko theory [30, 31], the effective time of photon propagation on a distance l is equal [6]

$$\tau_{ef}(l) \approx 2.6\tau_o\sqrt{k_o l}, \qquad (11.23)$$

where τ_o is the radiative lifetime of an isolated excited atom, and in the argon case $\tau_o \approx 2\,\mathrm{ns}$. As it follows from formula (11.23), a typical distance L of electron formation as a result of the reabsorption process depends on the rate constant k_{as} of associative ionization. Unfortunately, we have only a rough estimation for this rate constant. Nevertheless, formula (11.23) gives for a typical reabsorption distance $L = 1\,\mu\mathrm{m}$, if the rate constant is $k_{as} = 5 \times 10^{-12}\,\mathrm{cm}^3/\mathrm{s}$; it equals to $L = 10\,\mu\mathrm{m}$ at $k_{as} = 2 \times 10^{-12}\,\mathrm{cm}^3/\mathrm{s}$, and $L = 100\,\mu\mathrm{m}$, if $k_{as} = 5 \times 10^{-13}\,\mathrm{cm}^3/\mathrm{s}$. These rate constants are possible in reality that testifies about the possibility of the Firsov mechanism for propagation of the ionization wave in atmospheric air.

11.4 Luminous Phenomena in Upper Atmosphere

According to the above scheme of atmospheric electricity, a negative charge is collected at the lower edge of a cumulus cloud. This charge goes to the Earth as a result of precipitation of charged particles and lightning, whereas the positive charge on the upper cloud edge is conserved. The cloud charge creates an electric field both in lower atmospheric layers where this field determines the development of lightning and also in upper atmospheric layers below the ionosphere of a high conductivity. It should be noted that the electron behavior in gases in an external electric field is described by the parameter E/N, where E is the electric field strength, and N is the number density of atoms or molecules of this gas. Since the number density of atmospheric molecules decreases with an increasing altitude, electric processes proceed in the upper atmosphere at relatively low electric field strengths E. For example, according to Fig. 2.3, the number density of air molecules at altitudes of 80 km is fife orders of magnitude less than that at the sea level. Correspondingly, the breakdown electric field strength at such altitude is about 0.3 V/cm that corresponds to the electric potential 3 kV which may create electric breakdown in the layer of the depth of 10 km. Hence, electric processes of high amplitudes may be caused in an upper atmosphere by not high electric fields.

We below consider briefly luminous phenomena in an upper atmosphere under the action of electric fields. The possibility of electric phenomena in upper atmosphere was predicted by Wilson in 1925 [32] on the basis of measurements of electric fields in the stratosphere. Sprites as the most spread luminous objects in the upper atmosphere were observed [33] through many years after the Wilson hypothesis. Sprites are large luminous atmospheric discharges [19, 34–39] which are observed at altitudes 40–90 km and are similar by its nature to positive streamers [40, 41]. They propagate in atmospheric air with velocities $(10^8 - 10^9)$ cm/s. Due to the character of sprite propagation in a non-uniform air, sprites consist of a lower highly structured region and an upper diffuse tail [38, 42, 43]. In considering a sprite as a streamer at a low air pressure, one can use the similarity laws [36, 44] to this phenomenon. According to these laws, typical streamer times t vary with a pressure p as $t \sim p$, a streamer size r varies as $r \sim 1/p$, and the number density of electrons N_e at the front of the ionization wave varies as $N_e \sim p^2$ [36]. Hence, in contrast to streamers propagated in lower atmospheric layers, a transverse size of sprites is large and varies from ten up to several hundreds of meters.

In reality, sprites are a seldom phenomenon. The frequency of its appearance in the upper atmosphere is estimated as approximately 1000 flashes per day over all the globe [45]. For comparison, this number of lightning flashes is observed approximately through 10 s. Spectroscopic measurements exhibit a branched structure of sprites [38, 46, 47] that is typical for fractal objects. According to their properties, sprites are analogous to streamers if to take into account that these physical objects are characterized by different scales.

Along with sprites, other luminous objects are observed in the upper atmosphere, as elves [48], blue jets [49–52], halos which accompany sprites [53], gigantic jets [39, 54, 55] etc. In particular, the nature of these objects was analyzed in [56–59]. In this context we note that in contrast to processes at atmospheric pressure where the rate of quenching of excited molecular states in collisions with other molecules is large compared with the rate of the photoionizaton process, this relation is not valid at high altitudes. Indeed, these rates become equal at the air pressure about 30 Torr [15] that corresponds to the atmosphere altitude of approximately 24 km. Hence the collision quenching is not essential for sprites which develop at altitudes 40–90 km, and the efficiency of photoprocesses is higher than that at low altitudes. An estimation [60] for the electric field strength gives $E = (1 - 2)\,\mathrm{V/cm}$ at the altitude of 70 km that is compared with the breakdown one and therefore may cause some electrical phenomena. From another standpoint, air is excited lightly in upper atmosphere due to a low air density. Various structures and motions may arise in a forming plasma, as a self-consistent system, if it is located in external fields. This leads to creation of local electric fields due to collective interactions, and we observe this in the upper atmosphere in the form of various luminous objects.

References

1. H.A. Kramers, Phil. Mag. **46**, 836 (1923)
2. V.P. Krainov, H.R. Reiss, B.M. Smirnov, *Radiative Processes in Atomic Physics* (Wiley, New York, 1997)
3. L.D. Landau, E.M. Lifshitz, *Classical Theory of Fields* (Pergamon Press, Oxford, 1979)
4. B.M. Smirnov, *Introduction to Plasma Physics* (Mir, Moscow, 1977)
5. I.I. Sobelman, *Atomic Spectra and Radiative Transitions* (Springer, Berlin, 1979)
6. B.M. Smirnov, *Theory of Gas Discharge Plasma* (Springer, Heidelberg, 2014)
7. H. Rishbeth, O.K. Carriott, *Introduction to Ionospheric Physics* (Acadamic, New York, 1969)
8. S.R. Seshardi, *Fundamentals of Plasma Physics* (American Elsevier, New York, 1973)
9. B.M. Smirnov, *Physics of Weakly Ionized Gases* (Mir, Moscow, 1981)
10. A.F. Alexandrov, L.S. Bogdankevich, A.A. Rukhadze, *Principles of Plasma Electrodynamics* (Springer, Berlin, 1984)
11. T.H. Stix, *Waves in Plasmas* (American Institute of Physics, New York, 1992)
12. D.G. Swanson, *Plasma Waves* (Institute of Physics Publishing, Bristol, 2003)
13. B.M. Smirnov, *Plasma Processes and Plasma Kinetics* (Wiley, Berlin, 2007)
14. D. Bilitza, Radio Sci. **36**, 261 (2002)
15. M.B. Zheleznyak, A.K. Mnatsakanyan, S.V. Sizykh, High Temp. **20**, 357 (1982)
16. A.X. Zheleznyak, G.V. Naidis, S.V. Sizykh, High Temp. **20**, 423 (1982)
17. P.M. Dehmer, W.A. Chupka, J. Chem. Phys. **62**, 4525 (1975)
18. V.L. Carter, J. Chem. Phys. **56**, 4195 (1972)
19. N. Liu, V.P. Pasko, J. Geophys. Res. **109**, A04301 (2004)
20. C.W. Walter, P.C. Cosby, H. Helm, Phys. Rev. **50A**, 2930 (1994)
21. K. Bartschat, V. Zeman, Phys. Rev. **59A**, R2552 (1999)
22. A. Dasgupta, M. Blaha, J.L. Giuliani, Phys. Rev. **61A**, 012703 (1999)
23. D.R. Lide (ed.), *Handbook of Chemistry and Physics*, 86th edn. (CRC Press, London, 2003–2004)
24. E.D. Lozanskii, O.B. Firsov, *Theory of Spark* (Atomizdat, Moscow, 1975, in Russian)

25. O.B. Firsov, *About the Spark Theory*. Dissertation of Leningrad Physical-Technical Institute, 1947, in Russian
26. E.D. Lozanskii, J. Tech. Phys. **38**, 1563 (1968)
27. M.F. Jarrold, L. Misov, T. Bowers, J. Chem. Phys. **81**, 4369 (1984)
28. N.B. Kolokolov, A.B. Blagoev, **163**, 152 (1993)
29. B.M. Smirnov, *Properties of Gas Discharge Plasma* (Izdat. Politekhn. Inst., Petersburg, 2010, in Russian)
30. B.A. Veklenko, ZhETF **33**, 629 (1957)
31. B.A. Veklenko, ZhETF **33**, 817 (1957)
32. C.T.R. Wilson, Proc. Cambr. Philos. Soc. **22**, 534 (1925)
33. R.C. Franz, R.J. Nemzek, J.R. Winkler, Science **249**, 48 (1990)
34. D.D. Sentman, E.M. Wescott, D.L. Osborne, D.L. Hampton, M.J. Heavner, Geophys. Res. Lett. **22**, 1205 (1995)
35. W.A. Lyons, J. Geophys. Res. **101**, 29641 (1995)
36. V.P. Pasko, U.S. Inan, T.F. Bell, Geophys. Res. Lett. **25**, 2123 (1998)
37. T. Neubert, Geophys. Res. Lett. **22**, 12 (2003)
38. E.A. Gerken, U.S. Inan, J. Atmos. Solar Terr. Phys. **65**, 567 (2003)
39. H.T. Su, R.R. Hsu, A.B. Chen, Nature (e.a) **423**, 974 (2003)
40. M. Fillekrug, E.A. Mareev, M.J. Rycroft, *Sprites, Elves and Intense Lightning Discharges* (Springer, Heidelberg, 2006)
41. X. Lu, G.V. Naidis, M. Laroussi, K. Ostrikov, Phys. Rep. **540**, 123 (2014)
42. H.C. Stenbaek-Nielsen et al., Geophys. Res. Lett. **27**, 3829 (2000)
43. V.P. Pasko, H.C. Stenbaek-Nielsen, Geophys. Res. Lett. **29**, 1440 (2002)
44. N.I. Petrov, G.N. Petrova, Tech. Phys. **44**, 472 (1999)
45. C.L. Kuo, A.B. Chen, J.K. Chou e.a. J. Phys. **41D**, 234014 (2008)
46. T.M.P. Briels, E.M. van Veldhuizen, U. Ebert, I.E.E.E. Trans, Plasma Sci. **33**, 264 (2005)
47. R.A. Marshall, U.S. Inan, Radio Sci. **41**, RS6S43 (2006)
48. W.L. Boeck, O.H. Vaughan, R.J. Blakeslee et al., Geophys. Res. Lett. **19**, 99 (1992)
49. E.M. Wescott, D. Sentman, D. Osborne et al., Geophys. Res. Lett. **22**, 1209 (1995)
50. W.L. Boeck, O.H. Vaughan, R.J. Blakeslee et al., J. Geophys. Res. **100**, 1465 (1995)
51. E.M. Wescott, D. Sentman, D.L. Heavner et al., Geophys. Res. Lett. **23**, 2153 (1996)
52. W.A. Lyons, T.E. Nelson, R.A. Amstrong et al., Bull. Am. Meteorolog. Soc. **84**, 445 (2003)
53. C.P. Barrington-Leigh, U.S. Inan, M. Stanley, J. Geophys. Res. **106**, 1741 (2001)
54. V.P. Pasko, Nature **423**, 927 (2003)
55. O.A. van der Velde, W.A. Lyons, T.E. Nelson et al., J. Geophys. Res. **112**, D20104 (2007)
56. V.P. Pasko, J. Geophys. Res. Lett. **115**, A00E35 (2010)
57. U.S. Inan, S.A. Cummer, R.A. Marshall, J. Geophys. Res. Lett. **115**, A00E36 (2010)
58. V.V. Surkov, M. Hayakawa, Ann. Gephys. **30**, 1185 (2012)
59. D. Singh, R.P. Singh, A.K. Singh et al., Space Sci. Rev. **169**, 73 (2012)
60. B.D. Green et al., Geophys. Res. Lett. **23**, 2161 (1996)

Chapter 12
Conclusion

To sum up the above analysis, we note two principles of this consideration: the first is a detailed analysis of elementary processes in atmospheric air involving electrons, ions, atoms, and aerosols. In particular, one can emphasize that the ionosphere physics was created in the middle of the twentieth century as a result of measurements of the rate constants of appropriate elementary processes. This is reflected by a citation in the second part of the book by H. S. W. Massey, one of the classics of physics of excited atmospheres. The second principle is the physical and chemical kinetics of these processes and the analysis of some atmospheric phenomena such as atmospheric electricity, the greenhouse effect, and ionosphere properties. The material of this book is an addition to existing atmospheric books that allow one to understand the fundamentals of atmospheric physics.

It is necessary to mark out methods used in the book. Along with some areas of physics, mostly quantum mechanics, gas dynamics, electrodynamics, and statistical physics, we are based on global models including the global circuit model by Wilson, the standard atmospheric model, and the like, which are intended for global aspects of the atmosphere. Therefore this analysis has a qualitative character. But there are precise methods with a more accurate description. As an example, one can consider atmospheric dynamics from this standpoint. In particular, weather prediction is based on atmospheric dynamics that describe air transport using definite boundary conditions and equations of gas dynamics. As a result, evolution of the temperature, moisture, and other atmospheric parameters in a certain locality follows from the solution of these equations. Within the framework of the standard atmosphere model horizontal motion of air is excluded. But representing convective air motion as a sum of vortices, one can find the altitude distribution of particles and aerosols due to convective motion (Formulas (2.28), (2.29)). As a consequence, a transition between free water molecules and aerosols leads to a small water concentration in the stratosphere which is less by three orders of magnitude than that near the Earth's surface (see Fig. 2.9). Thus, the global and accurate methods relate to different aspects of the problem and add to each other.

B.M. Smirnov, *Microphysics of Atmospheric Phenomena*,
Springer Atmospheric Sciences, DOI 10.1007/978-3-319-30813-5_12

One more peculiarity of the above approach is the usage of data for elementary and other atmospheric processes with its global parameters that opens the possibility for a deep understanding of the problem under consideration. This understanding may be a basis for the subsequent precise solution of the problem or give additional information. In particular, the analysis of the evolution of cumulus clouds allows one to determine an average aerosol charge (Formula (8.14)). This approach allows us to check some conclusions in a simple manner. As an example, we show in Sect. 5.5 that an electric charge of 5 C which is referred to the leader in classical lightning books, may be transported only on the next stage of lightning development when a conductive lightning channel is formed. In addition, the analysis of the atmospheric greenhouse effect exhibits the dominant role of atmospheric water in this effect, rather than greenhouse gases. In particular, an increase of the atmospheric water amount by 10 % gives the same action on the greenhouse effect as doubling the amount of atmospheric carbon dioxide. Note that a deeper understanding of some problems on the basis of a given analysis may pose new problems. For example, the analysis of the atmospheric greenhouse effect exhibits the transition between atmospheric water molecules and aerosols to be of importance. j Of course, the possibilities of the used and accurate ways for atmosphere analysis are restricted. For example, an accurate solution of equations of gas dynamics and electrodynamics does not predict a point of the next lightning strike during a thunderstorm. In the same manner, it is impossible to estimate the contribution of deforestation or forest burning to the greenhouse effect, although this contribution may be comparable with that due to carbon dioxide eruptions due to contemporary energetics. Nevertheless, a detailed analysis on the basis of the used method can change our viewpoint of the role of some processes in the atmospheric greenhouse effect.

To emphasize the complexity of atmospheric physics, one can consider the role of cosmic rays in atmospheric properties. Although typical energies for processes involving cosmic rays and for atmospheric processes differ by approximately 10 orders of magnitude, cosmic rays are responsible for discharging of a dry atmosphere, for the origin of electrical breakdown in the atmosphere in lightning form, and the intensity of cosmic rays may be correlated with Earth's covering by clouds. Note that the method used in this book and based on estimations is capacious and especially is convenient for the analysis of complex systems. In spite of the possibilities of this method, it is based on a deep knowledge of various physical laws and is connected with a creative approach. Therefore the method of physical estimation allows us to analyze many problems of atmospheric physics and to obtain various results.

The represented approaches and results may be useful in the analysis of global phenomena, in particular, for climatology. Unfortunately, contemporary climatology is found at a low level and operates with groundless statements or unjustified values. In contrast to this, the physics uses founded estimations that are valid under certain conditions. In particular, there is an accepted change of the average Earth temperature from 1880 up to now, which according to NASA data is approximately 0.8 °C [1], whereas according to the above analysis (Chap. 10) the global Earth's heating is approximately 0.2 °C due to an increased amount of atmospheric carbon dioxide during this time. The indicated temperature increase of 0.8 °C may also be reached by an increase of the amount of atmospheric water vapor by 1/4 or by the transition 0.4 % of the atmospheric water vapor in aerosols. These estimations are based on the assumptions that other atmospheric parameters are invariable and estimated changes of parameters are relatively small. As a result, the approach used on the basis of physics and chemistry allows one to analyze separate aspects of the problem under certain conditions and to understand some questions of atmospheric physics, where contemporary climatology does not give any reliable and grounded result. It should be noted that physical methods are used widely in atmospheric analysis and are represented in a few tens of books indicated in the bibliography.

Reference

1. http://climate.nasa.gov

Appendix A
Appendices

A.1 Fundamental Physical Constants

Electron mass	$m_e = 9.10939 \times 10^{-28} g$
Proton mass	$m_p = 1.67262 \times 10^{-24} g$
Atomic unit of mass	$m_a = \frac{1}{12} m(^{12}C) = 1.66054 \times 10^{-24} g$
Ratio of masses	$m_p/m_e = 1836.15$, $m_a/m_e = 1822.89$
Electron charge	$e = 1.602177 * 10^{-19}\text{C} = 4.8032 \times 10^{-10}\, CGSE$
	$e^2 = 2.3071 \times 10^{-19}\, \text{erg} \cdot \text{cm}$
Planck constant	$h = 6.62607 \times 10^{-27}\, \text{erg} \cdot \text{s}$, $\hbar = 1.05457 \times 10^{-27}\, \text{erg} \cdot \text{s}$
Light velocity	$c = 2.99792 \times 10^{10}\, \text{cm/s}$
Fine-structure constant	$\alpha = e^2/(\hbar c) = 0.07295$
Inverse fine-structure constant	$1/\alpha = \hbar c/e^2 = 137.03599$
Bohr radius	$a_o = \hbar^2/(m_e e^2) = 0.529177\, \text{Å}$
Rydberg constant	$R = m_e e^4/(2\hbar^2) = 13.6057\, \text{eV} = 2.17987 * 10^{-18}\, \text{J}$
Bohr magneton	$\mu_B = e\hbar/(2m_e c) = 9.27402 * 10^{-24}\text{J/T} = 9.27402 * 10^{-21}\, \text{erg/Gs}$
Avogadro number	$N_A = 6.02214 \times 10^{23}\, \text{mol}^{-1}$
Stephan–Boltzmann constant	$\sigma = \pi^2/(60\hbar^3 c^2) = 5.669 \times 10^{-12}\, \text{W/(cm}^2\text{K}^4)$
Molar volume	$R = 22.4141/\text{mol}$
Loschmidt number	$L = N_A/R = 2.6867 * 10^{19}\, \text{cm}^{-3}$
Faraday constant	$F = N_A e = 96485.3\, \text{C/mol}$

A.2 Parameters of Atoms and Molecules of Atmospheric Air

See Figs. A.1, A.2, A.3, A.4, A.5, A.6 and A.7 and Tables A.1, A.2, A.3 and A.4.

B.M. Smirnov, *Microphysics of Atmospheric Phenomena*,
Springer Atmospheric Sciences, DOI 10.1007/978-3-319-30813-5

Table A.1 Potential ionization of nitrogen- and oxygen-contained atomic particles*

Atom, Molecule	Ionization potential, eV
N	14.534
O	13.618
Ar	15.760
N_2	15.58
O_2	12.07
NO	9.26
NH	13.49
OH	13.02
NO_2	9.59
N_2O	12.89
O_3	12.43

**Handbook of Chemistry and Physics.* 84th edition, ed. by D.R. Lide. (London, CRC Press, 2003–2004)

Table A.2 Electron affinity to nitrogen-contained and oxygen-contained atomic particles*

Atom, Molecule	Electron affinity, eV
O	1.461
O_2	0.45
NO	0.026
NH	0.37
OH	1.828
NO_2	2.27
O_3	2.10
NO_3	3.94

**Handbook of Chemistry and Physics.* 84th edition, ed. by D.R. Lide. (London, CRC Press, 2003–2004)

Table A.3 Binding energies for bonds of nitrogen- and oxygen-contained molecules*

Bond	Binding Energy, eV
$N–N$	9.78
$O–O$	5.17
$N–O$	6.54
$N–H$	3.51
$O–H$	4.46
$H–OH$	5.15
$O–NO$	3.16
$O–CO$	5.51
$O–N_2$	1.73
$H–NO$	2.02
$H–NO_2$	3.40

**Handbook of Chemistry and Physics.* 84th edition, ed. by D.R. Lide. (London, CRC Press, 2003–2004)

Table A.4 Parameters of diatomics for nitrogen, oxygen, and their radicals and ions*

Diatomic	T_e, cm^{-1}	$\hbar\omega_e$, cm^{-1}	$\hbar\omega_e x_e$, cm^{-1}	B_e, cm^{-1}	r_e, Å
$N_2(X^1\Sigma_g^+)$	0	2358.6	14.32	1.998	1.098
$N_2(A^3\Sigma_u^+)$	50203.6	1460.6	13.87	1.455	1.287
$N_2(B^3\Pi_g)$	59618.7	1734.0	14.39	1.638	1.212
$O_2(X^3\Sigma_g^-)$	0	1580.2	19	1.446	1.208
$O_2(a^1\Delta_g)$	7918.1	1483.5	12	1.418	1.216
$O_2(b^1\Sigma_g^+)$	13195	1432.8	13	1.400	1.227
$NO(X^2\Pi_{1/2})$	0	1904.2	14.07	1.672	1.151
$NO(X^2\Pi_{3/2})$	119.8	1904.0	14.10	1.720	1.151
$NO(A^2\Sigma^+)$	43966	2374.3	16.11	1.996	1.063
$NH(X^3\Sigma^-)$	0	3282	78.3	16.70	1.034
$NH(a^1\Delta)$	12570	3320	70	16.44	1.036
$OH((X^2\Pi_{3/2}$	0	3737.8	84.9	18.91	0.970
$OH((X^2\Pi_{1/2}$	126.2	3737.8	84.9	18.91	0.970
$OH(A^2\Sigma^+)$	32684	3178.9	92.9	17.36	1.012
$N_2^+(X^2\Sigma_g^+)$	0	2207.3	16.26	1.932	1.116
$N_2^+(A^2\Pi_u)$	9167.5	1903.5	15.02	1.744	1.116
$N_2^+(B^2\Sigma_u^+)$	25461.1	2421	24.1	2.085	1.075
$O_2^+(X^2\Pi_{g,3/2})$	0	1904.8	16.3	1.691	1.116
$O_2^+(X^2\Pi_{g,1/2})$	197.3	1904.8	16.3	1.691	1.116
$O_2^+(a^4\Pi_u$	32960	1035.7	10.4	1.105	1.381
$O_2^-(X^2\Pi_{1/2})$	0	1090	8	1.16	1.35
$O_2^-(X^2\Pi_{3/2})$	160	1090	8	1.16	1.35
$NO^+(X^1\Sigma^+)$	0	2376.7	16.3	1.997	1.063
$NO^+(a^3\Sigma^+)$	52150	1300	15	1.37	1.28
$NO^-(X^3\Sigma^-)$	0	1360	8	1.43	1.26
$NH^+(X^2\Pi)$	0	2922	–	15.3	1.07
$NH^+(A^2\Sigma^-)$	22200	1707	61	11.46	1.251
$OH^+(X^3\Sigma^-)$	0	3113.4	78.5	16.79	1.029
$OH^+(A^3\Pi)$	28439	2133.6	79.5	13.79	1.135
$OH^+(b^1\Sigma^+)$	29050	2980	–	16.32	1.03
$OH^-(X^1\Sigma^+)$	0	3700	–	19.13	0.964

*K.P. Huber, G. Herzberg. *Constants of Diatomic Molecules.* (New York, Van Nostrand Reinhold, 1979)

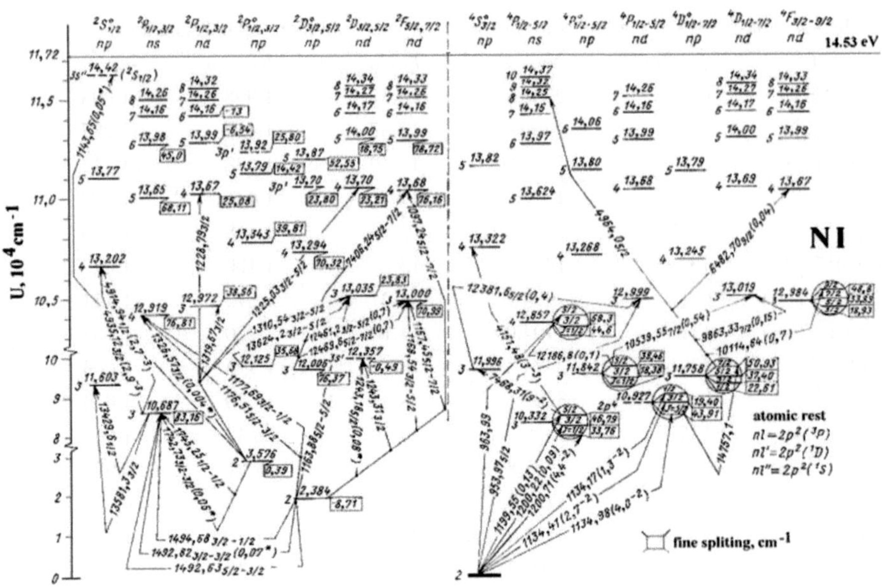

Fig. A.1 Spectrum of lower states of the nitrogen atom

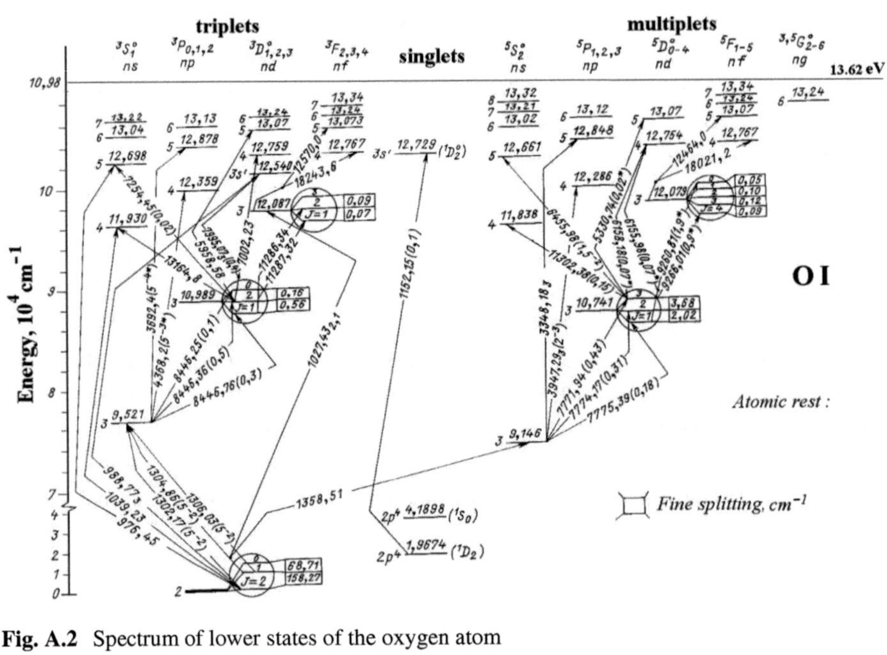

Fig. A.2 Spectrum of lower states of the oxygen atom

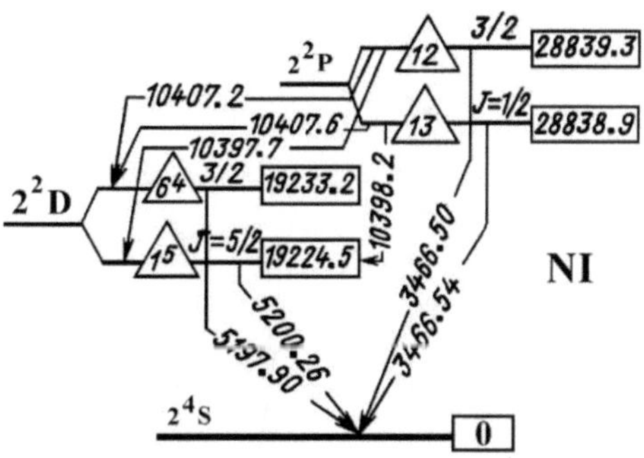

Fig. A.3 Lowest states of the oxygen atoms. Excitation energies of these states are given in rectangular boxes and are expressed in cm^{-1}, the radiative lifetimes are given in triangles and are expressed in seconds; for example, 1^5 is 1×10^5, the wavelengths of transitions are inside arrows

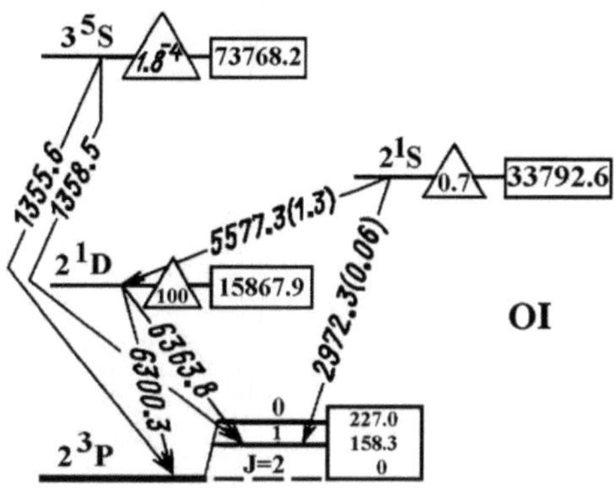

Fig. A.4 Lowest states of the oxygen atoms. Excitation energies of these states are given in rectangular boxes and are expressed in cm^{-1}, the radiative lifetimes are given in triangles and are expressed in seconds; for example, 1^5 is 1×10^5, the wavelengths of transitions are inside arrows

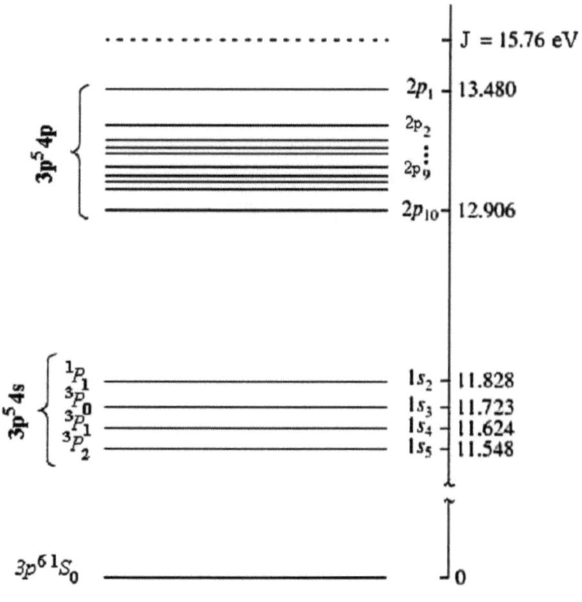

Fig. A.5 Potential curves of lower states for oxygen molecule and its ion

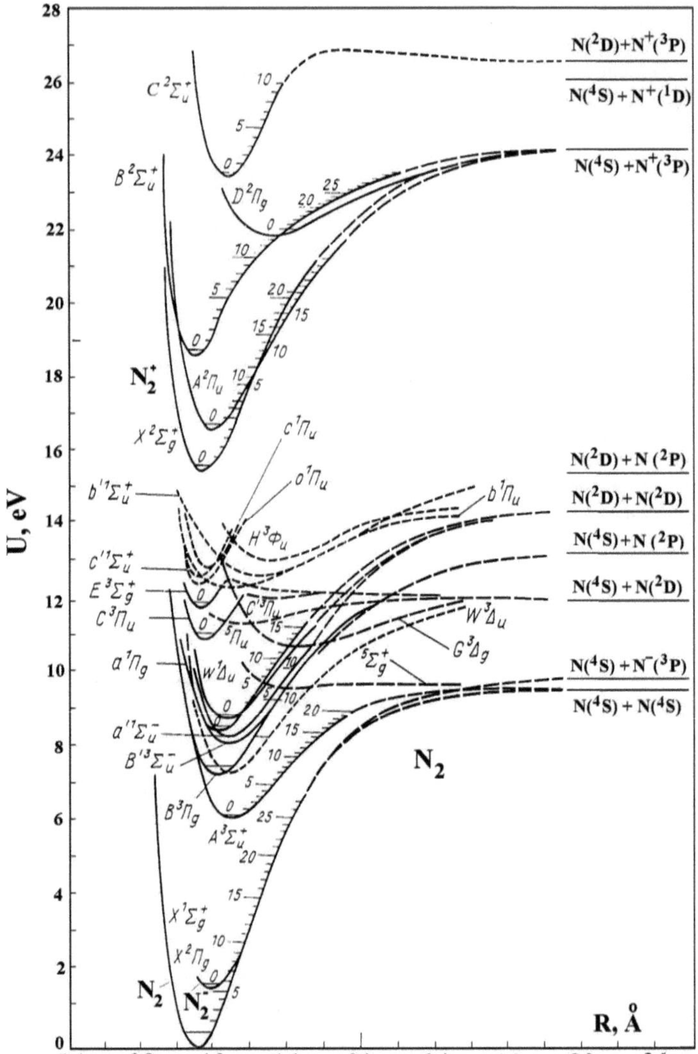

Fig. A.6 Potential curves of lower states for oxygen molecule and its ion

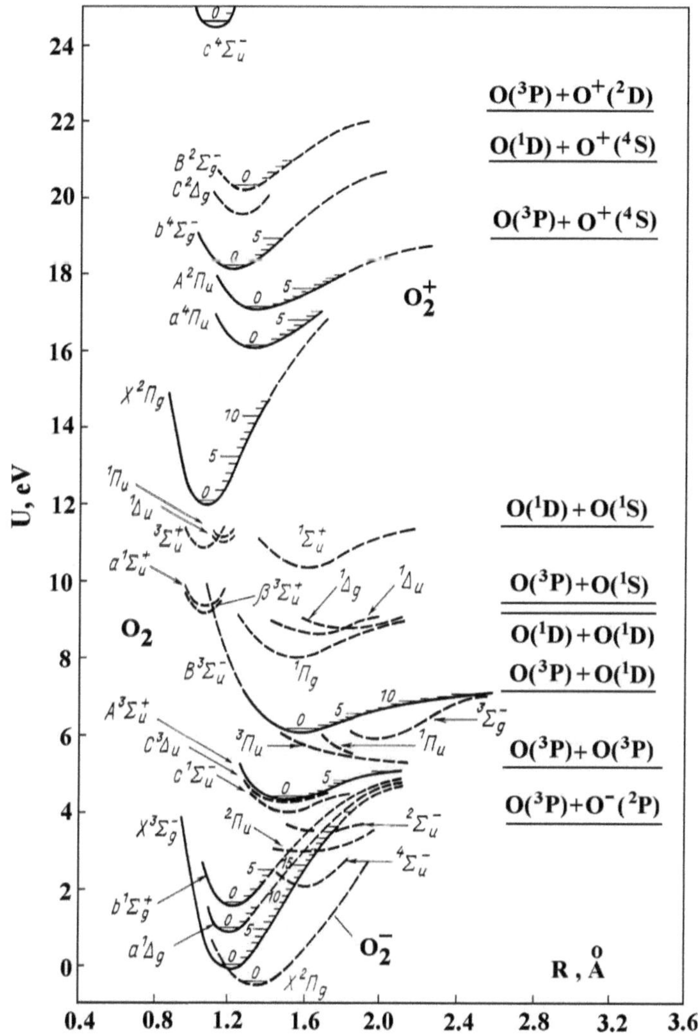

Fig. A.7 Potential curves of lower states for oxygen molecule and its ion

A.3 Conversional Factors for Units

Units of Energy

	1 J	1 erg	1 eV	1 K	1 cm^{-1}	1 MHz	1 kcal/mol	1 kJ/mol
1 J	1	10^{7}	6.2415×10^{18}	7.2429×10^{22}	5.0341×10^{22}	1.5092×10^{27}	1.4393×10^{20}	6.0221×10^{20}
1 erg	10^{-7}	1	6.2415×10^{11}	7.2429×10^{15}	5.0341×10^{15}	1.5092×10^{20}	1.4393×10^{13}	6.0221×10^{13}
1 eV	1.6022×10^{-19}	1.6022×10^{-12}	1	11604	8065.5	2.4180×10^{8}	23.045	96.485
1 K	1.3807×10^{-23}	1.3807×10^{-16}	8.6174×10^{-5}	1	0.69504	2.0837×10^{4}	1.9872×10^{-3}	8.3145×10^{-3}
1 cm^{-1}	1.9864×10^{-23}	1.9864×10^{-16}	1.2398×10^{-4}	1.4388	1	2.9979×10^{4}	2.8591×10^{-3}	1.1963×10^{-2}
1 MHz	6.6261×10^{-28}	6.6261×10^{-21}	4.1357×10^{-9}	4.7992×10^{-5}	3.3356×10^{-5}	1	9.5371×10^{-9}	3.9903×10^{-7}
1 kcal/mol	6.9477×10^{-21}	6.9477×10^{-28}	4.3364×10^{-2}	503.22	349.76	1.0485×10^{7}	1	4.184
1 kJ/mol	1.6605×10^{-21}	1.6605×10^{-28}	1.0364×10^{-2}	120.27	83.594	2.5061×10^{6}	0.23901	1

Units of Pressure

	1 Pa = 1 N/m^2	1 dyn/cm^2	1 Torr	1 atm*	1*at**	1 bar
1 Pa = 1 N/m^2	1	10	7.5001×10^{-3}	9.8693×10^{-6}	1.0197×10^{-5}	10^{-5}
1 dyn/cm^2	0.1	1	7.5001×10^{-4}	9.8693×10^{-7}	1.0197×10^{-6}	10^{-6}
1 Torr	133.332	1333.32	1	1.3158×10^{-3}	1.3595×10^{-3}	1.33332×10^{-3}
1 atm*	1.01325×10^5	1.01325×10^6	760	1	1.01332	1.01325
1*at**	9.80665×10^4	9.80665×10^5	735.56	0.96785	1	0.980665
1 bar	10^5	10^6	750.01	0.98693	1.0197	1

*atm—physical atmosphere, **at* = kg/cm^2—technical atmosphere

Units of Electric Charge

	1 e	1 *CGSE*	1 C
1 e	1	4.8032×10^{-10}	1.60218×10^{-19}
1 *CGSE*	2.0819×10^9	1	3.33564×10^{-10}
1 C	6.2415×10^{18}	2.99792×10^9	1

Units of Electric Voltage

	1 V	1 *CGSE*	1C/m
1 V	1	3.33564×10^{-3}	1.113×10^{-10}
1 *CGSE*	299.792	1	3.3364×10^{-8}
1 C/m	8.9875×10^9	2.99792×10^7	1

Units of Electric Field Strength

	1 V/cm	1 *CGSE*	1 C/m^2
1 V/cm	1	3.33564×10^{-3}	1.113×10^{-8}
1 *CGSE*	299.792	1	3.3364×10^{-6}
1 C/m^2	8.9875×10^7	2.99792×10^5	1

Units of Specific Electric Field Strength

	1Td*	1 V/(cm · Torr)
1Td*	1	2.829
1 V/(cm · Torr)	0.3535	1

*1 Td = 1×10^{-17} V · cm^2

Units of conductivity

	S/m	$1/(\Omega \cdot cm)$	1/s
S/m	1	0.01	1.11265×10^{-14}
$1/(\Omega \cdot cm)$	100	1	1.11265×10^{-12}
1/s	8.98755×10^{13}	8.98755×10^{11}	1

Units of Electric Resistance

	1 Ω	1 *CGSE*	1 *CGSM*
1 Ω	1	1.11265×10^{-12}	10^9
1 *CGSE*	8.98755×10^{11}	1	8.98755×10^{20}
1 *CGSM*	10^{-9}	1.11265×10^{-21}	1

Units of Current Density

	1 $e/(cm^2 \cdot s)$	1 *CGSE*	1 A/m^2
1 $e/(cm^2 \cdot s)$	1	2.99792 4.8032×10^{-10}	1.60218×10^{-15}
1 *CGSE*	2.0819×10^9	1	3.3356×10^{-6}
1 A/m^2	6.2415×10^{14}	2.9979×10^5	1

Units of Magnetic Field Strength

	1 Oe	1 *CGSE*	1 A/m
1 Oe	1	2.99792×10^{10}	79.5775
1 *CGSE*	3.33564×10^{-11}	1	2.65442×10^{-9}
1 A/m	0.012566	1.11265×10^{-21}	1

Units of Magnetic Induction

	1 *CGSE*	$1\ T = 1\ TWb/m^2$	1 Gs
1 *CGSE*	1	2.99792×10^6	2.99792×10^{10}
$1\ T = 1\ Wb/m^2$	3.33564×10^{-7}	1	10^4
1 Gs	3.33564×10^{-11}	10^{-4}	1

A.4 Conversional Factors in Formulas

Formulas of General Physics

Number	Formula	Factor C	Units used
1.	$v = C\sqrt{\varepsilon/m}$	5.931×10^7 cm/s	ε in eV, m in m_e^*
		1.389×10^6 cm/s	ε in eV, m in m_a^*
		5.506×10^5 cm/s	ε in K, m in m_e^*
		1.289×10^4 cm/s	ε in K, m in m_a^*
2.	$v = C\sqrt{T/m}$	6.692×10^7 cm/s	T in eV, m in m_e^*
		6.212×10^5 cm/s	T in K, m in m_e^*
		1.567×10^6 cm/s	T in eV, m in m_a^*
		1.455×10^4 cm/s	T in K, m in m_a^*
3.	$\varepsilon = Cv^2$	$3.299 * 10^{-12}$ K	v in cm/s, m in m_e^*
		6.014×10^{-9} K	v in cm/s, m in m_a^*
		$2.843 * 10^{-16}$ eV	v in cm/s, m in m_e^*
		5.182×10^{-13} eV	v in cm/s, m in m_a^*
4.	$\omega_H = CH/m$	$1.759 \times 10^7\ s^{-1}$	H in Gs, m in m_e^*
		$9648\ s^{-1}$	H in Gs, m in m_a^*
5.	$v = CE/H$	1×10^8 cm/s	E in V/cm, H in Gs
6.	$r_H = C\sqrt{\varepsilon m}/H$	3.372 cm	ε in eV, m in m_e^*, H in Gs
		143.9 cm	ε in eV, m in m_a^*, H in Gs
		3.128×10^{-2} cm	ε in K, m in m_e^*, H in Gs
		1.336 cm	ε in K, m in m_a^*, H in Gs
7.	$p = CH^2$	$4.000 \times 10^{-3} Pa = 0.04\ \mathrm{erg/cm^3}$	H in Gs

$^*m_e = 9.108 \times 10^{-28} g$ is the electron mass, $m_a = 1.6605 \times 10^{-24} g$ is the atomic mass unit

Explanations of Table

1. The particle velocity is $v = \sqrt{2\varepsilon/m}$, where ε is the energy and m is the particle mass.

2. The average particle velocity is $v = \sqrt{8T/(\pi m)}$ with the Maxwell distribution function over velocities; T is the temperature expressed in energetic units and m is the particle mass.

3. The particle energy is $\varepsilon = mv^2/2$, where m is the particle mass and v is the particle velocity.

4. The Larmor frequency is $\omega_H = eH/(mc)$ for a charged particle of a mass m in a magnetic field of a strength H.

5. The drift velocity of a singly charged particle in crossed electric and magnetic fields of strengths E and H, correspondingly.

6. The Larmor radius of a charged particle is $r_H = \sqrt{2\varepsilon/m}/\omega_H$, where ε is the energy of a charged particle, m is its mass, and ω_H is the Larmor frequency.

7. The magnetic pressure is $p_m = H^2/(8\pi)$.

Formulas of Gas and Plasma Physics

Number	Formula	Proportionality factor C	Units
1.	$\alpha = CN_e/T^3$	2.998×10^{-21}	N_e in cm^{-3}, T in eV
		4.685×10^{-9}	N_e in cm^{-3}, T in K
2.	$f = Cm^{3/2}T^{3/2}$	$2.415 \times 10^{15}\,cm^{-3}$	T in K, m in m_e^*
		$3.019 \times 10^{21}\,cm^{-3}$	T in K, m in m_e^*
		$1.879 \times 10^{20}\,cm^{-3}$	T in K, m in m_a^*
		$2.349 \times 10^{26}\,cm^{-3}$	T in eV, m in m_a^*
3.	$K = C/T_e^{9/2}$	$2.406 \times 10^{-22}\,cm^6/s$	T_e in 1000 K
		$3.894 \times 10^{-27}\,cm^6/s$	T_e in eV
4.	$\omega_p = C\sqrt{N_e/m}$	$5.642 \times 10^4\,s^{-1}$	N_e in cm^{-3}, m in m_e^*
		$1322\,s^{-1}$	N_e in cm^{-3}, m in m_a^*
5.	$r_D = C\sqrt{T/N_e}$	525.3 cm	N_e in cm^{-3}, T in eV
		4.876 cm	N_e in cm^{-3}, T in K

$^*m_e = 9.108 \times 10^{-28} g$ is the electron mass, $m_a = 1.6605 \times 10^{-24} g$ is the atomic mass unit

Explanations of Table

1. The ideality plasma parameter is $\alpha = N_e e^6/T_e^3$, where e is the electron charge, N_e is the number density of electrons, and T_e is the electron temperature.

2. Pre-exponent of the Saha formula is $\xi = [mT/(2\pi\hbar^2)]^{3/2}$.

3. The rate constant of three-body electron ion recombination $(2e + A^+ \rightarrow e + A)$ is $K = 1.5e^{10}/(m_e^{1/2}T_e^{9/2})$.

4. The plasma frequency is $\omega_p = \sqrt{4\pi N_e e^2/m_e}$.

5. The Debye–Hückel radius is $r_D = \sqrt{T/(8\pi N_e e^2)}$.

Radiative Transition Between Atom States

Number	Formula	Conversional factor C	Used units
1.	$\varepsilon = C\omega$	4.1347×10^{-15} eV	ω in s^{-1}
		6.6261×10^{-34} J	ω in s^{-1}
2.	$\omega = C\varepsilon$	$1.519 \times 10^{15}\,s^{-1}$	ε in eV
		$1.309 \times 10^{11}\,s^{-1}$	ε in K
3.	$\omega = C/\lambda$	$1.884 \times 10^{15}\,s^{-1}$	λ in μm
4.	$\varepsilon = C/\lambda$	1.2398 eV	λ in μm
5.	$f_{o*} = C\omega d^2 g_*$	1.6126×10^{-17}	ω in s^{-1}, d in D^*
		0.02450	$\Delta\varepsilon = \hbar\omega$ in eV, d in D^*
6.	$f_{o*} = Cd^2 g_*/\lambda$	0.03038	λ in μm, d in D*
7.	$1/\tau_{*o} = C\omega^3 d^2 g_o$	$3.0316 \times 10^{-40}\,s^{-1}$	ω, s^{-1}, d, D*
		$1.06312 \times 10^6\,s^{-1}$	$\Delta\varepsilon = \hbar\omega$ in eV, d in D*
8.	$1/\tau_{*o} = Cd^2 g_o/\lambda^3$	$2.0261 \times 10^6\,s^{-1}$	λ in μm, d in D*
9.	$1/\tau_{*o} = C\omega^2 g_o f_{o*}/g_*$	$1.8799 \times 10^{-23}\,s^{-1}$	ω in s^{-1}, d in D*
		$4.3393 \times 10^7\,s^{-1}$	$\Delta\varepsilon = \hbar\omega$ in eV; d in D*
10.	$1/\tau_{*o} = Cf_{o*}g_o/(g_*\lambda^2)$	$6.6703 \times 10^7\,s^{-1}$	λ in μm, d in D*

*D is Debye, 1 $D = ea_o = 2.5418 \times 10^{-18}$ *CGSE*

Explanation of Table

1. The photon energy $\varepsilon = \hbar\omega$, where ω is the photon frequency.
2. The photon frequency is $\omega = \varepsilon/\hbar$.
3. The photon frequency is $\omega = 2\pi c/\lambda$, where λ is the wavelength and c is the light speed.

4. The photon energy is $\varepsilon = 2\pi\hbar c/\lambda$.

5, 6. The oscillator strength for a radiative transition from the lower o to the upper $*$ state of an atomic particle that is averaged over lower states o and summed over upper states $*$ is equal to

$$f_{o*} = \frac{2m_e\omega}{3\hbar e^2}\left|\langle o\left|\mathbf{D}\right|*\rangle\right|^2 g_* = \frac{2m_e\omega}{3\hbar e^2}d^2 g_* = \frac{4\pi c m_e}{3\hbar e^2\lambda}d^2 g_* \,,$$

where $\mathbf{d} = \langle o|\mathbf{D}|*\rangle$ is the matrix element for the operator of the dipole moment of an atomic particle taken between transition states. Here $m_e,\ \hbar$ are atomic parameters, g_* is the statistical weight of the upper state, and $\omega = (\varepsilon_* - \varepsilon_o)/\hbar$ is the transition frequency, where $\varepsilon_o, \varepsilon_*$ are the energies of transition states and λ is the transition wavelength.

7–10. The rate of the radiative transition is

$$\frac{1}{\tau_{*o}} = B_{*o} = \frac{4\omega^3}{3\hbar c^3}d^2 g_o = \frac{32\pi^3}{3\hbar\lambda^3}d^2 g_o = \frac{2\omega^2 e^2 g_o}{m_e c^3 g_*}f_{o*} = \frac{8\pi^2 g_o}{\hbar g_*\lambda^2 c}f_{o*}$$

Here B is the Einstein coefficient; other notations are the same as above.

Transport Coefficients

Number	Formula	Coefficient C	Units
1.	$D = CKT$	8.617×10^{-5} cm^2/s	K in cm^2/(V · s), T in K
		1 cm^2/s	K in cm^2/(V · s), T in eV
2.	$K = CD/T$	11604 cm^2/(V · s)	D in cm^2/s, T in K
		1 cm^2/(V · s),	D in cm^2/s, T in eV
3.	$D = C\sqrt{T/\mu}/(N\overline{\sigma_1})$	4.617×10^{21}cm^2/s	$\overline{\sigma_1}$ in Å^2, N in cm^{-3}, T in K, μ in m_a^*
		1.595 cm^2/s	$\overline{\sigma_1}$ in Å^2, $N = 2.687 \times 10^{19}$cm^{-3}, T in K, μ in m_a^*
		171.8 cm^2/s	$\overline{\sigma_1}$ in Å^2, $N = 2.687 \times 10^{19}$ cm^{-3}, T in eV μ in m_a^*
		68.1115 cm^2/s	$\overline{\sigma_1}$ in Å^2, $N = 2.687 \times 10^{19}$ cm^{-3}, T in K, μ in m_e^*
		7338 cm^2/s	$\overline{\sigma_1}$ in Å^2, $N = 2.687 \times 10^{19}$ cm^{-3}, T in eV, μ in m_e^*
4.	$K = C(\sqrt{T\mu}N\overline{\sigma_1})^{-1}$	1.851×10^4 cm^2/(V · s)	$\overline{\sigma_1}$ in Å^2, $N = 2.687 \times 10^{19}$ cm^{-3}, T in K, μ in m_e^*
		171.8 cm^2/(V · s)	$\overline{\sigma_1}$ in Å^2, $N = 2.687 \times 10^{19}$ cm^{-3}, T in eV, μ in m_a^*
		7.904×10^5 cm^2/(V · s)	$\overline{\sigma_1}$ in Å^2, $N = 2.687 \times 10^{19}$ cm^{-3}, T in K, μ in m_e^*
		7338 cm^2/(V · s)	$\overline{\sigma_1}$ in Å^2, $N = 2.687 \times 10^{19}$ cm^{-3}, T in eV, μ in m_e^*
5.	$\kappa = C\sqrt{T/m}/\overline{\sigma_2}$	1.743×10^4 W/(cm · K)	T in K, m in m_a^*, $\overline{\sigma_2}$ in Å^2
		7.443×10^5 W/(cm · K)	T in K, m in m_e^*, $\overline{\sigma_2}$ in Å^2
6.	$\eta = C\sqrt{Tm}/\overline{\sigma_2}$	5.591×10^{-5} g/(cm · s)	T in K, m in m_a^*, $\overline{\sigma_2}$ in Å^2
7.	$\xi = CE/(TN\sigma)$	1.160×10^{20}	E in V/cm, T in K, σ in Å^2, N in cm^{-3}
		1×10^{16}	E in V/cm, T in eV, σ in Å^2, N in cm^{-3}

*$m_e = 9.108 \times 10^{-28}g$ is the electron mass, $m_a = 1.6605 \times 10^{-24}$ g is the atomic mass

Explanation of Table

1. The Einstein relation for the diffusion coefficient of a charged particle in a gas $D = KT/e$, where $D,\ K$ are the diffusion coefficient and mobility of a charged particle and T is the gas temperature.

2. The Einstein relation for the mobility of a charged particle in a gas $K = eD/T$.

3. The diffusion coefficient of an atomic particle in a gas in the first Chapman–Enskog approximation $D = 3\sqrt{2\pi T/\mu}/(16N\overline{\sigma_1})$, where T is the gas temperature, N is the number density of gas atoms or molecules, μ is the reduced mass of a colliding particle and gas atom or molecule, and $\overline{\sigma_1}$ is the average cross section of collision.

4. The mobility of a charged particle in a gas in the first Chapman–Enskog approximation $K = 3e\sqrt{2\pi/(T\mu)}/(16N\overline{\sigma_1})$; notations are the same as above.

5. The gas thermal conductivity in the first Chapman–Enskog approximation $\kappa = 25\sqrt{\pi T}/(32\sqrt{m}\overline{\sigma_2})$, where m is the atom or molecule mass and $\overline{\sigma_2}$ is the average cross section of collision between gas atoms or molecules; other notations are the same as above.

6. The gas viscosity in the first Chapman–Enskog approximation $\eta = 5\sqrt{\pi Tm}/(24\overline{\sigma_2})$; notations are the same as above.

7. The parameter of ion drift in a gas in a constant electric field $\xi = eE/(TN\sigma)$, where E is the electric field strength, T is the gas temperature, N is the number density of atoms or molecules, and σ is the cross section of collision.

Clusters or Small Particles

Number	Formula	Coefficient C	Used units
1.	$r_W = C(m/\rho)^{1/3}$	0.7346 Å	m in m_a^*, ρ in g/cm^3
2.	$n = C(r_o/r_W)^3$	4.189	r_o and r_W in Å
3.	$k_o = Cr_W^2\sqrt{T/m}$	4.5714×10^{-12} cm^3/s	r_W in Å, T in K, m in m_a^*
4.	$w = C\rho r^2/\eta$	0.2179 cm/s	r in µm, ρ in g/cm^3, η in 10^{-5}g/(cm · s)
	$w = Cr^2$	0.01178 cm/s	r in µm, ρ, η relate to air at $p = 1$ atm, $T = 300$ K
5.	$D = CKT$	8.617×10^{-5} cm^2/s	K in cm^2/(V · s), T in K
		1 cm^2/s	K in cm^2/(V · s), T in eV
6.	$K = CD/T$	11604 cm^2/(V · s)	D in cm^2/s, T in K
		1 cm^2/(V · s),	D in cm^2/s T in eV
7.	$D_o = C\sqrt{T/m}/(N_a r_W^2)$	1.469×10^{21}cm^2/s	r_W in Å, N_a in cm^{-3}, T in K, m in m_a^*
		0.508 cm^2/s	r_W in Å, $N_a = N_o$, T in K, m in m_a^*
		54.69 cm^2/s	r_W in Å, $N_a = N_o$, T in eV, m in m_a^*
8.	$K_o = C(\sqrt{Tm}\,Nr_W^2)^{-1}$	1.364×10^{19} cm^2/(V · s)	r_W in Å, N_a in cm^{-3}, T in K, m in m_a^*
		0.508 cm^2/(V · s)	r_W in Å, $N_a = N_o$, T in K, m in m_a^*
		54.69 cm^2/(V · s)	r_W in Å, $N = N_o$, T in K, m in m_a^*
9.	$D_o = CT/(r_W\eta)$	7.32×10^{-5} cm^2/s	r_W in Å, T in K, η in 10^{-5}g/(cm · s)
10.	$K_o = C/(r_W\eta)$	0.085 cm^2/(V · s)	r_W in Å, η in 10^{-5}g/(cm · s)

$^*m_a = 1.66054 \times 10^{-24} g$ is the atomic mass

Explanations of Table

1, 2. A number of atoms in a cluster or spherical particle consisting of n atoms and having a radius r is $n = (r/r_W)^3$ within the framework of the liquid drop model, where r_W is the Wigner–Seits radius of a cluster material. A cluster mass is $nm_a = 4\pi r^3 \rho/3$, where m_a is the atom mass and ρ is the mass density of a cluster material.

3. The reduced rate constant for collisions involving clusters is $k_o = \pi r^2 \cdot \sqrt{8T/(\pi m_a)}$.

4. The freefall velocity w in the gravitational field for a spherical particle of a radius r_o: $w = 2\rho g r_o^2/(9\eta)$, where g is the freefall acceleration, ρ is the mass density for a particle material, and η is the viscosity of a medium where the particle moves.

5. The Einstein relation for a charged particle located in a gas $D = KT/e$, where D, K are the diffusion coefficient and mobility of a charged particle and T is the gas temperature.

6. The Einstein relation $K = eD/T$ for the mobility K of a charged particle moved in a gas.

7. The reduced diffusion coefficient D_o of a particle in a gas in the kinetic regime, so that the particle diffusion coefficient D of a particle consisting of n atoms at the normal number density $N_a = N_o = 2.687 \times 10^{19}\text{cm}^{-3}$ of gas molecules, that is equal to $D_n = D_o/n^{2/3}$, where $D_o = 3\sqrt{2T/\pi m}/(16N_o r_W^2)$, T is the gas temperature, m is the gas atom mass, and r_W is the Wigner–Seits radius.

8. The reduced zero-field mobility of a spherical particle K_o in the kinetic regime, so that K is the mobility of a particle which consisting of n atoms is $K_n = K_o/n^{2/3}$, where $K_o = 3e/(8N_o r_W^2 \sqrt{2\pi mT})$ at the normal number density of gas atoms $N_a = N_o = 2.687 \times 10^{19}\text{cm}^{-3}$, and other notations are indicated in the previous point.

9. The reduced diffusion coefficient $d_o = T/(6\pi r_W \eta)$ of a particle in a gas in the diffusion regime at the normal number density $N_a = N_o = 2.687 \times 10^{19}\text{cm}^{-3}$ of gas molecules, so that the particle diffusion coefficient D_n of a particle consisting of n atoms is equal to $D_n = d_o/n^{1/3}$. Here, η is the gas viscosity and r_W is the Wigner–Seits radius; other notations are given above.

10. The reduced zero-field mobility of a spherical particle is $K_o = e/(6\pi r_W \eta)$ in the diffusion regime at the normal number density of gas atoms $N_a = N_o = 2.687 \times 10^{19}\text{cm}^{-3}$, so that the mobility of a particle consisting of n atoms is equal to $K_n = K_o/n^{1/3}$, and the notations used are indicated above.

Index

B.M. Smirnov, *Microphysics of Atmospheric Phenomena*,
Springer Atmospheric Sciences, DOI 10.1007/978-3-319-30813-5

MIX
Papier aus verantwortungsvollen Quellen
Paper from responsible sources
FSC® C105338

If you have any concerns about our products,
you can contact us on
ProductSafety@springernature.com

In case Publisher is established outside the EU,
the EU authorized representative is:
Springer Nature Customer Service Center GmbH
Europaplatz 3, 69115 Heidelberg, Germany

Printed by Libri Plureos GmbH
in Hamburg, Germany